Transgen

CELL AND MOLECULAR BIOLOGY IN ACTION SERIES

Transgenic Mammals

John Bishop

Pearson Education Limited
Edinburgh Gate, Harlow
Essex CM20 2JE
England

and Associated Companies throughout the World

First published 1999

ISBN 0 582 35730 6

British Library Cataloguing-in-Publication Data
A catalogue record for this book is
available from the British Library.

Library of Congress Cataloging-in-Publication Data
A catalog entry for this title is
available from the Library of Congress.

Set by 30 in Concorde BE
Printed in Great Britain by Henry Ling Ltd., at the Dorset Press,
Dorchester, Dorset.

Contents

Preface

The transgenic approach is now important in most if not all branches of mammalian biology. As a consequence students and workers come to it from a wide variety of backgrounds. The huge expansion in our understanding during the last century has made it quite impossible for anyone to have a comprehensive knowledge even of fundamental biology, never mind its ramifications. Thus, this book attempts both to summarise the subject area of transgenic mammals and also to supply the bare bones of the background information required to make the account comprehensible without recourse to other texts. The first chapter, *Overview*, is an attempt to describe the subject area in the broadest terms with the minimum of jargon and to clarify at the outset what constitutes a transgenic mammal, in contrast say to a clone or a product of surrogacy.

The treatment of transgenic mammals focuses on what might be called the theoretical aspects of the subject, all of which ultimately bear on the question of transgene expression. Thus to comprehensively understand transgene expression requires an understanding of alternative transgene configurations, of the processes that integrate foreign DNA into the chromosomes and of the various routes by which DNA is introduced into the germ line. The undoubted skills that are required to achieve these ends are only referred to in passing. The surgical skills are of course generally learned from an experienced practitioner; in the UK at least, this is a legal requirement.

The book is intended for more advanced undergraduates as well as for graduate students and postdoctoral scientists. A reasonably large number of references is given, with the aim of easing access to the literature for those who wish to pursue the subject more deeply. I hope that they may not prove necessary for those who do not. I also wish to apologise to authors whose contributions I may inadvertently have underestimated, misinterpreted or overlooked.

I have benefitted from generous assistance of various sorts from colleagues and acquaintances both here at the Centre for Genome Research and elsewhere, namely Austin and Andrew Smith, Yacine Cherifi, Tom Burdon, Chris Ferrard, Bob Wall, Ian Mather, Bill Ritchie and Ian Wilmut. Needless to say, I am responsible for all opinions expressed and any errors of fact or interpretation.

John Bishop

PART ONE

Introduction

CHAPTER ONE

Overview

Over the last fifteen years transgenic animals have been recognised to be an essential tool in biological research, particularly research into complex systems such as development, behaviour and memory that involve interactions between tissues or different types of cells. Transgenic mammals are particularly important because of their application to medicine. They give new insights into normal functioning, and they provide models of human disease that can be employed as test-beds for promising therapies. More than 99% of the transgenic mammals produced for these purposes are in fact laboratory mice. There are several reasons for this. First, the cost of husbandry increases roughly in proportion to the weight of the animal. Also (and no doubt driven over the years by the same cost considerations) the mouse resource and database are vastly greater than those available for any other mammal. These include a large number of different inbred mouse strains, the availability of an enormous number of genetic variants that have been accumulated over the better part of 100 years of study, a detailed genetic map, extensive DNA sequence data and a wealth of information about mouse physiology, behaviour and disease. Lastly, the most precise method for making a transgenic animal, the ES cell route, can so far be applied only in the mouse.

Although the physiology and behaviour of laboratory rats is better understood and they are essential for toxicology trials of pharmaceuticals, rather few genetic variants are available. Cost considerations and the unavailability of the ES cell route contribute to the moderate use of the rat in transgenic studies.

Transgenic plants are already commercially significant and will certainly become more so. There is also money to be made from transgenic mammals. Two areas are particularly promising. One is 'improvement', analogous to successes already recorded with plants. Examples are the genetic alteration of cows to make the milk that they produce more suitable for babies, or a basis for more efficient cheese making. The second is the use of animals such as sheep and cows as 'bioreactors' capable of producing, for example in their milk, vast amounts of a valuable foreign protein.

What is a transgenic mammal?

All transgenic animals are altered in a way that is heritable. A formal definition of a transgenic mammal is unlikely to be helpful. Instead, here is an exhaustive list of the types of heritable change that would lead to the inclusion of mammals carrying them within the catch-all designation 'transgenic'. By 'foreign DNA' we mean any piece of DNA, *including DNA from the same species*, introduced into the genome of another animal.

- integration of a fragment of foreign DNA into at least one chromosome
- alteration of any gene resulting from the introduction of foreign DNA, and excluding alterations produced by the action of chemicals or radiation on the DNA *in situ*
- rearrangement of chromosomes resulting from the introduction of foreign DNA, and again excluding effects due to chemicals and radiation
- any deliberately introduced persistent supernumerary genetic entity, for example an entirely artificial chromosome or an extrachromosomal DNA element that is replicated and transmitted to daughter cells

The common features of these changes are that they affect the genetic complement of the cells, they are heritable and they result from the introduction of foreign DNA into an embryo or a cell.

Procedures that do not involve transgenesis

Certain processes or procedures are sometimes erroneously thought to embody an element of transgenesis or genetic engineering. Those listed below do not *of themselves* involve transgenesis, although any of them might be incorporated in a protocol designed to generate transgenic animals:

- surrogacy
- artificial insemination
- *in vitro* fertilisation
- transfer of somatic cell nuclei into *oocytes* (unfertilised eggs), as in the sheep Dolly
- gene therapy

The foreign DNA is genetically manipulated

The **genome** of an organism is essentially all of its DNA[1], which is its total complement of genes, often plus additional 'non-gene' DNA. The genomes of all free-living organisms are large; that of a simple bacterium, for example, contains about 6000 genes, that of a mammal perhaps 100 000. Bacterial genes are small and compact while many mammalian genes are very large indeed. In addition, the genomes of mammals (as of other multicellular organisms) contain significant amounts of 'non-gene' DNA. As a result of these factors, the amount of DNA in a mammalian genome is about 500 times more than in a bacterial genome, rather than 16 times more as suggested by the gene count.

A particular gene, whether bacterial or mammalian, can be closely studied and manipulated only if it can be isolated away from the rest of the genome. In practice this is normally achieved by **molecular cloning**. Molecular cloning most frequently utilises the DNA of a bacterial **vector**. Vectors are 'parasites' of two types that can proliferate within but not outside cells. Some are bacterial viruses whose DNA can escape from cells packaged inside a protein coat, while others are circular (in the sense of a ring) DNA molecules called **plasmids** which cannot escape in this way. A crucial feature of a suitable vector is that its DNA can readily be separated from that of the host and purified to be free of other macromolecules, especially proteins. Most vectors have a very small DNA genome (5 to 50 genes; 3000 to 50 000 nucleotide pairs) which can be manipulated *in vitro* with relative ease. The end result of a molecular cloning exercise is usually the isolation of such a vector into which a foreign gene has been incorporated.

The first molecular clones of this sort occurred naturally as a result of rare spontaneous deviations from the normal functioning of bacterial viruses and plasmids. These deviant events led to the integration of a segment of the bacterial host genome into the genome of the vector. Examples are the F'*lac* derivatives of the plasmid sex factor F and the λ-*gal* derivatives of bacteriophage λ (a virus), which carry the host β-galactosidase (*lac*) and galactose (*gal*) genes respectively. The rare variant forms could be identified and isolated because they can supply to a bacterial host that lacks it the function of the gene that they carry. They were used, for example, in quantifying the expression of the bacterial gene and to study the affinity between the DNA and protein regulators of transcription. However, real progress in molecular cloning awaited the discovery and isolation of a number of enzymes that are essential for the manipulation of DNA *in vitro*, and the development of techniques for screening so-called DNA **libraries**.

A DNA library is a mixed set of molecular clones derived from one particular source, such as the entire genome of a mammal, produced by cutting the genomic DNA into small fragments in a controlled way and incorporating them at random into a large number of vector genomes, one fragment into each. The products of the proliferation of each vector–fragment combi-

[1] Organelles such as mitochondria contain small amounts of DNA, not classified as genomic.

nation constitute a molecular clone. Given the generation of a sufficient number of molecular clones (usually more than 10^6), all of the DNA sequences present in the mammalian genome will be represented in the library, some more frequently than others because of the random nature of the integration process. Using a suitable screening method a particular class of molecular clones, for example those that contain a particular gene, can be identified, isolated and propagated separately from the other clones in the library, providing an unlimited supply of the DNA fragments that they carry. These can then be analysed in detail and manipulated in a variety of ways to prepare a DNA fragment for use as a transgene.

Generation of transgenic mammals

Two completely different approaches to the generation of transgenic mammals have proved highly successful. The first to be developed was the direct microinjection of DNA into the nucleus of a 1-cell embryo. A few years later it became possible to propagate early embryonic mouse cells *in vitro* and to use them subsequently to generate lines of mice. An alteration to the genome of these cells *in vitro* is transformed into an alteration to the genome of the mice derived from them.

Direct injection into embryo pronuclei (see Figure 7.1)

In the mouse, mating usually occurs during the night, and next morning **1-cell embryos** (fertilised eggs) can be recovered from the oviducts. The 1-cell embryos contain two nuclei (**pronuclei**), one derived from the egg and the other from the sperm cell that fertilised it. Each of the pronuclei contains a single set of chromosomes rather than the two sets present in nuclei of normal cells. With the help of micromanipulators, foreign DNA can be injected into one of the pronuclei through an extremely fine glass needle. During overnight incubation the embryos divide and next day can be introduced into the oviducts of a surrogate dam. Many of the embryos develop normally and are born alive. This approach was first shown to be feasible in 1980 (Gordon *et al.* 1980) and hereditary transmission of transgenes was demonstrated during the following year (Gordon and Ruddle 1981; Costantini and Lacy 1981).

Depending on circumstances, between 5 and 25% of injected embryos turn out to have incorporated (**integrated**) some of the foreign DNA into their chromosomes. These are the **transgenic founders**. In most cases integration occurs at a single site on one chromosome in a given embryo, but integration at two or three sites also occurs more rarely. Generally, multiple copies of the transgene become integrated at each site (Figure 8.3). The term **transgene array** is used to refer collectively to the copies of a transgene incorporated at a single site.

About 75% of transgenic founder mice are **mosaic**[2], which is to say that they contain both normal cells and cells that carry the transgene. This is generally taken to reflect the fact that the transgene became integrated into part of a chromosome that had already replicated at least once. The remaining 25% of founders are uniformly transgenic animals.

Random insertion of foreign DNA into any genome is liable to inactivate a gene and, as exemplified by the mouse, mammals are no exception (Chapter 8). Interest has focused mainly on genes that cause pre-natal death when both copies are inactivated, primarily with a view to isolating and identifying the affected gene. The approach has been less successful than was hoped, seemingly because transgene integration is sometimes accompanied by alterations in the gross structure of the chromosome in the vicinity of the integration site. These alterations naturally tend to correlate with gene inactivation events, and for technical reasons they also make it much more difficult to isolate the gene responsible for pre-natal death.

In most cases the first objective is the expression of the transgene, perhaps followed by an investigation of any developmental or physiological perturbations that ensue. Less obvious approaches have also been adopted (Chapter 10). For example, by joining DNA fragments together it is often possible to direct the synthesis of a protein to a particular tissue or to have it synthesised at a particular time. With this approach the protein can be native to any source, even a virus or a bacterium. In an extension of the same concept, the synthesis of a lethal protein can be directed to a particular cell type which is killed after synthesis of the protein is initiated. For example, mice that lack growth hormone were produced by inducing the death of the small set of highly specialised cells that synthesise it. Another approach is to gain experimental control of the expression of a gene by arranging things in such a way that the synthesis of its protein product occurs only when some unnatural but innocuous substance is administered to the transgenic animal.

Applications

Transgenic mice produced by DNA injection have been employed in a vast range of studies. For instance the method allows evaluation of the effects of gene mutations thought to be implicated in the inception of cancer. It has also contributed greatly to our understanding of the immune defence system. Indeed there are few areas of biology which have not been illuminated by the study of transgenic mice.

Similar procedures are also used to generate transgenic rats, rabbits, farm animals, etc. with variations dictated by the differences between the species in details of husbandry, reproduction and early development. Much effort is presently being applied to producing transgenic animals with commercial potential. One very obvious target is milk. Secreted abundantly by

[2] A mosaic animal is one made up of at least two types of cell, differing in their genomes, descended from a single 1-cell embryo.

healthy animals in a semi-sterile state and conveniently collected with established technology, milk is an obvious vehicle for the production of pharmaceutical proteins. Alterations to the food properties of milk also have commercial potential. Changes in the quality of meat and the efficiency of feed conversion are in prospect.

Introduction of DNA into embryonic stem cells (see Figure 7.2)

The alternative method of generating transgenic animals is so far confined to mice, and was made possible by a notable advance in the study of mouse embryogenesis, namely the isolation of cells called **embryonic stem cells** or **ES cells**, that proliferate indefinitely in culture (Martin 1981; Evans and Kaufman 1981) and are able to **colonise** early embryos (Bradley *et al.* 1984). This is to say that they multiply alongside the cells of the embryo and participate in developmental processes, with the result that all parts of the mature animal contain descendants both of cells that originated in the embryo and of those that were introduced from the cell culture.

Cells that proliferate in culture by cell division, whether bacterial or animal cells, can be said to proliferate **clonally**, meaning that the precise genome of any cell is replicated in its descendants through successive divisions. One division produces two identical cells, two divisions produce four and n divisions produce 2^n identical cells. The division products of each initial cell can be kept together, by adherence to a surface or by isolating non-adherent cells in different containers, e.g. in the 96 separate wells of multiwell culture dishes. Under these circumstances the clone descended from each cell forms a **colony** of identical cells. This makes **colony selection** possible. Selection is generally applied by modifying the culture medium in some way, leading to the death of most of the cells and the survival of rare colonies, each derived from one of a minority of cells that display a particular survival property.

Molecularly cloned foreign DNA is usually introduced into ES cells by subjecting them briefly to an electrical potential that partially disrupts the cell membrane, allowing some exchange between the contents of the cell and the surrounding liquid, which contains the DNA. The outcome depends crucially upon the structure of this DNA, which is designed so as to achieve a specific purpose. Frequently this is the inactivation of a particular gene already present in the mouse genome (a **resident gene**) by replacing all or part of it with the foreign DNA. The foreign DNA must carry a selective **marker gene**, preferably one configured in such a way that it will be expressed only when substituted into the resident gene. The marker gene confers on the cell the ability to survive colony selection. The selective conditions kill those cells that do not express it and as a result only the cells that have achieved the replacement of resident with foreign DNA survive, each forming a colony. The DNA must also carry two regions of identity with the resident gene or its immediate chromosomal environment (Figure 9.1). These lie on either side of the marker gene, arranged in such a way as to allow them to pair with the identical chromosomal regions[3]. Cellular mechanisms bring about the substitution. As a rule only one of the two resident **alleles** will be substituted in this way[4]. The inactivated gene is a **null allele**, or **gene knockout**, meaning that its normal product is absent or non-functional, and the overall procedure is commonly referred to as **gene targeting**. In practice, the colonies that survive selection generally include a substantial majority that have integrated the foreign DNA into their chromosome complement in the 'wrong' way. The minority of surviving colonies that have integrated it in the desired way are identified by analysis of DNA samples. These necessarily still carry the marker gene that was substituted for a region of the gene, and this might in some cases affect the phenotype of the animals that will subsequently be produced. A method is available to eliminate the marker gene, if this is desirable. To make this possible the transgene is constructed in a special way and an extra round of cell treatment and selection is required.

The next step employs early multicellular embryos that differ genetically from the ES cells, including a difference in hair colour genes. A few cells from an ES cell colony that has incorporated the foreign DNA as intended are injected into each embryo, and the embryos are introduced into surrogate dams (Gossler *et al.* 1986). Embryos successfully colonised by the ES cells develop into **chimeric**[5] pups distinguished by having sizeable patches of hair of two colours, corresponding to cells derived from the embryo on the one hand and from the ES cells on the other. If chimerism extends to the germ line these animals will produce gametes derived from the embryo and gametes from the ES cells. About half of the latter will carry the altered allele and half the normal allele. Some of the offspring of a mating between this chimeric mouse and a normal mouse will therefore carry one set of ES cell chromosomes and one set from the normal mouse, and usually half of these pups will carry the altered gene on the ES cell chromosome. Additional matings between such mice will produce some **homozygous null** offspring, i.e. offspring carrying two inactivated copies of the gene and no active copies.

Mice like this that completely lack the expression of a gene are intended to allow its primary function to be identified and to explore the ramifications of its effects *in vivo*. If the homozygous null condition is lethal, the first abnormalities to appear give a pointer to the cause of death. Other homozygous null mice are viable but functionally abnormal. However, probably the most significant general result to have emerged is that in a surprisingly high number of cases the homozygous null is normal

[3] The pairing of identical regions of chromosomes is a normal process, occurring invariably for example during the meiotic divisions that precede the formation of mature germ cells..

[4] Most of the chromosomes are present in two copies, and the two copies of a given gene are termed **alleles**. Generally only one X chromosome will be present since the ES cells commonly employed have the male XY chromosome complement.

[5] A chimera is an animal made up of at least two types of cells with different genomes and descended from cells differing in origin (*cf.* mosaic, footnote 2).

or nearly so. The prevalent view is that this is due to redundancy, a given product or function being supplied by more than one protein or biochemical pathway, and in some cases there is evidence to support this view.

More subtle approaches to gene action can be brought into play. For example, an indicator (**reporter**) function is commonly substituted for the normal function of a gene while leaving in place the regulatory DNA regions required for its correct expression. The bacterial *lacZ* gene is often employed in this way. The expectation is that the gene product, an enzyme, will be synthesised only in those cells in which the normal product of the gene is synthesised, and at the appropriate times. Cells that express the *lacZ* reporter can be identified by incubating tissue sections with a chromogenic substrate. Another reporter causes cells that express it to fluoresce under ultraviolet light. This procedure can be carried out with mice that carry one modified and one normal copy of the gene, overcoming most potential viability problems. To take one example, a number of genes have been shown to be transiently active in specific groups of cells during development, and their expression correlated with developmental defects that arise in homozygous null embryos.

Applications

The ES cell route has usually been employed to inactivate a gene, alter it or replace its protein coding region with a reporter, and less often to substitute its coding region with a different functional gene, although this is likely to be done more and more frequently. The main applications of ES cell transgenic mice are to medicine and 'pure science'. The medical applications include improved understanding of all aspects of the healthy animal and of therapies. The scientific applications, not really so pure because they invariably impinge upon both infectious and hereditary diseases, as well as disease models for the development of medicine, again range widely over biochemistry and physiology, but are particularly effective in the areas of mammalian development, neurobiology, learning and memory.

Nuclear transfer (see Figure 7.3)

The 'cloning' technique pioneered at the Roslin Institute in Scotland and the laboratories of Granada Inc. in Texas culminated in the generation of a lamb ('Dolly') from an enucleated egg into which the nucleus from a cultured **somatic** cell[6] of a mature sheep had been introduced. Comparison of the sheep and the mouse gives an indication of how frustrating the subtle differences in the developmental pathways of different mammals can be. On the one hand ES cells that colonise the germ line have until now been

[6] Somatic cells are all those cells that are not germ line cells, and make up the body (*soma*) of the animal. The cell employed in making Dolly was derived from mammary tissue. The somatic and germ line cell lineages diverge at an early stage in development. Somatic cells are said to be diploid, i.e. to contain two sets of chromosomes, one from each parent, although if they are proliferating they contain the equivalent of four sets in the period immediately prior to cell division.

derived only from certain inbred lines of mice; on the other, successful transfer of nuclei from cultured somatic cells has not been reported in the mouse as of December 1998. This development may not be far away, however, since transfer of nuclei from differentiated (but not cultured) mouse cells has very recently been successful.

The method is conceptually simple. First the nucleus is removed manually from an unfertilised oocyte. Next a non-dividing somatic cell is placed in contact with the oocyte and the two are fused by applying an electrical pulse, which also activates the egg, mimicking the process of natural fertilisation. Although the cytoplasm of the somatic cell mingles with that of the egg, the cell is 50–100 times smaller in volume and the egg macromolecules must largely dictate immediate developments. If the somatic cell is at an inappropriate stage in its division cycle the incoming nucleus becomes disorganised and development of the embryo comes to a halt. Success depends on the use of a somatic cell brought to a non-dividing state in culture by depriving it of external stimuli that provoke growth.

The result of this procedure is an activated oocyte with two chromosome sets, equivalent to the sum of the two nuclei of the 1-cell embryo, each of which contains a single set of chromosomes. The cytoplasm of a normal oocyte is primed with proteins and RNA molecules that are required during the very early stages of development. The somatic nucleus contains a full complement of genes which are now reprogrammed to take over the developmental program in the same way as the genes of a normal embryo. This is perhaps not surprising. The chromosomes of sperm are very highly condensed and their DNA is largely bound up with a few unique proteins not found in any other cell type. Upon entry of the sperm nucleus into the oocyte cytoplasm, these special proteins are replaced by the normal ones, in a process dictated by molecules already present in the egg. That a similar, and possibly less radical, reprogramming of a somatic nucleus should occur is thus not unexpected.

Applications

Nuclear transfer has applications outside the field of transgenesis, such as the propagation of an animal with a particularly desirable complement of genes, but we will concentrate here on transgenic applications. These are two in number:

Herds of transgenic farm animals

Unlike inbred lines of laboratory mice, sheep and cows are not uniform and variation in the genetic makeup of individuals produced from two germ cells, even when they come from animals of the same breed, is likely to make for variation in expression of the transgene (see below). For this reason the propagation of transgenic founders by breeding is not without risk. Nuclear transfer would make possible the production of large numbers of more or less identical copies of any sheep or cow that has a proven high performance, by the transfer of its somatic nuclei into oocytes.

However the application of the procedure to livestock is expensive, and the number of animals produced per treated oocyte is not yet sufficiently high to make this an economically attractive route, although this situation may change rapidly.

An alternative to ES cells

The power of the ES cell transgenic route is mainly due to the opportunity it offers to select those cells in which foreign DNA has become integrated in a predetermined way. Nuclear transfer promises the same advantages in those animals (i.e. all but mice) from which ES cells that colonise the germ line have not yet been derived. The method employed to bring about DNA integration would be the same as described above for ES cells, after which nuclear transfer into oocytes would be carried out with cells from the selected colonies. Cultured foetal cells, which give higher success rates than cells from an adult, would presumably be employed. This procedure may already be economically sound. One obvious application is to modify natural products. For instance, the protein coding regions of bovine milk protein genes could be replaced with equivalent human gene regions to 'humanise' cow milk. Another is to target foreign genes to chromosomal sites that are known to permit high-level expression.

Transgene expression (see Chapter 10)

Quite frequently the expected expression of a transgene is not realised in practice. Expression may be lower than anticipated or not confined to the expected tissues or may occur only in unexpected tissues. There may be high variation in expression between transgenic lines, or between animals within a line or from one generation to the next. Such effects are at the least a great inconvenience, and at worst they can invalidate experiments or make prohibitive the cost of commercialisation. Some of the contributory causes are as follows:

- a lack of necessary DNA elements, which may lie outside the believed boundaries of the gene or within **introns**[7] that have been discarded
- effects of genetic background, due to a number of genes, each with a small effect, that differ from one animal to another, for example between different strains of laboratory mice
- chromosomal position effects, i.e. effects due in some way to the chromosome immediately surrounding the transgene

[7] Most mammalian genes consist of alternating exons and introns. The RNA produced with such a gene as the template is processed in the nucleus to remove the introns, joining up the exons end-to-end, before export to the cytoplasm. Thus the introns play no part when the RNA comes to be a template in its turn, this time for the synthesis of a protein.

- inactivation by **DNA methylation**[8], which may relate to the number of transgene copies present in the integrated array.

Missing elements

To a large extent this source of misexpression is trivial, due simply to inadequate understanding of the gene that is being manipulated, and in particular of the location of the regulatory DNA regions required for its correct expression, which usually lie outside the coding region.

Some genes are very large and therefore difficult to manipulate *in vitro*. In these and other cases it has been a common practice to substitute the **cDNA** (a DNA copied from the cytoplasmic messenger RNA) for most of the coding region, thus eliminating most or all of the introns which make up much of the length of large genes (see footnote 7). This can have disastrous effects upon expression. In some cases the basis of the effect is understood, at least in a broad way. In particular intron 1 of some genes, or occasionally another intron, contains a sequence that co-operates with the promoter to initiate RNA synthesis. Removal of the relevant intron from these genes therefore inevitably results in defective expression. In other genes, a so-called silencer element that normally restricts expression to one particular developmental subline is located in an intron. Introns have another effect: the introduction of a so-called **generic** intron into an intronless gene increases expression in some cases at least. This phenomenon has not been analysed in any detail *in vivo*. Experiments with cultured cells suggest that the stimulation relates to the processing of the RNA that follows transcription.

Genetic background

The genetic background effectively signifies all the genes apart from the gene or allele under study. Potential problems due to genetic background are amply documented in respect of both normal genes and transgenes. The genomes of different inbred strains of mouse differ significantly in perhaps 2000 genes. In view of the enormously complex interactions between gene products it is not surprising that an alteration to a gene sometimes produces different effects in different genetic backgrounds. This is not a problem as long as the effect is recognised or when an inbred strain is employed, for example when a normal gene and a mutant allele are viewed against the same background. The problem arises in strains of mixed ancestry, when different individuals differ in genetic background in unknown ways. It touches both routes by which transgenic mice are generated, and in both cases it has arisen essentially for technical reasons (see Chapter 10).

[8] DNA methylation in general is a dynamic process of addition of methyl (CH_3) residues to the nucleotide bases in the DNA chains. In mammals, of the four bases only a subset of the cytosine bases become methylated: those that are neighboured by a guanine base on one side (but not if the guanine is on the other side). As a general but not invariable rule, methylation of certain of these cytosine bases in a given gene correlates with transcriptional inactivity, and conversely demethylation is the rule when and where the gene is acting as a template for RNA synthesis.

Chromosomal position effect

The observation that leads to the concept of position effect derives from comparisons between transgenic lines that carry the same transgene. Recall that a transgenic line is the set of animals descended from a single founder animal. Gene integration into the chromosomes is random following pronuclear microinjection and, prior to integration, foreign DNA molecules are assembled into long linear **concatemers**[9] by processes that are also random (Figure 8.3). Consequently the foreign DNA integrated in each founder is unique with respect both to its chromosomal position and to the number and configuration of transgenes in the integrated array. Different mouse lines carrying the same transgene are found to differ widely in the level of transgene expression, in some cases up to 1000-fold. Furthermore, when the expression level is compared with the number of copies of the transgene present in the different lines, a positive correlation is not found (**copy-number independence**). Where the foreign gene exhibits a tissue-specific pattern of expression *in situ*, or a characteristic temporal pattern during development, this is often not replicated in the transgenic lines, or at least in some of them. Different lines may show quite different expression patterns and the transgene may be expressed in tissues in which it is normally silent (**ectopic expression**). When added to some transgenes, sequences called **matrix attachment regions** seem to insulate them against chromosomal position effects.

A significant minority of transgenes do not show these effects. Expression of these transgenes is substantially the same as that of the same genes *in situ*, more or less irrespective of their integration sites. They also exhibit **copy-number dependence**, i.e. the level of expression correlates positively with the number of transgene copies present in the array. Analysis of a few of these genes has identified regions that are responsible for their correct expression and for overriding the influence of chromosomal position.

Methylation

In the normal course of events much of mammalian DNA methylation relates to gene expression in a cause and effect way (more methylation, less expression) although there is no consensus as to whether methylation is ever the *primary* cause of repression. Transgenes appear to become methylated when they might be expected not to be, leading to the cessation of transgene expression. It is possible that the tandem arrays in which transgenes are commonly arranged attract methylation, and there is evidence of a positive correlation between the number of copies of a given transgene in an array and the level to which the array becomes methylated. This could of course provide one explanation for copy-number independence.

[9] The structures formed are many times longer than the DNA molecules that are introduced. Mostly these appear as though joined end-to-end, but partial molecules are also incorporated. For convenience we will call them **concatemers** prior to integration into a chromosome, and **arrays** following integration.

Some conclusions

Problems relating to their expression crop up much more frequently when transgenes are introduced by random insertion than when the route of integration is homologous recombination (gene targeting). In the latter case the question of chromosomal position effect does not arise, DNA elements outside the coding regions are less likely to be missing, multiple copies of the transgene become integrated only infrequently and, perhaps as a consequence, DNA methylation is less of a problem.

The microinjection route has generally been employed to introduce a new functional gene into the genome by random integration, in effect to add an extra gene to an already full complement. In the great majority of cases, the objective requires the gene to be expressed, in some cases at a high level. It is therefore unfortunate that many transgenes are susceptible to all the negative complications listed above. However, the problems can be reduced by careful practice, and indeed we can look forward to a time in the near future when all of them will be under control.

In contrast, the ES cell route has usually been employed to inactivate a resident gene, or alter it in some way. The principal problems arise from the use of hybrid mouse strains. One way to confirm that the phenotype is due to the mutation is **transgene rescue**. If the phenotype is lost when a normal gene is reintroduced into the genome (preferably by generating several independent lines) we can be confident that it resulted from the inactivation of the resident gene.

One of the main obstacles to transgenic research, and especially to its commercial application, has been the lack of ES cells derived from animals other than mice, resulting in the unavailability of the more controlled ES cell route to transgenesis. Thus, the pharmaceutical industry is so far denied transgenic rats engineered in this way for toxicology testing, and the most direct way of substituting a milk protein with an altered one is presently barred, to cite obvious examples. Two ways forward present themselves. First, ES cells may become available in other animals at any time. Secondly, and probably more immediately, the nuclear transfer procedure developed in sheep and cows may be combined with DNA integration and clonal selection and also adapted for use with other animals.

Ethical issues

The view of the present author is that transgenic animals and plants raise practical issues, but not ethical ones. Nature, after all, is prolific in experimentation. There may be justified public concern about potential damage to the environment through transfer of genetic material from transgenic to wild organisms, but that is a practical issue. Indeed, one's ethics are a question of personal belief, and there are as many valid ethical positions as there are individuals.

References

Bradley, A., Evans, M., Kaufman, M.H., and Robertson, E. (1984) Formation of germ-line chimaeras from embryo-derived teratocarcinoma cell lines. *Nature*, **309**, 255–256.

Costantini, F. and Lacy, E. (1981) Introduction of a rabbit beta-globin gene into the mouse germ line. *Nature*, **294**, 92–94.

Evans, M.J. and Kaufman, M.H. (1981) Establishment in culture of pluripotential cells from mouse embryos. *Nature*, **292**, 154–156.

Gordon, J.W., Scangos, G.A., Plotkin, D.J., Barbosa, J.A., and Ruddle, F.H. (1980) Genetic transformation of mouse embryos by microinjection of purified DNA. *Proceedings of the National Academy of Sciences of the United States of America*, **77**, 7380–7384.

Gordon, J.W. and Ruddle, F.H. (1981) Integration and stable germ line transmission of genes injected into mouse pronuclei. *Science*, **214**, 1244–1246.

Gossler, A., Doetschman, T., Korn, R., Serfling, E., and Kemler, R. (1986) Transgenesis by means of blastocyst-derived embryonic stem cell lines. *Proceedings of the National Academy of Sciences of the United States of America*, **83**, 9065–9069.

Martin, G.R. (1981) Isolation of a pluripotent cell line from early mouse embryos cultured in medium conditioned by teratocarcinoma stem cells. *Proceedings of the National Academy of Sciences of the United States of America*, **78**, 7634–7638.

CHAPTER TWO

Why employ transgenic animals?

The methods that must be employed to generate transgenic animals are costly and laborious and the time span of experiments is very long. What great advantages do transgenic animals bring to biological research that might justify their use?

The most important test of many concepts is to introduce a gene, usually a modified gene, into cells and to study its expression. A simple example is the frequently asked question as to which neighbouring DNA regions regulate the expression of a gene, and how they do this. How is tissue specificity of transcription determined, which regions are responsible for hormonal responses, and so on? Many tests have been carried out by transfecting the unmodified gene and structurally modified derivatives of it into cultured cells, and much useful information has been gained in this way. There are dangers in this approach however, and it has yielded many false results. One reason to believe this is that different results are frequently obtained with different cell lines (see for example Ott *et al.* 1984; Vassar *et al.* 1989). Both cannot be correct, but which is wrong? Or are both wrong, perhaps, and the correct result still unknown? And if we continue to work only with cultured cells, how will we know when we have arrived at the correct result?

Properties of many cell types change rapidly when cultured

Liver parenchyma cells, alternatively called hepatocytes, monitor the composition of blood and provide the primary barrier to toxins; they also synthesise many of the plasma proteins, secreting abundant quantities of several of these. This makes the activity of the corresponding genes easy to detect at the different levels of expression: rate of synthesis of RNA, amount of mRNA in the cytoplasm, amount of protein synthesised and secreted.

Note that the expression of these genes is restricted to one or a few tissues. The discussion that follows takes hepatocytes as an example and is mainly relevant to tissue-specific gene expression.

The hepatocytes make up about 90% of the liver cell mass, so that a culture of dissociated liver cells is primarily a hepatocyte culture. It is found that within about one day of conventional culture, the activity of most of the tissue-specific genes has effectively come to a halt (Friedman *et al.* 1986; Padgham *et al.* 1993; Paquereau and Le Cam 1992). Not only that, but some years ago it was shown that when liver cells were dissociated *in situ* by perfusion with collagenase, their activity fell off dramatically and could be partially restored by reversing the dissociation (Clayton *et al.* 1985). This is a strong indication that normal activity in these cells is dependent on contact with the intercellular matrix, and indeed it has been found that cultivation of the cells on matrices of various sorts slows their functional decay (Enat *et al.* 1984; Isom *et al.* 1985; Fujita *et al.* 1986; Spray *et al.* 1987; Ben Zeev *et al.* 1988; Guzelian *et al.* 1988). Seldom, however, do they maintain activities near to *in vivo* levels.

Without labouring the point, it should be remembered that there are many differences in growth conditions of cells *in vivo* and *in vitro*. Cells of solid tissues are fed by arterioles carrying plasma and red blood cells that closely control every aspect of their environment from CO_2 tension and pH to endocrine ligands. Furthermore, they are incorporated into well-ordered three-dimensional structures where they often make contact with cells of other types and where they are subject to paracrine stimulation. Reproducing all of this *in vitro* would be something of a task, to say the least, and even then, could we ever be sure that we had got everything right?

So far we have been discussing primary cell cultures, one step away from *in vivo* existence and not usually maintained through many subsequent cell divisions. When DNA is introduced into such cells, by whatever means, generally only a small, or even tiny proportion of the cells take it up. In addition to the problems discussed above there is the additional one that the cells are likely to be in the process of dying. This is obviously unsatisfactory from the viewpoint of the molecular biologist. Consequently most DNA transfection experiments have employed established cell lines, composed of cells that grow continuously in culture at a more or less steady rate. In fact most studies have employed cell lines that have become immortal, either spontaneously or after deliberate manipulation by mutagen treatment or virus infection. Such cells offer obvious practical advantages, not least that fresh cultures can be established from single cells, allowing special characteristics to be selected for: stable transfection with foreign DNA for instance.

The paradigms were bacteria and yeasts, which contributed so much to the foundation and development of molecular biology. However bacteria and yeasts are free-living, while mammalian cells are not. When mammalian cells are induced to proliferate as if they are bacteria or yeasts they undergo dramatic changes, which have been extensively documented.

Very often the chromosome number changes, generally increasing. Cultures quickly become heterogeneous, stabilising at what is called their mean chromosome number. The process is a dynamic one, chromosomes being lost and gained around the mean value. The spectrum of available transcription factors, upon which the regulation of cellular processes vitally depends, changes dramatically (Xanthopoulos *et al.* 1989; Pellicer *et al.* 1980; DiPersio *et al.* 1991). Some components of signal transduction pathways are lost and others gained, different isoforms become available, and so on.

One attempt to address the problems was to immortalise cells conditionally. This can be done by introducing the gene that codes for a temperature-sensitive mutant form of the large T antigen (Tag) of simian virus 40 (Georgoff *et al.* 1984; Isom and Georgoff 1984; Woodworth *et al.* 1986). Normally non-proliferative cells will proliferate steadily if they are producing Tag, activated by a mechanism that is well understood. During this time, DNA can be introduced into the cells and stable transformants can be isolated, if required. At elevated temperature the temperature-sensitive Tag is inactivated and proliferation stops. At this point it is hoped that the cells will return to their 'normal' differentiated state. Supposing that this does happen, the result will be a population of cells equivalent to a primary culture. The same basic problems of matrix attachment, cell–cell contacts and paracrine and humoral influences remain if our objective is to investigate the *in vivo* characteristics of the cells.

These observations apply to cells isolated from solid tissues. With cell types such as fibroblasts and lymphocytes there is much less cause for concern. Also, there is no intention to question all work with cultured cells; much of our understanding of cellular biochemistry stems from this source. Rather it is to emphasise the gulf that exists between cells *in vivo* and their cultured derivatives.

Some genes behave differently when transfected into cells and as mammalian transgenes

We have seen that the expression of *resident* genes can change when cells are explanted. To make the case for transgenic animals, we also have to show that transgenes are correctly expressed *in vivo*. This is a complicated issue, which will be dealt with in detail in Chapter 10. Meanwhile suffice it to say that the correct expression of transgenes has been demonstrated in many instances. At this point, however, it is germane to note cases in which transgenes are expressed differently in cultured cells and *in vivo*. The first to come to light were the α-fetoprotein and serum albumin genes. In the mouse these genes, which are structurally related and closely linked, are coordinately expressed in the embryonic yolk sac and the foetal liver and gut. Expression of the serum albumin gene continues in the liver in adult life, while the α-fetoprotein gene becomes silent shortly after birth. A relatively small promoter region from each of these genes was functional in hepatoma cells but not in fibroblasts or in hepatoma cells that did not express the corresponding resident gene, and this was taken to indicate tissue-specific expression (Ott *et al.* 1984; Widen and Papaconstantinou 1986). The same constructs were non-functional as transgenes *in vivo*, but both became active when coupled with nearby enhancer elements (Hammer *et al.* 1987; Pinkert *et al.* 1987) (for a description of enhancer elements, see Chapter 10).

It transpired that the serum albumin enhancer element is active (i.e. stimulates expression) in some but not all liver-derived cell lines. The α-fetoprotein gene, on the other hand, has three enhancer elements with equivalent and non-additive effects on expression in transfected liver-derived cell lines but additive effects in foetal liver *in vivo*. Interestingly, the three enhancer elements affect expression differently in the three tissues (Hammer *et al.* 1987).

There are potential problems inherent in making comparisons amongst the dozen or so studies summarised above. The α-fetoprotein and serum albumin genes were altered in various ways; the cultured cells were assayed during the transient phase of expression following transfection, i.e. while the foreign DNA was extrachromosomal; cell lines from at least three species were employed in different studies as well as primary hepatocyte cultures. Yet these are the realities of most work done with cultured cells. Overall, it is quite impossible to deduce the true *in vivo* characteristics of the two genes from the results obtained with the cell cultures.

Perhaps the most dramatic difference so far established between cultured cells and transgenic animals is a substantial stimulatory effect of introns on expression when genes are integrated into the mouse genome but not when the same genes are stably integrated in cultured cells (Brinster *et al.* 1988; Aronow *et al.* 1995). This is almost certainly due to the absence from the cultured cells of tissue-specific proteins that enhance expression when bound to specific sequences within the introns.

The literature contains quite a few examples of similar differences between transfected cells and transgenic mice (see Table 2.1), with the common theme that expression *in vivo* generally requires more DNA elements to be present than does expression in cultured cells. The entire spectrum of effects can be rationalised in terms of gains and losses of *trans*-acting transcription factors following explantation, during propagation and as a consequence of immortalisation; enhancers and matrix attachment elements can only function correctly in the presence of appropriate *trans*-acting factors (see for example Haynes *et al.* 1996). At first sight this is paradoxical in relation to the loss of tissue-specific functions that occurs when cells are explanted. The answer probably lies in the habitual use of selected cell lines that are permissive of gene expression.

Table 2.1 Some examples of genes that show differences in expression in transfected cells and transgenic mice.

Gene	*References*
Rat serum albumin	Pinkert *et al.* 1987
Rat α-fetoprotein	Hammer *et al.* 1987
Rat elastase I	Swift *et al.* 1989
Human keratin K14	Vassar *et al.* 1989
Mouse *N-myc*	Zimmerman *et al.* 1990
Rat collagen $I_{\alpha 1}$	Bogdanovic *et al.* 1994
Mouse hepatocyte nuclear factor 4	Zhong *et al.* 1994

Complex processes, cell–cell and tissue–tissue interactions

The overriding need for transgenic animals in research work is in the analysis of complex processes. Some of these are listed in Table 2.2. The basis of the complexity is not the same in every case. Oncogenesis involves progressive but diverse changes in cells which are largely autonomous. Development, differentiation and immune responses involve complex cell–cell and tissue–tissue interactions. Diseases are pleiotropic, the primary lesion often affecting several cell types and usually having consequential effects on others. Learning and behaviour are whole-organism phenomena, with the added complication that their most important manifestations need to be studied by observing living animals.

Aspects of these can of course be studied by traditional methods. For example, the structure of macromolecules can only be studied by reductionist methods. Again, parameters of interactions between macromolecules are best studied by biochemical methods. At the other extreme, it is obvious that the molecular basis of learning cannot be understood by employing reductionist methods alone.

The utility of transgenic animals in the study of complex processes is by no means restricted to transgenes that are expressed correctly. Advantage can be taken of ectopic expression, provided of course that it is well defined and characterised. It is also worth mentioning at this point that many contributions of transgenic animals to the understanding of complex processes to date have been through the inactivation or more rarely modification of resident genes (Chapter 9).

An example

A simple example of how results obtained *in vitro* can be misleading is the case of the muscle regulatory gene *MyoD*. This gene was initially identified as inducing the synthesis of skeletal muscle-specific proteins when intro-

Table 2.2 Complex processes.

Development	Oncogenesis
Differentiation	Many inherited diseases
Brain function	Many degenerative diseases
Behaviour	Some infectious diseases
Immune responses	Organism–environment interactions
Endocrine pathways	Therapeutics

duced into a variety of cell types in culture (Lassar *et al.* 1989; Weintraub *et al.* 1989). This led readily to the conclusion that *MyoD* is the crucial switch in muscle development. However, when mice were generated in which the gene was inactivated (by gene 'knockout' in ES cells) there was no obvious effect on skeletal musculature (Braun *et al.* 1992). *MyoD* is a member of a family of muscle-specific transcription factors (the *MyoD* family). Another member of the family, *Myf-5,* is similar to *MyoD*, and its inactivation also fails to prevent skeletal muscle development. However, when both genes were inactivated, skeletal muscle failed to develop altogether (Rudnicki *et al.* 1993). It appears that *MyoD* and *Myf-5* are expressed in different skeletal muscle precursor lineages, either of which is competent to generate the full complement of skeletal muscle when the other lineage is compromised. Any consequent impairment of function is not readily recognised. Another member of the same family, the myogenin gene (*Myog*), is essential for the development of skeletal muscle (Nabeshima *et al.* 1993; Hasty *et al.* 1993). It would be reasonable to suppose that the myogenin gene acts upstream of both *MyoD* and *Myf-5.* In fact, it acts downstream of both these genes in the two lineages (Figure 2.1).

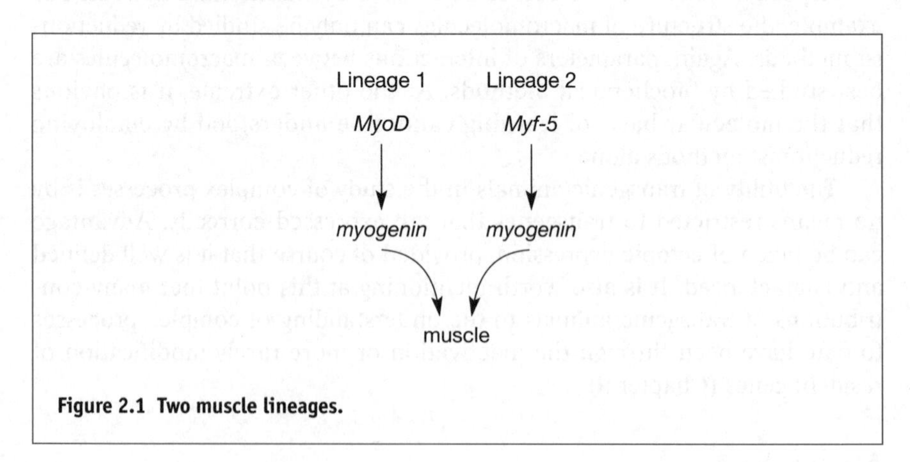

Figure 2.1 Two muscle lineages.

References

Aronow, B.J., Ebert, C.A., Valerius, M.T., Potter, S.S., Wiginton, D.A., Witte, D.P., and Hutton, J.J. (1995) Dissecting a locus control region: facilitation of enhancer function by extended enhancer-flanking sequences. *Molecular & Cellular Biology*, **15**, 1123–1135.

Ben Zeev, A., Robinson, G.S., Bucher, N.L., and Farmer, S.R. (1988) Cell–cell and cell–matrix interactions differentially regulate the expression of hepatic and cytoskeletal genes in primary cultures of rat hepatocytes. *Proceedings of the National Academy of Sciences of the United States of America*, **85**, 2161–2165.

Bogdanovic, Z., Bedalov, A., Krebsbach, P.H., Pavlin, D., Woody, C.O., Clark, S.H., Thomas, H.F., Rowe, D.W., Kream, B.E., and Lichtler, A.C. (1994) Upstream regulatory elements necessary for expression of the rat COL1A1 promoter in transgenic mice. *Journal of Bone & Mineral Research*, **9**, 285–292.

Braun, T., Rudnicki, M.A., Arnold, H.H., and Jaenisch, R. (1992) Targeted inactivation of the muscle regulatory gene *Myf-5* results in abnormal rib development and perinatal death. *Cell*, **71**, 369–382.

Brinster, R.L., Allen, J.M., Behringer, R.R., Gelinas, R.E., and Palmiter, R.D. (1988) Introns increase transcriptional efficiency in transgenic mice. *Proceedings of the National Academy of Sciences of the United States of America*, **85**, 836–840.

Clayton, D.F., Harrelson, A.L., and Darnell, J.E. (1985) Dependence of liver-specific transcription on tissue organization. *Molecular & Cellular Biology*, **5**, 2623–2632.

DiPersio, C.M., Jackson, D.A., and Zaret, K.S. (1991) The extracellular matrix coordinately modulates liver transcription factors and hepatocyte morphology. *Molecular & Cellular Biology*, **11**, 4405–4414.

Enat, R., Jefferson, D.M., Ruiz Opazo, N., Gatmaitan, Z., Leinwand, L.A., and Reid, L.M. (1984) Hepatocyte proliferation in vitro: its dependence on the use of serum-free hormonally defined medium and substrata of extracellular matrix. *Proceedings of the National Academy of Sciences of the United States of America*, **81**, 1411–1415.

Friedman, J.M., Babiss, L.E., Clayton, D.F., and Darnell, J.E. (1986) Cellular promoters incorporated into the adenovirus genome: cell specificity of albumin and immunoglobulin expression. *Molecular & Cellular Biology*, **6**, 3791–3797.

Fujita, M., Spray, D.C., Choi, H., Saez, J., Jefferson, D.M., Hertzberg, E., Rosenberg, L.C., and Reid, L.M. (1986) Extracellular matrix regulation of cell–cell communication and tissue-specific gene expression in primary liver cultures. *Progress in Clinical Biology Research*, **226**, 333–360.

Georgoff, I., Secott, T., and Isom, H.C. (1984) Effect of simian virus 40 infection on albumin production by hepatocytes cultured in chemically defined medium and plated on collagen and non-collagen attachment surfaces. *Journal of Biological Chemistry*, **259**, 9595–9602.

Guzelian, P.S., Li, D., Schuetz, E.G., Thomas, P., Levin, W., Mode, A., and Gustafsson, J.A. (1988) Sex change in cytochrome P-450 phenotype by growth hormone treatment of adult rat hepatocytes maintained in a culture system on matrigel. *Proceedings of the National Academy of Sciences of the United States of America*, **85**, 9783–9787.

Hammer, R.E., Krumlauf, R., Camper, S.A., Brinster, R.L., and Tilghman, S.M. (1987) Diversity of alpha-fetoprotein gene expression in mice is generated by a combination of separate enhancer elements. *Science*, **235**, 53–58.

Hasty, P., Bradley, A., Morris, J.H., Edmondson, D.G., Venuti, J.M., Olson, E.N., and Klein, W.H. (1993) Muscle deficiency and neonatal death in mice with a targeted mutation in the myogenin gene. *Nature*, **364**, 501–506.

Haynes, T.L., Thomas, M.B., Dusing, M.R., Valerius, M.T., Potter, S.S., and Wiginton, D.A. (1996) An enhancer LEF-1/TCF-1 site is essential for insertion site-independent transgene expression in thymus. *Nucleic Acids Research*, **24**, 5034–5044.

Isom, H.C., Secott, T., Georgoff, I., Woodworth, C., and Mummaw, J. (1985) Maintenance of differentiated rat hepatocytes in primary culture. *Proceedings of the National Academy of Sciences of the United States of America*, **82**, 3252–3256.

Isom, H.C. and Georgoff, I. (1984) Quantitative assay for albumin-producing liver cells after simian virus 40 transformation of rat hepatocytes maintained in chemically defined medium. *Proceedings of the National Academy of Sciences of the United States of America*, **81**, 6378–6382.

Lassar, A.B., Buskin, J.N., Lockshon, D., Davis, R.L., Apone, S., Hauschka, S.D., and Weintraub, H. (1989) MyoD is a sequence-specific DNA binding protein requiring a region of myc homology to bind to the muscle creatine kinase enhancer. *Cell*, **58**, 823–831.

Nabeshima, Y., Hanaoka, K., Hayasaka, M., Esumi, E., Li, S., and Nonaka, I. (1993) Myogenin gene disruption results in perinatal lethality because of severe muscle defect. *Nature*, **364**, 532–535.

Ott, M.O., Sperling, L., Herbomel, P., Yaniv, M., and Weiss, M.C. (1984) Tissue-specific expression is conferred by a sequence from the 5' end of the rat albumin gene. *EMBO Journal*, **3**, 2505–2510.

Padgham, C.R., Boyle, C.C., Wang, X.J., Raleigh, S.M., Wright, M.C., and Paine, A.J. (1993) Alteration of transcription factor mRNAs during the isolation and culture of rat hepatocytes suggests the activation of a proliferative mode underlies their de-differentiation. *Biochemical & Biophysical Research Communications*, **197**, 599–605.

Paquereau, L. and Le Cam, A. (1992) Electroporation-mediated gene transfer into hepatocytes: preservation of a growth hormone response. *Analytical Biochemistry*, **204**, 147–151.

Pellicer, A., Wagner, E.F., el Kareh, A., Dewey, M.J., Reuser, A.J., Silverstein, S., Axel, R., and Mintz, B. (1980) Introduction of a viral thymidine kinase gene and the human beta-globin gene into developmentally multipotential mouse teratocarcinoma cells. *Proceedings of the National Academy of Sciences of the United States of America*, **77**, 2098–2102.

Pinkert, C.A., Ornitz, D.M., Brinster, R.L., and Palmiter, R.D. (1987) An albumin enhancer located 10 kb upstream functions along with its promoter to direct efficient, liver-specific expression in transgenic mice. *Genes & Development*, **1**, 268–276.

Rudnicki, M.A., Schnegelsberg, P.N., Stead, R.H., Braun, T., Arnold, H.H., and Jaenisch, R. (1993) MyoD or Myf-5 is required for the formation of skeletal muscle. *Cell*, **75**, 1351–1359.

Spray, D.C., Fujita, M., Saez, J.C., Choi, H., Watanabe, T., Hertzberg, E., Rosenberg, L.C., and Reid, L.M. (1987) Proteoglycans and glycosaminoglycans induce gap junction synthesis and function in primary liver cultures. *Journal of Cell Biology*, **105**, 541–551.

Swift, G.H., Kruse, F., MacDonald, R.J., and Hammer, R.E. (1989) Differential requirements for cell-specific elastase I enhancer domains in transfected cells and transgenic mice. *Genes & Development*, **3**, 687–696.

Vassar, R., Rosenberg, M., Ross, S., Tyner, A., and Fuchs, E. (1989) Tissue-specific and differentiation-specific expression of a human K14 keratin gene in transgenic mice. *Proceedings of the National Academy of Sciences of the United States of America*, **86**, 1563–1567.

Weintraub, H., Tapscott, S.J., Davis, R.L., Thayer, M.J., Adam, M.A., Lassar, A.B., and Miller, A.D. (1989) Activation of muscle-specific genes in pigment, nerve, fat, liver, and fibroblast cell lines by forced expression of MyoD. *Proceedings of the National Academy of Sciences of the United States of America*, **86**, 5434–5438.

Widen, S.G. and Papaconstantinou, J. (1986) Liver-specific expression of the mouse alpha-fetoprotein gene is mediated by *cis*-acting DNA elements. *Proceedings of the National Academy of Sciences of the United States of America*, **83**, 8196–8200.

Woodworth, C., Secott, T., and Isom, H.C. (1986) Transformation of rat hepatocytes by transfection with simian virus 40 DNA to yield proliferating differentiated cells. *Cancer Research*, **46**, 4018–4026.

Xanthopoulos, K.G., Mirkovitch, J., Decker, T., Kuo, C.F., and Darnell, J.E. (1989) Cell-specific transcriptional control of the mouse DNA-binding protein mC/EBP. *Proceedings of the National Academy of Sciences of the United States of America*, **86**, 4117–4121.

Zhong, W., Mirkovitch, J., and Darnell, J.E., Jr. (1994) Tissue-specific regulation of mouse hepatocyte nuclear factor 4 expression. *Molecular & Cellular Biology*, **14**, 7276–7284.

Zimmerman, K., Legouy, E., Stewart, V., Depinho, R., and Alt, F.W. (1990) Differential regulation of the N-myc gene in transfected cells and transgenic mice. *Molecular & Cellular Biology*, **10**, 2096–2103.

PART TWO

Background

CHAPTER THREE

Mouse reproduction and development

The procedures employed to generate transgenic animals involve intervention at an early stage in development. Fertilised eggs or early embryos are removed from donors, manipulated, usually cultured *in vitro* for a time and finally introduced into a recipient, effectively a surrogate mother, or dam in animal parlance. It is therefore useful to have a broad understanding of reproduction and early development in mammals. Here, as elsewhere in the book, the mouse is presented as the model animal. Although there are considerable differences in detail, particularly in the timing of events, from one mammal to another, the broader picture applies generally to mammals and the principles that emerge are widely applicable.

Meiosis

The common feature linking male and female gametogenesis is that both involve a meiotic division on the pathway to the mature gamete. In other respects gametogenesis is astonishingly different in the two sexes. Like premitotic cells, the premeiotic cell contains four genomes. Meiosis, however, involves two rounds of cell division without further DNA synthesis, so that the grand-daughter cells each contain a single haploid genome. The salient features of meiosis are as follows (see Figure 3.1):

- At the first meiotic division homologous duplicated chromosomes are paired on the spindle. Each duplicated chromosome is said to have two chromatids. During the pairing of the duplicated chromosomes crossing-over occurs between the chromatids.
- Normally sister centromeres pass to the same daughter nucleus at this first division. Segregation is random, in the sense that a given chromatid pair may pass to either daughter. Thus it is as a result of this first

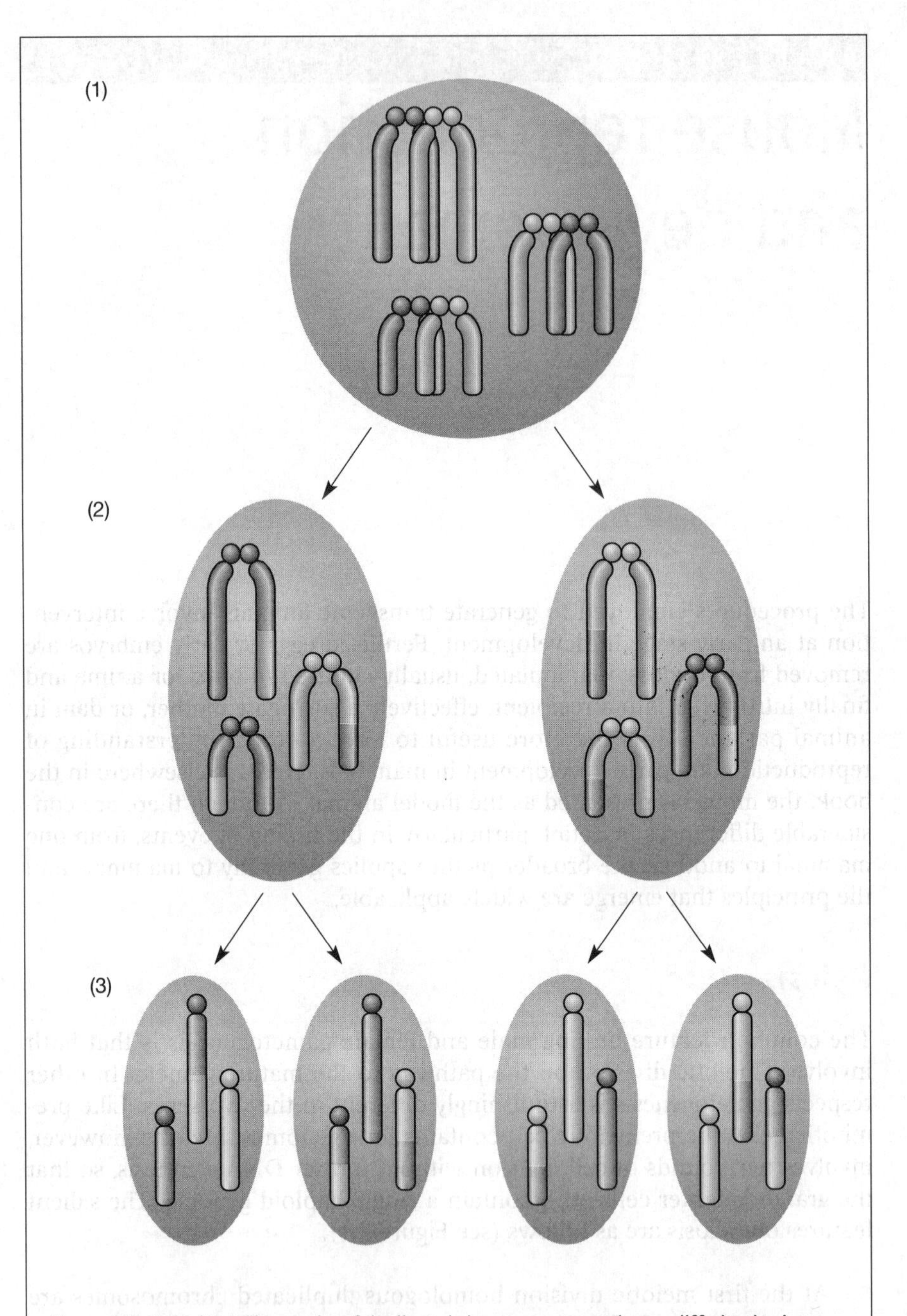

Figure 3.1 Meiosis. 1. Three pairs of duplicated chromosomes are shown, differing in size; paternal and maternal chromatids are black and blue. 2. Segregation of like centromeres to daughter cells at the first meiotic division, following exchange of distal parts of chromatids. In the female cytokinesis is very unequal; the smaller daughter cell becomes the first polar body and the larger the oocyte. 3. Segregation of haploid sets of chromatids to four gametes at the second meiotic division in the male. In the female only the oocyte continues into second meiosis, cytokinesis is again unequal, and the smaller cell becomes the second polar body.

segregation that the greater part of the variety among the gametes is generated. Each daughter cell contains two genomes and two sets of chromatids. Usually one chromatid of each pair will be recombinant (following crossing-over). Because of the co-segregation of sister centromeres, each pair of chromatids will be identical proximal to the crossover and will differ distally. In the female only, one of the daughter cells of each first meiotic division is effectively discarded (Figure 3.3).

- At the second meiotic division the sister centromeres segregate to the two daughter nuclei. Each grand-daughter cell is haploid. Again in the female only, one of the grand-daughter cells is discarded.

Oogenesis

Shortly after birth oocyte numbers are at their maximum. At this stage the tiny oocytes are arrested in the prophase of the first meiotic division and destined to remain so until such time as they are activated prior to ovulation. Each oocyte lies within a primordial follicle defined by a layer of epithelial granulosa cells. During much of adult life small numbers of primordial follicles begin to mature, a process that takes about 15 days (Figure 3.2). During this time the oocyte increases enormously in size and the granulosa cells proliferate and are now called follicle cells. The maturing follicle is defined by a basement membrane lined with a substantial layer of follicle cells enclosing a large fluid-filled space (the antrum). Within this lies the oocyte, now enclosed by a thick glycoprotein layer, the zona pellucida, around which clusters a second thick layer of follicle cells, sometimes called the cumulus. Cellular processes from the adjacent cumulus cells penetrate the zona pellucida and make contact with the oocyte through gap junctions. At this stage substances found in the antral fluid, which include cAMP, inosine/hypoxanthine and possibly also the protein ligand of the c-kit receptor, are required to suppress further maturation of the oocyte. The follicles now become susceptible to stimulation by the pituitary polypeptide gonadotrophins, follicle stimulating hormone (FSH) and luteinising hormone (LH). The level of circulating FSH rises and falls with a 4-day periodicity in the mouse. Some of the more mature follicles respond to elevated FSH by further growth and proliferation of the follicle cells. Ovulation is finally triggered by a sharp peak in the level of circulating LH. In response to this the arrested first meiotic division is completed within about 10 hours (Figure 3.3). The division of the cytoplasm into two cells (cytokinesis) is very unbalanced, with the result that one of the post-meiotic nuclei remains in the oocyte while the other is pinched off with a small amount of cytoplasm to form a small cell called the first polar body. The oocytes are now arrested prior to the second meiotic division. The follicles open, releasing the oocytes, still with numbers of associated cumulus cells, and these pass into the mouth of the oviduct. They settle down in a transient distension of the oviduct called the atrium and await the arrival of spermatozoa.

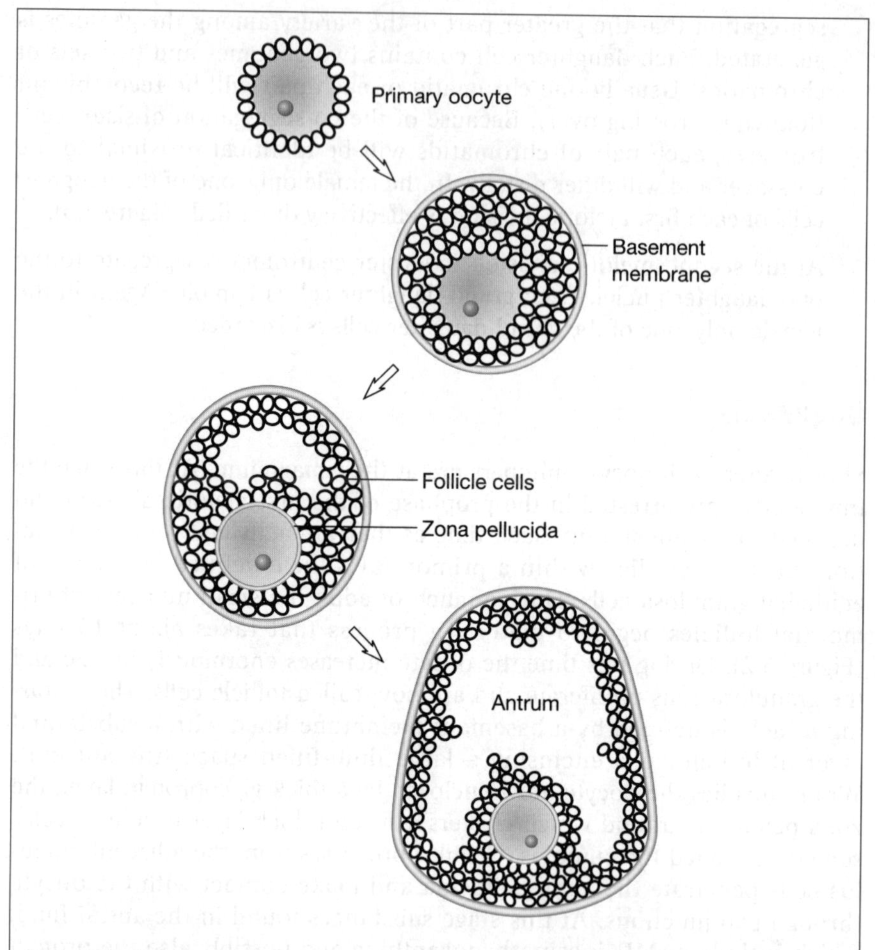

Figure 3.2 Oocyte maturation, showing the increase in size and number of follicle cells and the development of the antrum. The scale of the last diagram is less than the others (compare the sizes of the oocyte).

The number of follicles that respond to FSH and LH is smaller than the number that might be expected to do so, and in the normal course of events the non-responding oocytes die. Many of these can be artificially entrained by administering gonadotrophins. Although numerous different preparations could presumably be employed, the accepted procedure is to inject the animals with gonadotrophin from pregnant mares' serum (PMSG) and then, 48 hours later, with human chorionic gonadotrophin (HCG). Ovulation occurs about 12 hours later and during this night each female is housed with a proven stud male. The PMSG presumably simulates FSH, while the HCG simulates the LH peak. The responses of individual mice to the regime vary considerably, presumably in relation to the status of their natural cycles at the time of treatment. When the treatment is successful it is not unusual to obtain over 50 fertilised oocytes from a single dam.

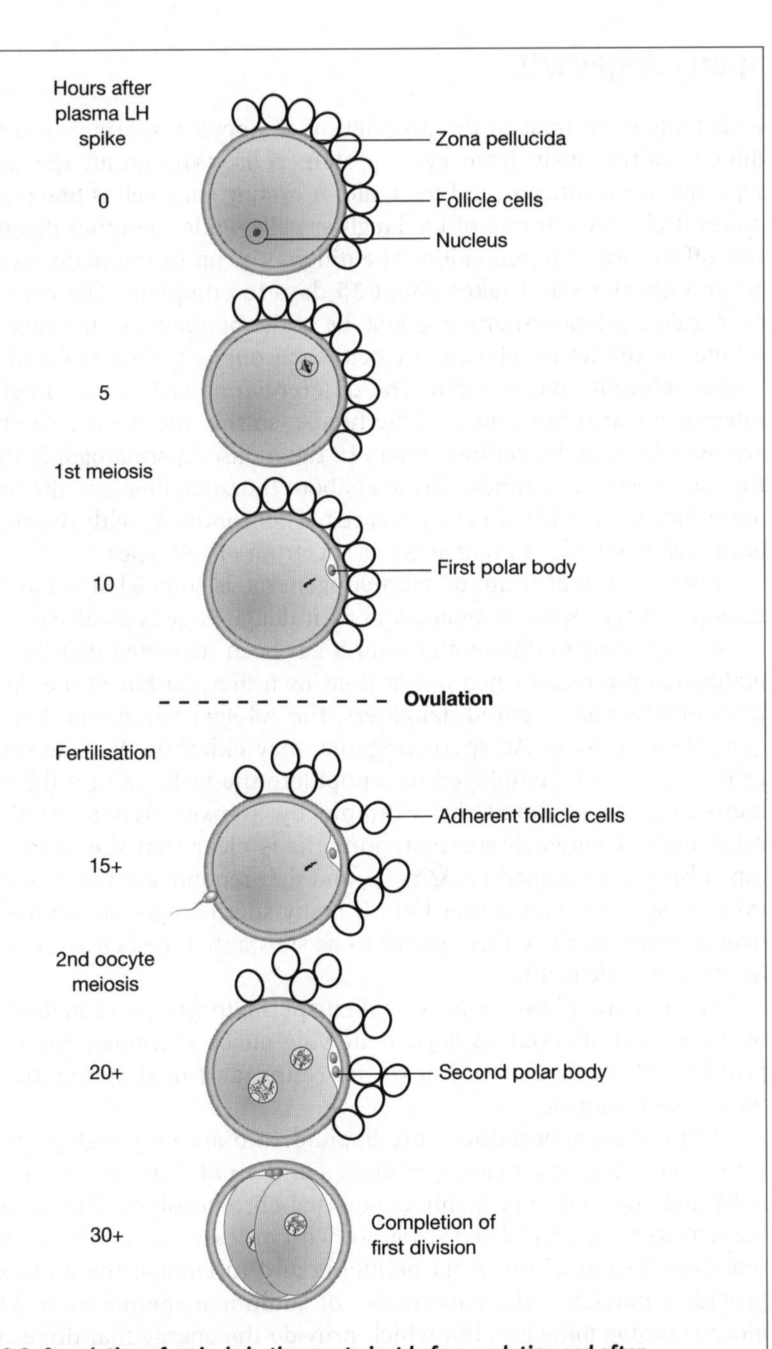

Figure 3.3 Completion of meiosis in the oocyte just before ovulation and after fertilisation. Follicle cells actually surround the oocyte completely, as in Figure 3.2. The first meiotic division is triggered by the plasma LH spike, the second by fertilisation. Following second meiosis the chromosomes of the maternal and paternal nuclei are duplicated and, following breakdown of the nuclear membranes, the two sets of paired chromosomes form up on a single spindle. The completion of the first division of the zygote is then normal and cytokinesis follows.

Spermatogenesis

In complete contrast to the production of oocytes, spermatozoa are produced continuously from cycling stem cells, throughout the period of reproductive competence of the male. A cycling stem cell is one that regenerates itself through one of its daughter cells while the other daughter cell sets off towards differentiation. The differentiation of spermatozoa involves several divisions and takes about 35 days to complete. The cycling stem cells, called primary spermatogonia, lie at the periphery of the seminiferous tubules in the testes. Having divided, each one is quiescent for about two weeks before it divides again. The differentiating cells move progressively inwards towards the centre of the tubule, so that the mature spermatozoa are released into the central lumen. In the mouse, spermatogonia that lie in the same region of a tubule divide at about the same time and the differentiation of the committed cells proceeds synchronously, with the result that each region exhibits a homogeneous spectrum of cell types.

Figure 3.4, a diagram of spermatogenesis, is somewhat simplified. For example, A-type spermatogonia and their daughter cells divide several times before beginning to differentiate and it has been suggested that the A1 spermatogonia are replenished not at their own division but at the division of their putative great grand-daughters, the A4 spermatogonia. Yet another type, the rare As or A0 spermatogonia, may either be the true progenitor cells or a 'reserve' employed to repopulate the testis should the A1 spermatogonia be depleted, for example by a toxin. Whatever the exact relationships between spermatogonia, it is clear that the stem cells are replenished during spermatogenesis, and that repopulation occurs following even drastic treatments that kill off many spermatogonia. Replenishment may be ongoing, since there seems to be substantial wastage of spermatogonia by apoptotic death.

The meiotic-phase cells are called spermatocytes, and meiosis occurs over a period of about 13 days. Following meiosis terminal differentiation proceeds through phases known as round and oval spermatids to the mature spermatozoa.

The mature spermatozoa are haploid, and are very highly specialised cells, consisting of a head, a midpiece and a tail. The nucleus lies in the head and contains very highly condensed chromosomes. The head carries the acrosome, a specialised structure with at least two functions, to facilitate penetration of the zona pellucida and to change the zona so as to provide a barrier to the penetration of additional spermatozoa. The midpiece contains mitochondria which provide the energy that drives the tail, providing motility.

Among the more dramatic events in spermatogenesis is the replacement of most of the histones, which bind to the DNA in all somatic cells, with smaller and even more basic polypeptides called protamines which are uniquely present in spermatozoa. The protamine mRNAs are generated in the spermatocytes, but are initially sequestered in nucleoprotein complexes

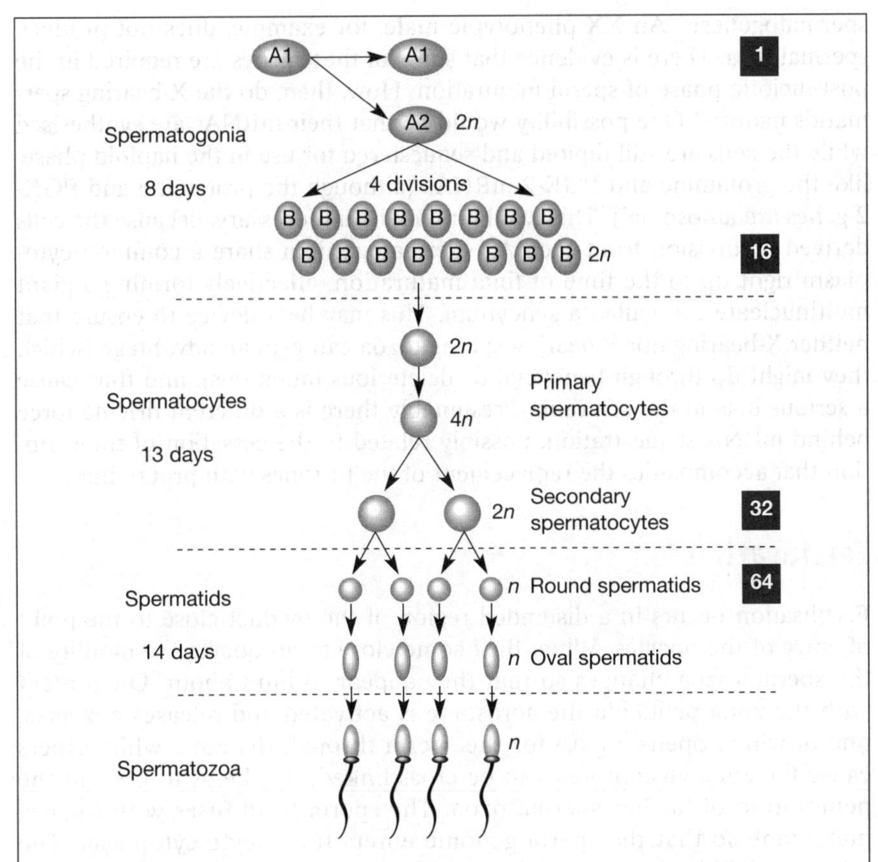

Figure 3.4 Schematic of the development of spermatozoa from a single primary spermatogonium. *n*, 2*n* and 4*n* indicate 1, 2 and 4 complements of chromosomes (or chromatids). The boxed numbers show the numbers of cells descended from a single committed spermatogonium at each stage. At an early division the cycling primary spermatogonium is regenerated. Altogether about 16 secondary spermatogonia are derived by mitotic divisions in about eight days, before meiosis. The first meiotic division generates 32 secondary spermatocytes from these and the second 64 haploid round spermatids. During the following two weeks differentiation into mature spermatozoa takes place. For most of the 35 days the cells form a syncytium, which breaks down in the later stages of maturation.

and translated only later at the spermatid stage. In addition to proteins unique to the testis, like the protamines, there are testis-specific isotypes of several proteins found in somatic cells. These include testis-specific isozymes of lactate dehydrogenase (LDH-X) and phosphoglycerate kinase (PGK-2), and isotypes of α-tubulin and cytochrome c. It is possible that these isotypes are specially tuned to function effectively in the testis.

Syncytial development of spermatozoa

The part of the Y chromosome that is not homologous to part of the X contains rather few genes. Some of them, however, are necessary for spermatogenesis. An XX phenotypic male, for example, does not produce spermatozoa. There is evidence that some of these genes are required in the post-meiotic phase of sperm maturation. How, then, do the X-bearing spermatids mature? One possibility would be that their mRNAs are synthesised while the cells are still diploid and sequestered for use in the haploid phase, like the protamine and PGK-2 mRNAs (although the protamine and PGK-2 genes are autosomal). This would in fact be unnecessary, because the cells derived by division from each A2 spermatogonium share a common cytoplasm right up to the time of final maturation, effectively forming a giant multinucleate cell called a syncytium. This may be a device to ensure that neither X-bearing nor Y-bearing spermatozoa can gain an advantage (which they might do through beneficial or deleterious mutations), and thus cause a serious bias in the sex ratio. Presumably there is a different driving force behind mRNA sequestration, possibly related to the cessation of transcription that accompanies the replacement of the histones with protamines.

Fertilisation

Fertilisation occurs in a distended region of the oviduct close to the point of entry of the oocytes. When they come close to an oocyte the motility of the spermatozoa changes so that they appear to hunt about. On contact with the zona pellucida the acrosome is activated and releases enzymes, one of which opens a path for the sperm through the zona while others cause the zona glycoproteins to be cross-linked; this helps to prevent the penetration of further spermatozoa. The sperm head fuses with the egg membrane so that the sperm genome enters the oocyte cytoplasm. The mitochondria in the sperm midpiece do not enter the oocyte; thus mitochondria are maternally inherited.

Fertilisation initiates the second oocyte meiotic division (Figure 3.3), during which sister centromeres normally separate to the daughter cells. Cytokinesis is again very unequal. In this case the very small daughter cell is called the second polar body, and this nestles against the first. A single haploid set of chromosomes (chromosome number n) remains within the oocyte, and a nuclear membrane forms around it.

Following fertilisation the male chromosomes undergo changes. The very basic protamines, the principal protein component of the sperm chromosomes, are phosphorylated and exchanged for histones from the egg cytoplasm between 4 and 12 hours after fertilisation, and during the first part of this time period a nuclear membrane is formed. The fertilised oocyte (or zygote or 1-cell embryo) now has two haploid nuclei (pronuclei). About 10 hours after fertilisation DNA synthesis begins in both nuclei, which become diploid after another 7 hours. The nuclear membranes break down and the two sets of chromosomes form up on a normal mitotic spindle. Separation of two diploid sets of chromosomes, reconstitution of nuclear membranes and cytokinesis are complete about 16 hours after fertilisation.

From the time of fertilisation until the first cell division there is no detectable RNA synthesis, although protein synthesis occurs utilising maternal (oocyte) mRNA. However changes occur in the availability of different mRNAs for translation, and these are reflected in changes in the pattern of protein synthesis.

From about 10 hours after fertilisation there is a period of approximately 6 hours when two pronuclei are present and certain types of manipulation can be carried out. These include removal and replacement of haploid nuclei and microinjection of substances (proteins, DNA, chemicals) into one or other pronucleus. In contrast, when the objective is to introduce a diploid somatic nucleus (Chapter 7) an unfertilised oocyte that has been mechanically enucleated is employed.

Early indications that mammals require both a maternal and a paternal chromosome complement came from experiments in which pronuclei were transplanted from one fertilised oocyte to another. Figure 3.5 shows an explanatory diagram and the results of these experiments are discussed in Chapter 6.

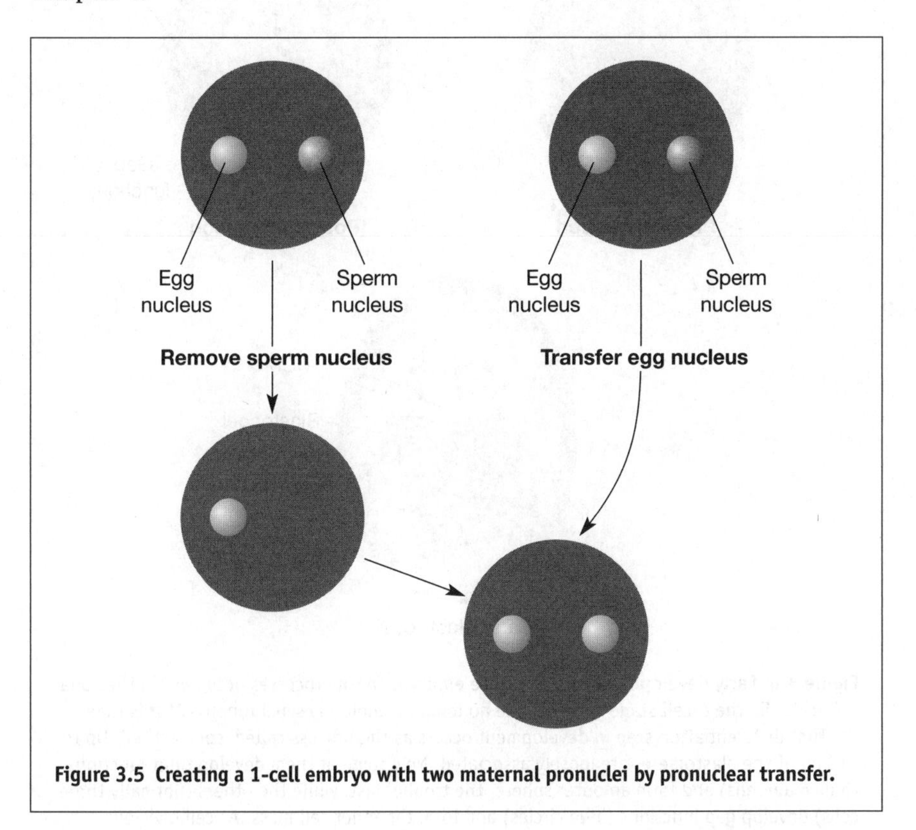

Figure 3.5 Creating a 1-cell embryo with two maternal pronuclei by pronuclear transfer.

Early development

Early cleavage and the blastocyst

The 2-cell embryo divides to produce four cells and then again to produce eight, by which time cell divisions have become less synchronous. At this stage the cells are very loosely aggregated. The synthesis of mRNA and of new proteins directed by the embryo nuclei become marked during the latter part of the 2-cell stage. Some time between there being eight and 16 cells (called a morula) the external cells form a shell linked together by tight junctions, while a small number remain within (Figure 3.6). This

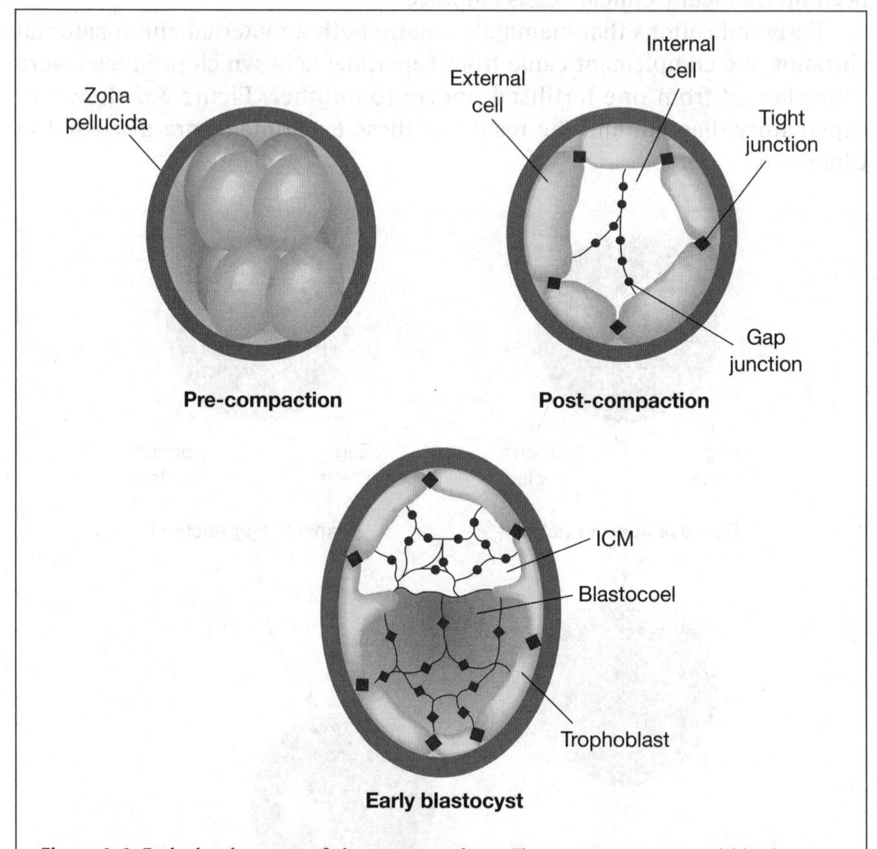

Figure 3.6 Early development of the mouse embryo. These processes occur within the zona pellucida. By the 8-cell stage divisions are no longer completely synchronous. At this time the first differentiation step in development occurs as the process called 'compaction'. Up to this point the blastomeres are loosely associated. Now some of them develop tight junctions (blue diamonds) and form an outer sphere, the trophoblast, while the others (normally three cells) develop gap junctions (blue circles) and form the inner cell mass. As cell divisions continue the cells become progressively smaller and the blastocoel cavity develops in what is now called the blastocyst.

process is called compaction because the embryo becomes more compact. The outer trophoblast cells are destined to form extraembryonic structures exclusively (e.g. the placenta) while the cells within, known as the inner cell mass (ICM), will form both the embryo and further extraembryonic structures (e.g. the amnion and the yolk sac). However, the cellular junctions break down transiently at the following cell division and at this time the trophoblast cells will still contribute to the embryo if injected into a blastocyst.

Divisions continue and by 4–5 days post coitum (dpc) the blastocoel cavity has opened up and the ICM has spun off another class of (at least partly) differentiated cells, the primitive endoderm, also destined to be extraembryonic. The remainder of the ICM forms the epiblast which will become the embryo proper. All this has taken place within the zona pellucida, but the embryo now emerges from the zona ('hatches') and proceeds rapidly to implantation in the uterine wall through the action of the trophoblast cells (the trophectoderm).

Two stages of this early development are of particular interest in relation to transgenic animals, namely the pre-compaction morula and the early blastocyst.

Embryo fusion

If the zona pellucida is removed from pre-compaction embryos and two are pushed close together and cultivated appropriately *in vitro*, they fuse to form a single embryo called a chimera, or chimeric embryo. Three embryos can be induced to fuse in the same way. In the latter case, as an example, the fused embryos are said to be hexaparental, reflecting the fact that each of the original embryos was produced by the fusion of gametes from two parents, making six in all; similarly, the three sets of diploid cells together contain six haploid genomes. When such chimeric embryos are introduced into suitably prepared dams many of them develop normally and are delivered as chimeric mice (also called allophenic mice). Chimeras are usually generated by fusing together pre-compaction embryos from two strains of mice that differ in a number of characteristics, coat colour in particular. Chimeras can then be identified visually, and in fact measuring the areas of coat of the different colours provides a good estimate of the degree of chimerism. When this was done many times it was found that on average 75% of the pups born were chimeras (Mintz 1970). This is the predicted proportion if, at some point in development, three cells are designated to be the progenitors of the embryo (see Table 3.1). It is reasonable to suppose that these correspond to the few cells that are internalised at the time of compaction. The fact that the trophoblast cells are still totipotent at this time would only affect the argument if they mix with the inner cells during the next (and last) synchronous division. Hexaparental mice also demonstrate that *at least* three cells are the embryo progenitors (Markert and Petters 1978).

Table 3.1 Proportion of animals produced by aggregation of two pre-compaction embryos that are predicted to be chimeric if different numbers of cells contribute to the embryo.

Number of cells contributing to the embryo	1	2	**3**	4	5
Predicted chimeras (%)	0	50	**75**	88	94

Segregation of somatic and germ line cells

Cells taken at the 2-cell to 8-cell stages form chimeras when fused with an 8-cell morula or alternatively when injected into a blastocyst, and will contribute to the germ line. This is also true of ICM cells and primitive ectoderm cells from blastocysts. These cells are consequently said to retain totipotency. At a later stage in the development of the embryo there is a segregation of germ line cells, which are destined to form the new generation of gametes, and somatic cells, which are destined to form the remainder of the embryo. Cells destined to be somatic lose the capacity to contribute to the germ line and are said to be pluripotent, because they will contribute to all the somatic tissues. The germ line cells are first recognisable as primordial germ cells (PGCs) at about eight days. PGCs from 8–12 day embryos are not totipotent, but become so if cultivated in a special way (see below). (Note that the word pluripotent is sometimes used to mean totipotent.)

Embryonic carcinoma cells (EC cells)

It was discovered many years ago that if mouse blastocysts are placed inside mouse kidney capsules (or indeed at some other ectopic sites) some of them will develop into teratomas, which are non-malignant growths consisting almost entirely of differentiated cell types. Others develop into teratocarcinomas, malignant growths consisting of a mixture of undifferentiated stem cells and differentiated cells. When a teratocarcinoma is explanted in culture dishes, undifferentiated cells proliferate and in due course clonal lines can be isolated. In some cases teratocarcinomas were explanted on top of mitotically arrested feeder cells. The feeder cells were initially chosen from among the large number of available fibroblast and epithelial cell lines on just this basis, i.e. that they support proliferation as against differentiation. They are grown to confluence and then treated with mitomycin-c or γ-irradiated. This allows cytoplasmic metabolism to continue but prevents any further cell division, avoiding overgrowth or contamination of the teratocarcinoma cells. In retrospect, the reason that the chosen cell lines support undifferentiated proliferation is that the cells secrete growth factors that stimulate the teratocarcinoma cells appropriately.

The clones that arise are called EC cells. These cells differentiate spontaneously, especially if removed from contact with the feeder cells. Their differentiation can be provoked and also directed down a particular pathway, for example by adding retinoic acid to the culture medium.

If cells from some EC cell lines are introduced into a blastocyst a chimera will result. These EC cells consistently contribute to the somatic tissues but too seldom to the germ line to be of general utility. Thus they are pluripotent. This means of course that the EC cell genome cannot be propagated by mating the chimeric mice. However their pluripotency is taken to mean that they derive ultimately from the ICM cells of the blastocyst that was inserted into the kidney capsule in the first place.

Embryonic stem cells (ES cells)

When mouse blastocysts are teased out onto a feeder cell layer in culture, outgrowths of cells occur (Chapter 7). These are mainly of two types, ICM cells and trophectoderm cells from the trophoblast. ICM cell colonies can be physically separated from trophectoderm cells and continuously growing cultures established which are then called ES cells. These cells share many of the properties of EC cells, with the very important difference that when they are injected into a blastocyst they exhibit the full potential of the ICM cells from which they were derived and contribute both to the somatic tissues and to the germ line of the resulting chimera. That is to say, they are totipotent. Consequently, by breeding the chimeras, mice can be generated that carry a haploid chromosome set derived from the ES cells. By inbreeding, mice can then be produced that are homozygous for particular alleles that originated in the ES cells.

The properties of ES cells that have made them probably the most powerful tool in transgenic science to date are as follows:

- they can be propagated and manipulated in culture
- transgenes can be introduced into the cells and selection can be applied to isolate rare clones of cells that have integrated a transgene in a particular predesigned way
- subsequently, the manipulated cells can be introduced into chimeras and, by breeding the chimeras, mice can be generated that are heterozygous for the transgene
- by mating together the heterozygous transgenic mice, transgenic homozygotes can be obtained

In an interesting development of ES cell technology the cells are introduced into tetraploid embryos by aggregation (Nagy *et al.* 1990; Nagy *et al.* 1993) or by blastocyst injection (Wang *et al.* 1997). The diploid cells outgrow the tetraploid cells in the growing embryo, with the result that the pups that develop are mainly composed of descendants of the ES cells. In some instances the pups died neonatally, but in others they have survived (Nagy *et al.* 1993; Wang *et al.* 1997).

In one of the two initial successful derivations the ES cells were generated from random bred ICR × inbred SWR/J F_1 embryos (Martin 1981). In the other, embryos from a subline of line 129 were employed (Evans and

Kaufman 1981). There is a belief that the line 129 sublines, which are also a particularly good source of teratocarcinomas, are a particularly favourable source. However, more recently ES cell lines have been produced from C57BL embryos (Ledermann and Burki 1991) and embryos from other mouse strains (see Chapter 7). Attempts to generate ES cells from other species, even rats, have so far been unsuccessful. Most reports of ES-like stem cells have not been followed by the successful generation of chimeras. Recently the isolation of pluripotent stem cells from rabbits, marmosets and rhesus monkeys has been reported, and stem-like cells have been isolated from bovine embryos following nuclear transfer (see Chapter 7) but as yet there have been no reports of germ line transmission from ES cells of any species other than the mouse.

Relationship between EC cells and ES cells

The supposition is that ES cells and EC cells represent the same cell type. Similarities between ES and EC cells in morphology, cell surface antigens and protein synthesis profiles support this view. Under appropriate conditions *in vitro* both cell types form embryoid bodies (complex structures containing differentiated cells) and as pointed out above, some EC cell lines, like ES cells, will contribute to somatic tissues *in vivo*. The differences between the two are rather slight: many EC cell lines do not contribute to chimeras; those that do tend to throw up teratocarcinomas; EC cells effectively do not produce gametes *in vivo*. The likely reason for these differences is that during the *in vivo* tumour phase of EC cell production, mutations and other genome alterations occur that confer a growth advantage while reducing their developmental potential. Certainly, EC cell lines tend to exhibit chromosomal abnormalities. The supposition is that this is largely avoided in ES cells because they are cultured from the outset under controlled conditions, with regular observation and subcultivation.

Embryonic germ cells (EG cells)

Primordial germ cells can be isolated from 8–8.5 day embryos and after migration to the germinal ridge at 12 days. Upon cultivation in the presence of Steel factor, the ligand of the *c-kit* receptor, basic fibroblast growth factor and LIF (see below), these give rise to cells that resemble ES cells, contribute to chimeras when injected into blastocysts and populate the germ line (Matsui *et al.* 1992; Stewart *et al.* 1994; Labosky *et al.* 1994). Unlike ES cells, but like the PGCs from which they are derived (Szabo and Mann 1995), the EG cells do not preserve parental methylation imprints (see Chapter 6).

Leukaemia inhibitory factor (LIF)

LIF (at that time known as differentiation inhibitory activity, DIA) was isolated as a factor secreted by feeder cells and active in promoting ES cell proliferation (Smith *et al.* 1988). When added to the culture medium it supports the proliferation of mouse ES cells in the absence of a feeder layer. In addition to inhibiting ES cell differentiation, LIF affects various adult tissues in different ways, for example stimulating acute-phase protein synthesis in hepatocytes and stimulating osteoclasts, and is also required for uterine blastocyst implantation. LIF is produced in a soluble form and a form that associates with the extracellular matrix, the two forms differing only in a few N-terminal amino acids. Their synthesis seems not to be coordinately regulated. Undifferentiated ES cells contain small amounts of the mRNA for matrix-associated LIF. Upon differentiation they produce larger amounts of mRNA for both forms.

The LIF receptor is a transmembrane receptor of the class II cytokine receptor superfamily. It has no intrinsic tyrosine kinase activity and associates with another transmembrane protein called gp130. Both the LIF receptor and gp130 are components of the receptors for the mitogens oncostatin-M, ciliary neurotrophic factor and cardiotrophin-1, while gp130 also forms part of the receptors for interleukins 6 and 11 (IL-6 and IL-11). Like other receptors of the superfamily, LIF receptor is produced in membrane-bound and soluble forms by differential splicing of the same transcription unit. The latter form lacks transmembrane and cytoplasmic receptor domains. The LIF–gp130 complex associates with the Janus kinases, Jak1 and Jak2 (Lutticken *et al.* 1994), which activate genes *via* phosphorylation of Stat3 (Narazaki *et al.* 1994). Stat3 is required for the undifferentiated propagation (self-renewal) of ES cells (Niwa *et al.* 1998). A second signal transduction pathway originating with LIF–gp130 acts *via* ras and the MAP kinase cascade. It is not yet clear which genes are activated by these pathways. gp130 can also be activated in other ways to exert the same inhibitory effect on ES cell differentiation. Both ciliary neurotrophic factor and oncostatin-M activate gp130, as does a mixture of IL-6 and the soluble form of the IL-6 receptor.

ES cell lines can be produced and maintained without employing feeder cells if LIF is added to the medium. Such ES cells differentiate normally *in vitro* if LIF is withdrawn, and they colonise both somatic and germ lines when introduced into blastocysts.

At later stages of development (8 to 12 dpc) the cells destined to develop into gametes, the progenitor germ cells (PGCs), if cultivated for a time *in vitro* in the presence of basic fibroblast growth factor (bFGF) and Steel factor (*c-kit* ligand) as well as LIF, are converted to embryonic germ cells (EGs). EGs can subsequently be cultivated in LIF alone and will contribute to embryos in the same way as ES cells.

Gene targeting experiments show that LIF (Stewart *et al.* 1992), the LIF receptor (Ware *et al.* 1995; Li *et al.* 1995) and gp130 (Kawasaki *et al.* 1997) are dispensable for early embryonic development. This serves to underline the fact that ES cells, despite their totipotency, are not precisely equivalent to inner cell mass cells and indicates that a different pathway maintains totipotent stem cells *in vivo*.

References

Evans, M.J. and Kaufman, M.H. (1981) Establishment in culture of pluripotential cells from mouse embryos. *Nature*, **292**, 154–156.

Kawasaki, K., Gao, Y.H., Yokose, S., Kaji, Y., Nakamura, T., Suda, T., Yoshida, K., Taga, T., Kishimoto, T., Kataoka, H., Yuasa, T., Norimatsu, H., and Yamaguchi, A. (1997) Osteoclasts are present in gp130-deficient mice. *Endocrinology*, **138**, 4959–4965.

Labosky, P.A., Barlow, D.P., and Hogan, B.L. (1994) Mouse embryonic germ (EG) cell lines: transmission through the germline and differences in the methylation imprint of insulin-like growth factor 2 receptor (Igf2r) gene compared with embryonic stem (ES) cell lines. *Development*, **120**, 3197–3204.

Ledermann, B. and Burki, K. (1991) Establishment of a germ-line competent C57BL/6 embryonic stem cell line. *Experimental Cell Research*, **197**, 254–258.

Li, M., Sendtner, M., and Smith, A. (1995) Essential function of LIF receptor in motor neurons. *Nature*, **378**, 724-727.

Lutticken, C., Wegenka, U.M., Yuan, J., Buschmann, J., Schindler, C., Ziemiecki, A., Harpur, A.G., Wilks, A.F., Yasukawa, K., Taga, T., *et al.* (1994) Association of transcription factor APRF and protein kinase Jak1 with the interleukin-6 signal transducer gp130. *Science*, **263**, 89–92.

Markert, C.L. and Petters, R.M. (1978) Manufactured hexaparental mice show that adults are derived from three embryonic cells. *Science*, **202**, 56–58.

Martin, G.R. (1981) Isolation of a pluripotent cell line from early mouse embryos cultured in medium conditioned by teratocarcinoma stem cells. *Proceedings of the National Academy of Sciences of the United States of America*, **78**, 7634–7638.

Matsui, Y., Zsebo, K., and Hogan, B.L. (1992) Derivation of pluripotential embryonic stem cells from murine primordial germ cells in culture. *Cell*, **70**, 841–847.

Mintz, B. (1970) Clonal expression in allophenic mice. *Symp. Int. Soc. Cell. Biol.*, **9**, 15.

Nagy, A., Gocza, E., Diaz, E.M., Prideaux, V.R., Ivanyi, E., Markkula, M., and Rossant, J. (1990) Embryonic stem cells alone are able to support fetal development in the mouse. *Development*, **110**, 815–821.

Nagy, A., Rossant, J., Nagy, R., Abramow-Newerly, W., and Roder, J.C. (1993) Derivation of completely cell culture-derived mice from early-passage embryonic stem cells. *Proceedings of the National Academy of Sciences of the United States of America*, **90**, 8424–8428.

Narazaki, M., Witthuhn, B.A., Yoshida, K., Silvennoinen, O., Yasukawa, K., Ihle, J.N., Kishimoto, T., and Taga, T. (1994) Activation of JAK2 kinase mediated by the interleukin 6 signal transducer gp130. *Proceedings of the National Academy of Sciences of the United States of America*, **91**, 2285–2289.

Niwa, H., Burdon, T., and Chambers, I. (1998) Self-renewal of pluripotent embryonic stem cells is mediated via activation of STAT3. *Genes & Development*, **12**, 2048–2060.

Smith, A.G., Heath, J.K., Donaldson, D.D., Wong, G.G., Moreau, J., Stahl, M., and Rogers, D. (1988) Inhibition of pluripotential embryonic stem cell differentiation by purified polypeptides. *Nature*, **336**, 688–690.

Stewart, C.L., Kaspar, P., Brunet, L.J., Bhatt, H., Gadi, I., Kontgen, F., and Abbondanzo, S.J. (1992) Blastocyst implantation depends on maternal expression of leukaemia inhibitory factor. *Nature*, **359**, 76–79.

Stewart, C.L., Gadi, I., and Bhatt, H. (1994) Stem cells from primordial germ cells can reenter the germ line. *Developmental Biology*, **161**, 626–628.

Szabo, P.E. and Mann, J.R. (1995) Biallelic expression of imprinted genes in the mouse germ line: implications for erasure, establishment, and mechanisms of genomic imprinting. *Genes & Development*, **9**, 1857–1868.

Wang, Z.Q., Kiefer, F., Urbanek, P., and Wagner, E.F. (1997) Generation of completely embryonic stem cell-derived mutant mice using tetraploid blastocyst injection. *Mechanisms of Development*, **62**, 137–145.

Ware, C.B., Horowitz, M.C., Renshaw, B.R., Hunt, J.S., Liggitt, D., Koblar, S.A., Gliniak, B.C., McKenna, H.J., Papayannopoulou, T., Thoma, B., *et al.* (1995) Targeted disruption of the low-affinity leukemia inhibitory factor receptor gene causes placental, skeletal, neural and metabolic defects and results in perinatal death. *Development*, **121**, 1283–1299.

Further reading

General

Gilbert, S. F. (1997) *Developmental biology*, Sinauer Associates, Sunderland, Mass.

Spermatogenesis

Willison, K. and Ashworth, A. (1987) Mammalian spermatogenic gene expression. *Trends in Genetics*, **3**, 351–355.

Dym, M. (1994) Spermatogonial stem cells of the testis. *Proceedings of the National Academy of Sciences of the United States of America*, **91**, 11287–11289.

Gene expression during early development

Nothias, J.Y., Majumder, S., Kaneko, K.J., and DePamphilis, M.L. (1995) Regulation of gene expression at the beginning of mammalian development. *Journal of Biological Chemistry*, **270**, 22077–22080.

Forlani, S. and Nicolas, J. (1997) Gene activity in the preimplantation mouse embryo. In *Transgenic animals, generation and use*, L.M. Houdebine (ed), Harwood Academic Press, Amsterdam, pp 345–359.

EC cells and ES cells

Robertson, E.J. (1987) *Teratocarcinomas and embryonic stem cells – a practical approach*, IRL Press, Oxford.

Smith, A.G. (1992) Mouse embryo stem cells: their identification, propagation and manipulation. *Seminars in Cell Biology*, **3**, 385–399.

Leukemia inhibitory factor

Smith, A.G., Nichols, J., Robertson, M., and Rathjen, P.D. (1992) Differentiation inhibiting activity (DIA/LIF) and mouse development. *Developmental Biology*, **151**, 339–351.

CHAPTER FOUR

Mouse husbandry and genetics

Husbandry

The advantages of the mouse as an experimental animal are:

- small size and rapid, non-seasonal reproduction
- toleration of inbreeding
- the existence and availability of an enormous number of genetic variants
- a long history of careful husbandry and record keeping
- the large mouse database of strains, genes, nucleotide sequences and chromosomal markers

Table 4.1 lists the important time factors in relation to mouse husbandry.

Table 4.1 Parameters of the mouse life cycle.

Litter size	10
Gestation time	19–20 days (*ca.* 3 weeks)
Weaning	3 weeks pp, 6 weeks pc
Sexual maturity	5 weeks pp, 8 weeks pc
Oestrous cycle	3–4 days
Can become pregnant again immediately after parturition	

(pc = post coitus or post conception; pp = post partum)

In mammalian reproduction females are rate limiting. The enormous fecundity of mice is illustrated in Table 4.2 which shows the notional rate of

Table 4.2 Notional maximum rate of expansion of a mouse colony from a single pair over a period of 20 weeks, assuming a litter size of 10 and no mortality.

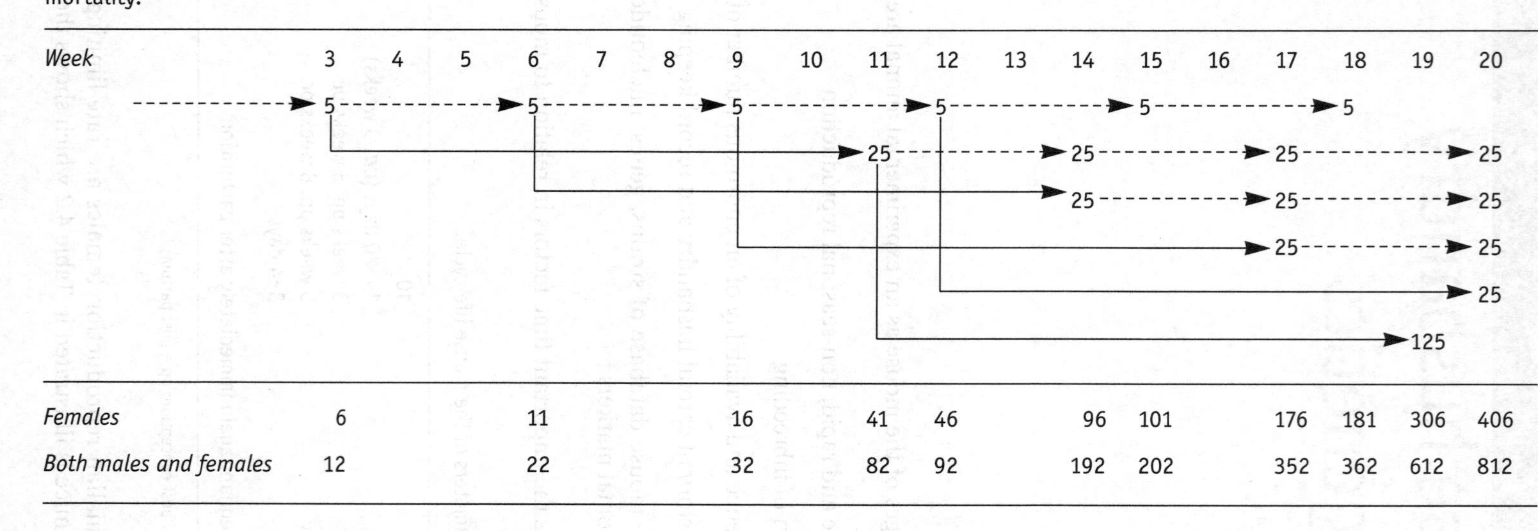

Week	3	4	5	6	7	8	9	10	11	12	13	14	15	16	17	18	19	20
	5			5			5			5			5			5		
									25			25			25			25
												25			25			25
															25			25
																		25
																	125	
Females	6			11			16		41	46		96	101		176	181	306	406
Both males and females	12			22			32		82	92		192	202		352	362	612	812

expansion of a mouse colony founded by one male and one female, with gestation time rounded up to 21 days for convenience. The internal part of the table, and the first summation row, list the total number of females in the colony. This is doubled to give the total of males and females in the last row. Row 2, for example, shows the founder female giving birth to five female offspring every three weeks. The continuous lines (with arrowheads) trace the time from birth of these females until they produce their first litters. Five females produce 25 female pups according to the model.

Although mice are the best mammalian model animal, their care nevertheless requires expensive facilities and is labour intensive. Table 4.3 lists some of the requirements.

Table 4.3 Factors affecting the cost of maintaining a mouse colony. Most are tightly regulated.

Facilities	*Labour*
Filtered air	Observation
Controlled temperature and humidity	Health monitoring
Controlled light/dark cycle	Feeding and watering
Cage cleaning and sterilisation	Cage cleaning and sterilisation
Security	Mating
	Weaning and segregation
	Marking for identification

Genetics

Like human genetics, mouse genetics is in the process of rapid change driven by technical advances in DNA manipulation, nucleotide sequencing and other molecular biological methods such as the polymerase chain reaction (PCR). These advances mainly affect the mapping of genes, at all levels from chromosomal assignments to single nucleotide polymorphisms. However, they also impinge on breeding programmes where their effect is to supplant to some extent classical methods based on the assumption of random recombination and segregation of chromosomes and employing probability calculations. While molecular methods give more control, eliminate the need for the assumption of randomness and significantly reduce the time span to reach some objectives, they are more labour intensive. This issue is taken up again later in the chapter. The classical methods are of course important from a historical point of view, and they also remain useful.

Laboratory mouse strains

Mouse breeding is reputed to have been recreational in China more than 2000 years ago. The practice is said to have been taken up in Japan and to have been imported into Europe from there. Effects of this history are still discernible; the Y chromosome of the C57BL/6 inbred strain, for example, is derived from an Asiatic subspecies (Nagamine *et al.* 1994). More recently mice were commonly kept as pets and 'fancy' strains were bred in 19th century Europe and America. Current strains date from that period; fancy strains were imported into laboratories from dealers and breeders in the first two decades of the 20th century and in many cases continuous records are available from that time.

Figure 4.1 shows the known relationships between some of the more commonly employed laboratory strains. All of the strains shown are highly inbred and effectively homozygous at all loci. In many cases strains have been kept apart continuously for about 100 years. In others, crosses between lines were made with some purpose in view, and some selection would have been applied, e.g. for coat colour, while a new inbred line was being established (see below). However, no control could be exerted over the vast majority of genes and for these the relative contributions of the parental lines to the eventual inbred product were largely a matter of chance. With the advent of molecular markers the contemporary strains can be extensively characterised at the DNA level and differences between them determined.

Inbred strains were passed from laboratory to laboratory and to suppliers of laboratory mice. It was recognised that in due course the populations maintained in different laboratories would come to differ from each other,

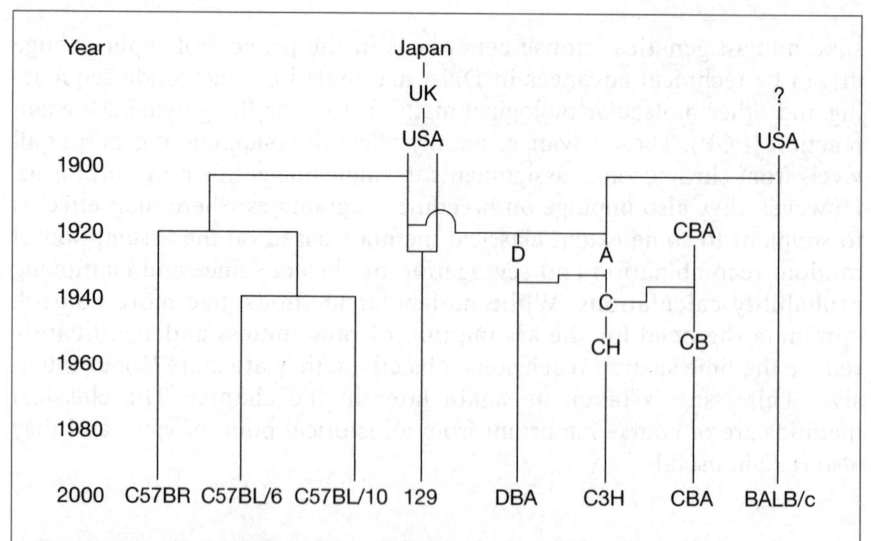

Figure 4.1 Relationships between some of the most frequently utilised inbred mouse strains. As shown, they were developed in the USA.

either subtly through the fixation of newly arisen spontaneous mutations or even, drastically, through inadvertent contamination by another strain. For this reason strain designations are given a laboratory identification. Thus C57BL/6J is maintained at the Jackson Laboratory while C57BL/6Ola is maintained by the Harlan-Olac company.

Genetic background

The genomes of different inbred strains of mouse differ in perhaps 0.1% of base pairs, which equates to approximately 2000 significant allelic differences (discussed in Chapter 10). Some of these are differences in known genes while others are unrecognised.

When we talk about differences in the genetic background of mouse strains, we are referring to the observation that a particular gene is expressed differently when present in different strains. Such differences are sometimes quantitative and sometimes qualitative, as when a gene is expressed in a different spectrum of tissues in the different strains. In some cases the differences can be mainly or entirely attributed to the effect of one or very few genes at other loci. The genes responsible will then be called modifiers of the gene in question. In other cases a simple picture is unavailable, either because no investigation has been undertaken, or because investigation has shown that the difference in expression is due to a large number of genes, each exerting a small effect. In these circumstances we have an unattributed difference in genetic background.

A new allelic variant of a gene is often discovered in a poorly characterised strain, or in a mouse captured from the wild. To harmonise the genetic background of alleles from different sources, they are introduced into a standard strain, more often than not C57BL/6, by backcrossing (see below). When this has been done, the strains carrying the different alleles are said to be congenic.

Breeding programmes: backcrossing and inbreeding

Because of the random segregation of chromosomes the effects of backcrossing and inbreeding are expressed in terms of probabilities.

Backcrossing

The objective is to bring a new allele (e.g. a newly discovered mutation or a transgene) into the same standard genetic background as the normal allele. Mice carrying the new allele are mated with mice carrying the normal allele. Some of the F_1 offspring are mated to the standard mice and the offspring are the first backcross generation (B_1). These mice are screened to determine which carry the new allele and some of these are again mated with mice carrying the normal allele, giving B_2 offspring. The process is repeated and as time goes on more and more of the genes come to be present in two copies of the standard allele.

The probability P that any given chromosome segment is homozygous after n generations is $1-0.5^n$ (or the homozygosity is $100\cdot(1-0.5^n)\%$). The probability values after different numbers of backcross generations are listed in Table 4.4. The number of generations required to produce a congenic line can be reduced by using a high-density molecular mapping technique (discussed below) to determine which backcross offspring have inherited two copies of the standard polymorphisms. In principle, this could be done at the B_1 generation, but with 20 chromosomes there are one million different combinations of homozygous standard and heterozygous chromosomes at this stage, even if there were no recombination in the heterozygous parent. However, screening limited numbers of animals at B_3 or B_4 could be advantageous.

Table 4.4 Progress towards the genetic background of the backcross strain at different backcross generations (B_1–B_7).

Generation	*P that a given chromosome segment remains heterozygous (0.5^n)*	*Homozygosity*
B_1	0.5	50%
B_2	0.25	75%
B_3	0.125	87.5%
B_5	0.0313	97.9%
B_7	0.0078	99+%

Because mice carrying the mutant allele or transgene are being selected at each generation for breeding the next, these estimates do not apply to genes that lie close to it on that particular chromosome. In general, the closer it is to the mutant allele, the less likely a gene is to become homozygous. This issue is explored further in the section dealing with the effect of background on transgenes (Chapter 10). Here in particular, more rapid progress can be made by screening polymorphisms in the relevant chromosome by molecular methods.

Full-sibling mating (inbreeding)

The objective is to make the members of each chromosome pair of a strain of mouse identical and all the genes homozygous. Two siblings are chosen as the first generation parents (P_0). Their offspring are the F_1 generation. Pairs of these are mated to give the F_2 generation and if several pairs are mated there will be several families of F_2 offspring. The rule is that the descendants of each family are kept as a separate line by making full-sibling matings at each generation. This is done because some of the lines are likely to become infertile and die out, or else become unprolific.

The probability that any given chromosome segment is homozygous after n generations is $1-0.75^n$ (or the homozygosity is $100\cdot(1-0.75^n)\%$).

Probability values after different numbers of generations of full-sib mating are listed in Table 4.5. We are assuming that all genes are heterozygous in the P_0 generation, which will seldom if ever be true, so that in practice the degrees of homozygosity will be higher than shown.

Table 4.5 Progress towards homozygosity at all loci after different numbers of generations of full-sibling mating (F_3–F_{15})

Generation	*P that a given chromosome segment remains* heterozygous *(0.75^n)*	*Homozygosity*
F_3	0.422	57.8%
F_5	0.237	76.3%
F_7	0.133	86.7%
F_{10}	0.056	94.4%
F_{15}	0.013	98.7%

Gene mapping

Four types of marker are commonly mapped, often in separate exercises which require later amalgamation. In the chronological order in which these came into use, and in fact in inverse order to their power, these are:

- Morphological markers, identified through defective mutant alleles both spontaneous and induced (e.g. by irradiation) that produce an altered phenotype.
- Isozyme loci, mainly detected by protein electrophoresis, often together with staining methods coupled to enzymatic conversion of specific substrates.
- Cloned genes, often linked to restriction fragment length polymorphisms (RFLPs, see below), but sometimes recognised by means of specific oligonucleotide probes.
- Anonymous and highly polymorphic short DNA regions.

Some of the methods employed to map genes are described in the sections that follow.

Recombinant inbred strains

Inbreeding is carried out by full-sibling mating (brother–sister mating). During inbreeding different genes (actually different segments of chromosomes) become homozygous after different numbers of generations of inbreeding. They are then said to be fixed within the line in question; the word refers graphically to the fact that, during inbreeding, once the line carries two copies of the same allele the other allele is of course lost and for ever excluded. When several lines are being carried forward in parallel the same gene will be fixed (at different generations) in some of the lines, while the alternative allele of the same gene will be fixed in all the other lines. In this way some lines come to carry two copies of one allele and some two copies of the other. This fact is used to advantage in generating recombinant inbred (RI) strains. Two inbred strains are mated and then several discrete strains are generated by inbreeding in this way; once fixation is universal each strain carries a different spectrum of chromosome segments from the two originating strains (Figure 4.2).

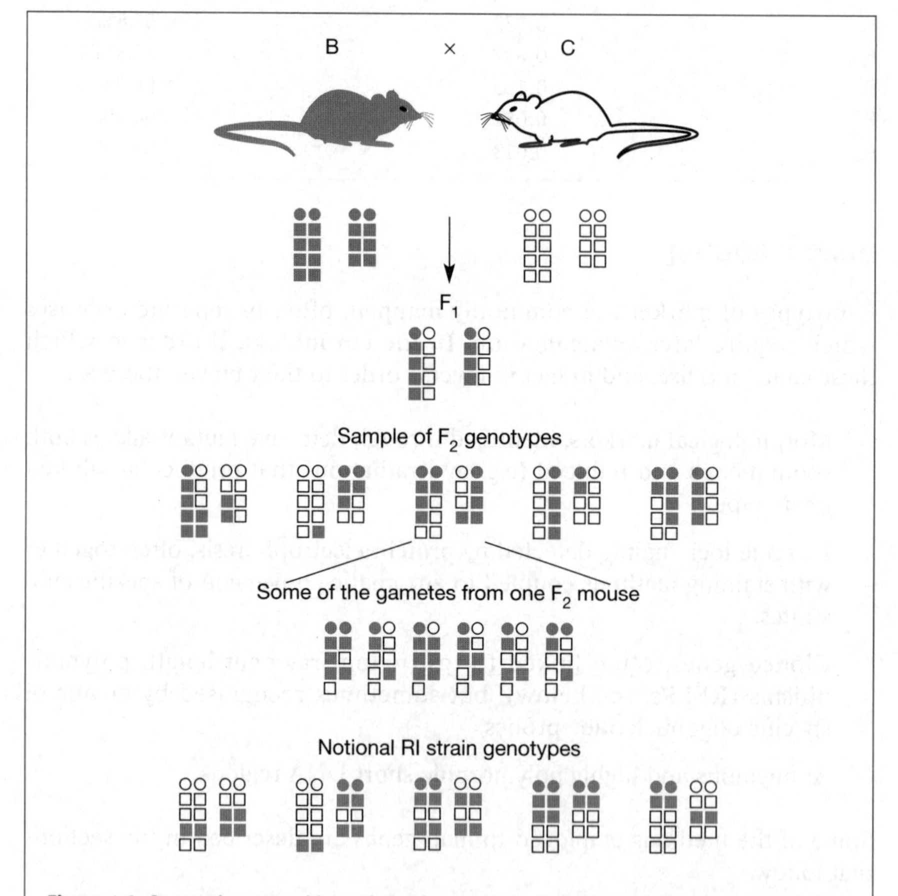

Figure 4.2 Generating recombinant inbred strains. The parental mouse strains are labelled B and C. The figure shows only two chromosome pairs, for the purpose of illustration, each with round centromeres and divided into four and three segments respectively. The F_1 generation of mice all carry the same set of autosomes, one of each pair from each parental strain. Due to segregation and recombination at meiosis during gametogenesis in the F_1, the F_2 generation is genetically very heterogeneous. Again due to segregation and recombination at meiosis, the chromosomes in the F_2 gametes are yet more heterogeneous. The last row shows a notional selection of possible genotypes following inbreeding of separate lines to the point of complete fixation.

The two inbred strains are chosen to be as dissimilar as possible within any constraints that may be relevant. These are crossed to yield F_1 offspring, which should be chromosomally identical (all heterozygous at all loci). F_1 animals are crossed to give F_2 progeny. Because of independent chromosome segregation the F_1 gametes are extremely diverse in chromosomal composition and those of the F_2 animals are even more so (Figure 4.2). In frequency terms the F_1 genotype will reappear about once in 10^6 F_2 mice and the parental types about once in 10^{12}.

Next, F_2 animals taken at random are crossed in pairs. This is the equivalent of the P_0 generation mentioned above, except that two different genomes are present rather than four. A large number of pairs (e.g. 30) are mated in this way, each pair being the founders of an RI strain. The strains are propagated by full-sibling mating, and the result after about 12 generations is that each has become homozygous over most of the genome. However, because of the diversity of F_2 animals and the random fixation of chromosomal regions, each of the strains carries a different complement of alleles from the original parental strains; for example, see Table 4.6. In this tiny sample U^a and V^a on chromosome I of strain A have remained together, as have U^b and V^b from strain B, while the alleles at the *w* locus have segregated independently of those at the *U* and *V* loci. The database for some sets of RI strains now extends to about 20 strains and several hundred markers.

Table 4.6 Recombinant inbred strains. Illustrates the fixation of different combinations of alleles from the two inbred parental strains in the derived recombinant inbred strains. Alleles at linked loci (*U* and *V*) may or may not become separated in some lines with a probability depending on the map distance between them. A small set of RI strains like that illustrated would be of little value.

Locus (Chromosome)	*U (I)*	*V (I)*	*w (II)*	*X (III)*	*y (IV)*	*z (V)*
Parental strain A	U^a	V^a	w^a	X^a	y^a	z^a
Parental strain B	U^b	V^b	w^b	X^b	y^b	z^b
RI strain 1	U^a	V^a	w^b	X^a	y^b	z^b
RI strain 2	U^b	V^b	w^a	X^a	y^a	z^b
RI strain 3	U^b	V^b	w^a	X^b	y^b	z^a
RI strain 4	U^a	V^a	w^a	X^a	y^b	z^a

During the long period of full-sib mating there is an opportunity at each generation for crossing-over to occur between homologous chromosomal regions wherever they have not yet been fixed. Quite a lot of recombination occurs and this is the basis for a method of mapping genes within as well as between chromosomes.

To be mapped a new gene has to be associated with a difference between the parental lines. This can be anything at all that can be scored, for example protein electrophoretic mobility or more commonly nowadays an associated DNA polymorphism, RFLP or minisatellite. All the strains in the set are then screened and identical segregation patterns (concordances) between the new and established markers are sought. In relation to the above example, one polymorphic form might concord with U^aV^a and the other with U^bV^b, indicating that the gene is located on chromosome I and linked to the *U* and *V* loci.

A considerable advantage of RI strains is that, once established, they are self-perpetuating and generate a continuously expanding database. They are particularly useful for mapping isozyme markers, because of the large amount of material available from each strain. However, because the number of strains generated from a cross is necessarily limited, so too is the resolution between markers that they provide.

Interspecific crosses

The utility of RI strains is limited by the frequency of polymorphisms among inbred laboratory strains of *Mus musculus*. *Mus spretus* is not completely isolated reproductively from *M. musculus* in that matings between the two species are successful and the female progeny, but not the male, are fertile. Because of the evolutionary distance between the two species, differences between them are much more numerous than between strains of *M. musculus*. However, the chromosomes have remained co-extensive. This permits the use of interspecific crosses in gene mapping.

The example (Figure 4.3) follows one chromosome pair through the procedure. A number of *musculus* × *spretus* F_1 females are backcrossed to *musculus* males. Recombination in the female germ line gives rise to chromosomes that are part *musculus* and part *spretus*. Thus some regions of the chromosome in the offspring are homozygous *musculus* while others are heterozygous. In this case the DNA of each of the N_2 backcross mice is the resource, and only DNA polymorphisms can be detected. On the other hand, since strains are not being maintained, large numbers of backcross progeny can be afforded.

The genomes of laboratory mice mainly originate from two subspecies, *M. musculus musculus* and *M. m. domesticus*, with overlapping geographical ranges. Less radical inter*sub*specific crosses can be performed between laboratory strains and either *M. m. castaneus* or *M. m. molossinus*. The genetic differences between laboratory mice and the latter two subspecies are about the same as that between *musculus* and *spretus*, and at the same time both sexes of each of the F_1 hybrids are fertile, with the result that both male and female meiosis can be scrutinised.

Cell hybrids

Different cells from the same or different species can be fused together by treating a mixture of the two cells, thrust into close proximity by centrifugation, with an inactivated Sendai virus, or more usually with polyethylene glycol. The initial products are heterokaryons, cells that contain more than

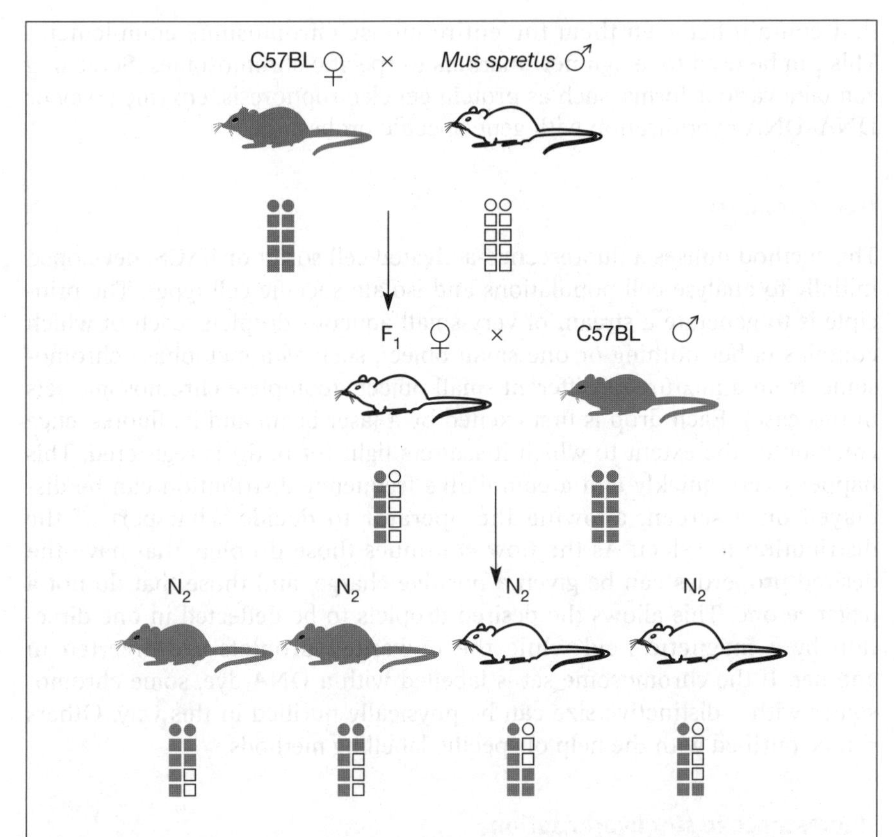

Figure 4.3 Interspecific cross. The fate of a single chromosome is illustrated, with the centromeres shown as circles and notional chromosome segments as squares. A C57BL female is mated to a *Mus spretus* male. F_1 female offspring, which are fertile, are mated to C57BL males. All individuals in the N_2 generation carry a haploid set of C57BL chromosomes from their fathers, together with a haploid set of (mainly) recombinant chromosomes from their mothers. The chromosomes in the N_2 generation illustrate how the different positions of the recombinant breakpoint in a given chromosome can facilitate the mapping of polymorphic genes. Note that the DNA of each individual mouse is the limited resource in this case.

one nucleus coexisting in the same cytoplasm. Some of these contain multiple identical nuclei while others contain nuclei of both types. When the heterokaryons proliferate the nuclei fuse. If the two cell types chosen carry different selective markers, growth in an appropriate medium will kill off unfused cells and fused cells that contain only one type of nucleus, leaving the cells that contain at least one chromosome set from each cell type. As the cells continue to proliferate they shed chromosomes. It happens that human–mouse hybrids preferentially shed human chromosomes, while mouse–Chinese hamster hybrids shed mouse chromosomes. By careful cloning and cytological observation it is possible to derive cell clones from a mouse–hamster hybrid that contain only one or two mouse chromosomes. By screening a number of these clones a set of clones can be chosen that contain between them the entire mouse chromosome complement. This can be used to assign new markers to specific chromosomes. Screening can take various forms, such as protein gel electrophoresis, enzyme assay or DNA–DNA hybridisation with gene-specific probes.

Flow cytometry

This method utilises a fluorescence-activated cell sorter or FACS, developed initially to analyse cell populations and isolate specific cell types. The principle is to generate a stream of very small aqueous droplets, each of which contains either nothing or one small object, such as a metaphase chromosome, from a mixture of different small objects (complete chromosome sets in this case). Each drop is first excited by a laser beam and its fluorescence emission or the extent to which it scatters light (or both) is registered. This happens very quickly and a cumulative frequency distribution can be displayed on a screen, allowing the operator to decide what part of the distribution to select. As the flow continues those droplets that have the desired properties can be given a positive charge, and those that do not a negative one. This allows the desired droplets to be deflected in one direction by a magnetic field while the unwanted droplets are diverted in another. If the chromosome set is labelled with a DNA dye, some chromosomes with a distinctive size can be physically purified in this way. Others can be purified with the help of specific labelling methods.

Fluorescence *in situ* hybridisation

Fluorescence *in situ* hybridisation or FISH allows the assignment of a specific DNA or RNA probe (see below) to a specific chromosome, and in some cases to a region of that chromosome. Both metaphase and interphase chromosomes can be examined in this way, and nascent gene transcripts can be detected by targeting RNA. The starting point is a cytological preparation, which is labelled by hybridisation with one or more specific DNA probes. A wide variety of visualisation techniques is available. The probe itself may be fluorescent or, for greater sensitivity, it may be labelled with biotin or digoxygenin. Biotin-labelled probes are typically detected using fluorescence labelled avidin or, to give more signal amplification, unlabelled avidin can be bound and then detected with fluorescence labelled anti-avidin. Digoxygenin is detected with an anti-digoxygenin antibody. To allow the assignment of a gene-specific signal to a particular chromosome, probes are available that stain a subset of chromosomes or only one chromosome. Signals are detected by fluorescence microscopy or laser-based microscopy and recorded by photography or with a television camera that records the image in a computer file.

It is outside the scope of this book to give a detailed account of molecular mapping methods. Only a brief overview is provided.

It is important to grasp the very large size of the mouse genome. As in other mammals, the haploid mouse genome contains about 3 000 000 000 base pairs, or just over half as many base pairs as there are people on earth. The size can alternatively be expressed as 3 million kilobases (kb) or 3000 megabases (Mb). About 10% of this consists of 'simple sequence' repetitive DNA which is mainly located in the region of the centromeres. Simple sequences are also associated with the telomeres and particularly useful classes of simple sequence, from the point of view of chromosome mapping, are scattered through the genome. In addition several classes of highly repetitive (but not 'simple') sequences, the SINEs (short interspersed elements, retroposons) and LINEs (long interspersed elements, related to retroviruses) are dispersed through the genome, the most notable in the mouse being the B1 family of functionless sequences, which are related to a small number of sequences with important functions. After accounting for these classes of repetitive DNA, about 75% of the genome, or 2.2×10^6 kb, remains and contains perhaps 100 000 genes.

Consideration of the fragmentation of the DNA by restriction enzymes gives some feeling for genome size. Two features of the DNA have to be considered:

- The G+C content (G+C)/(G+C+A+T) which is about 0.42 for the mouse.
- The low frequency of CpG neighbours. This is the result on an evolutionary scale of the specific methylation of C residues within CpG doublets and the deamination of 5-methyl-C to give thymidine (see also Chapter 10). As a result the frequency of the CpG doublet is about 0.01 rather than the 0.061 expected from the base composition alone.

We can calculate an expected frequency of the sequence GAATTC recognised by the *Eco*RI restriction enzyme to be 0.000312 per base pair, which would give an average fragment size of 3.2 kb. This means that *Eco*RI digestion will generate about 10^6 different fragments from the mouse genome. Similarly, the expected frequency of the sequence GGATCC recognised by the *Bam*HI restriction enzyme is 0.000164, with an average fragment size of 6.1 kb, generating 5×10^5 different fragments from the genome.

The restriction enzyme *Sal*I recognises the sequence GTCGAC. As a consequence of the CpG dimer at its centre the expected frequency of this sequence is actually 15% of 0.000164 or 0.0000246, with an average fragment size of 40 kb. *Not*I recognises the sequence GCGGCCGC, which is eight nucleotides long, consists entirely of Cs and Gs and contains 2 CpG dimers. This gives it an expected frequency of 8.5×10^{-8} and an average fragment size of 12 Mb (12 000 kb), and it is expected to generate only about 250 different fragments from the genome.

Restriction fragment length polymorphisms

It is obvious that the longer a nucleotide sequence, the less frequent will be its occurrence in 'random DNA'. Randomly chosen sequences of 16 nucleotides can be expected to occur on average once in the mouse genome and if a 17 nucleotide sequence does occur in the genome, it can reasonably be expected to be unique. This simple view is complicated by the occurrence of repeated sequences, gene duplications and tightly knit gene families in the genome, but very few DNA regions of 1 or 2 kb will not contain a unique sequence. Such a unique sequence can be called a sequence probe, and can be a molecularly cloned DNA fragment or can be synthesised chemically. A sequence probe has the property of being able to form DNA duplexes with the target DNA, the single complementary genomic DNA sequence. By the use of appropriate methods the probe can be used to detect the presence of the target DNA in such a way that we obtain information about it that can be used, for example, in gene mapping. The example given in Box 4.1 is by way of clarification.

BOX 4.1: DETECTING RFL POLYMORPHISMS

Suppose we isolate DNA from an inbred laboratory mouse and digest it with a restriction enzyme like *Eco*RI, generating approximately one million fragments with an average size of 3–4 kb. We can sort the fragments roughly on a size basis by placing the mixture at one location in a flat agarose gel and applying an electrical potential (e.g. 10 volts/cm) across it for a few hours. All the fragments move towards the anode, but smaller fragments move faster, and therefore further, because the gel matrix retards larger and larger fragments more and more. The matrix also prevents the DNA molecules from floating away. We immerse the gel in a bath of alkali to denature the DNA (i.e. separate the complementary strands) and then, by placing the gel surface in contact with a sheet of material that binds DNA (a filter), we can transfer the DNA out of the gel onto the filter rather in the manner of making a photocopy (except that we use up the original in the process). We now have an array of DNA fragments, with their complementary strands separated and bound to the filter with the shortest fragments towards one end and the longest towards the other.

We label the probe so that it can be detected in the presence of the filter-bound genomic DNA. The label can be a radioactive isotope or a fluorescent dye or can consist of a molecule such as biotin which can be incorporated into the DNA and act as a hook for another detection agent. The labelled probe is denatured and then annealed with the filter, i.e. held under conditions that favour the formation of DNA duplexes. The probe may be denatured double-stranded DNA, in which case each of its strands will form duplexes with the complementary target strand, or it may be a single-stranded polynucleotide which will form a duplex with only one strand of the target. Finally excess probe (i.e. probe that has not formed duplexes with the complementary sequences in the target) is washed away.

We now have probe associated with one of the million or so fragments attached to the filter (sometimes more than one, but ignore this possibility for the time being). The size of the target fragment was determined by the locations of the nearest sites, one to each side, recognised by whatever restriction enzyme we used to digest the DNA, and this size determined by applying the detection procedure to visualise the probe.

We can apply this procedure repeatedly employing a different restriction enzyme each time, and each time the target fragment will be of a different size (see Table). By simple extensions of the technique we can obtain the information required to construct a map showing the locations of different restriction enzyme sites around the target sequence (derived map). If we then repeat the process with a different inbred mouse strain, we are likely to derive a map that differs in one or more respects from the first (polymorphic variant). This is a restriction fragment length polymorphism, and it will usually be possible to distinguish the DNA of the two inbred strains after digestion with a single restriction enzyme or at most two (simultaneously). We will also be able to detect the two allelic forms in a heterozygote (e.g. offspring of a mating between mice of the two strains). This RFLP can then be used in different mapping strategies to fix its location in relation to other known markers.

	Fragment lengths, kb	
Enzyme	*Normal individual*	*Polymorphic variant*
*Pst*I	10	10
*Pvu*II	6 & 8	6 & 8
*Sac*I	7	14
*Pst*I & *Pvu*II	6 & 2	6 & 2
*Pst*I & *Sac*I	5	5
*Pvu*II & *Sac*I	3 & 4	3 & 8

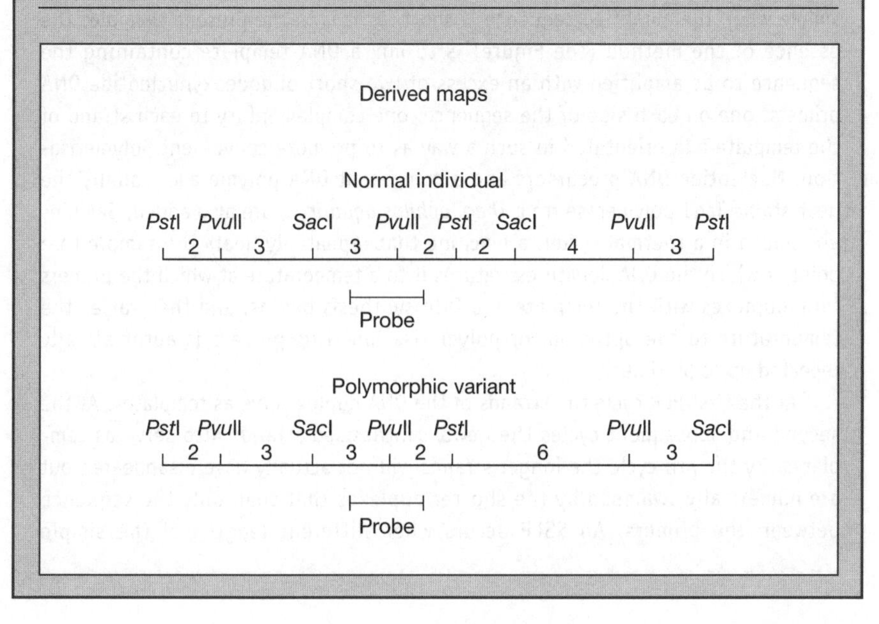

Minisatellites and microsatellites

The genome contains several different types of direct tandemly repeated DNA sequence, for example, the ribosomal RNA genes, the 5S RNA genes and the large blocks of simple-sequence (satellite) DNA in the pericentromeric regions. There are also small sets of tandem repeats dispersed through the genome, called minisatellites (small sets of short repeat sequences) and microsatellites (small sets of extremely short – dinucleotide, trinucleotide – repeat sequences). These share a strong tendency to polymorphic variation, at any given locus, in the *number* of the repeats in the array, either due to replication errors or to unequal crossing-over.

BOX 4.2: DETECTING SSLPs BY PCR

The use of SSLPs depends on prior work to identify them. This is done in two ways: by screening libraries of short genomic DNA fragments by hybridisation with an appropriately designed DNA oligonucleotide, and by scanning the database of published mouse nucleotide sequences. In the latter case the nucleotide sequences flanking each repeat will be available; in the former they require to be determined. A primer sequence is then chosen from each side of each repeat.

Since the PCR method was first shown to be practicable in 1985 (Saiki *et al.* 1985) and especially since heat-stable polymerases were introduced (Saiki *et al.* 1988) its use has mushroomed and literally hundreds of protocols have been published. The objective of the method is to selectively amplify a particular DNA sequence from among a complex mixture of sequences. It is sufficiently sensitive to amplify a single sequence of 100 to 10 000 nucleotides from a few copies of the entire 3×10^9 nucleotide mammalian genome to the point that it can be visualised on an agarose or acrylamide gel stained with eithidium bromide. More stringent conditions are required to amplify longer sequences, but the procedure is quite simple when the amplified sequence is short, as it is in the present example. The essence of the method (see Figure) is to mix a DNA template containing the sequence to be amplified with an excess of two short oligodeoxynucleotide DNA primers, one on each side of the sequence, one complementary to each strand of the template and orientated in such a way as to promote convergent polymerisation. Nucleotide DNA precursors and a competent DNA polymerase, usually the heat-stable Taq1 polymerase from *Thermophilus aquaticus*, are also added. Samples are placed in a thermal cycler, a machine that repeatedly heats the sample to a point at which the DNA denatures, returns it to a temperature at which the primers form duplexes with the template and DNA synthesis begins, and then raises the temperature to the optimum for polymerisation. This process is automatically repeated up to 50 times.

At the first PCR cycle the strands of the DNA duplex serve as templates. At the second and subsequent cycles the newly synthesised strands also serve as templates. By the *n*th cycle the longer strands will not actually have disappeared but are numerically swamped by the shorter duplexes that span only the sequence between the primers. An SSLP occurs when different lengths of the simple

▶

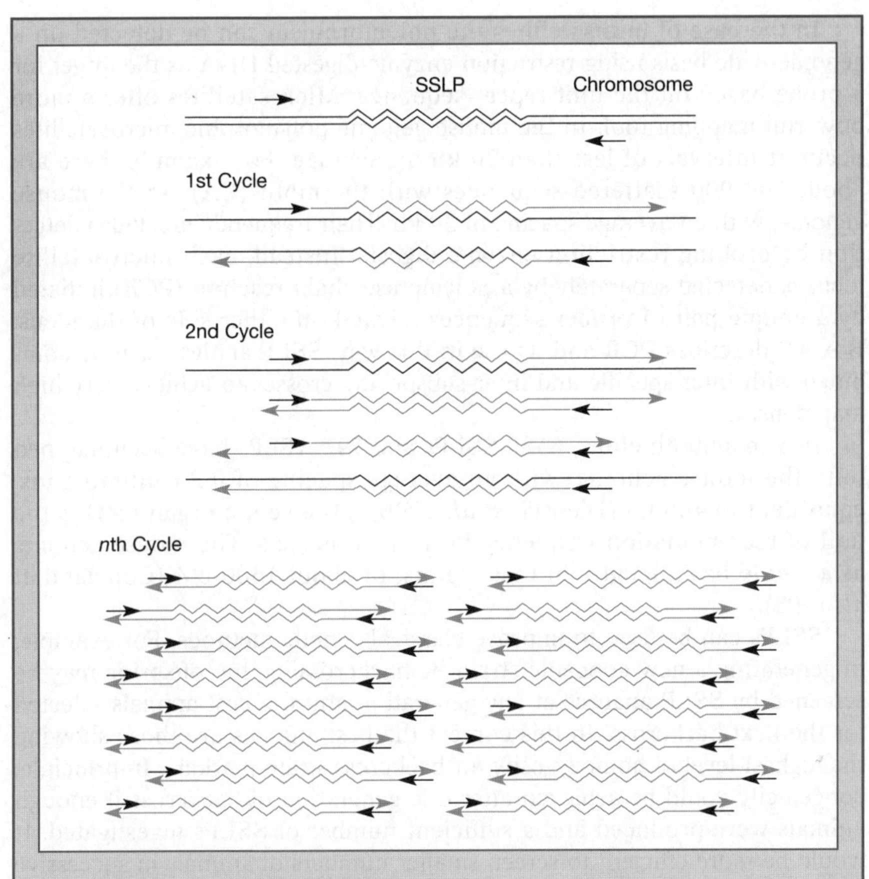

SSLP analysis by PCR. A simple sequence (wavy double line) is shown embedded in a chromosome (double line). The two lines represent the strands of the DNA duplex. Primers are shown as small black arrows, newly synthesised DNA in blue.

sequence occupy the same chromosomal position. Since they lie between the same flanking regions both are amplified from the same primers, while the amplification products are distinguishable by their different sizes.

Several computer programs are available that will select primers from a sequence to conform with predetermined requirements, such as optimal annealing temperature, G+C content, lack of self-complementarity, length of product, etc. Taq1 is a non-proof-reading polymerase with a significant error rate. If this is a problem, thermostable proof-reading DNA polymerases are also commercially available.

In the application of PCR to detecting SSLPs, the lengths of the sequences amplified with the same primer pair in different genomes are of course crucial. A procedure which maximises accuracy and minimises the work load is to employ fluorescence-labelled primers with templates of total genomic DNA and to detect the amplified bands with an automated nucleotide sequencing apparatus.

In the case of minisatellites the polymorphism can be detected on a genome-wide basis using restriction enzyme-digested DNA as the target for a probe based on the unit repeat sequence. Microsatellites offer a more powerful mapping tool. In the mouse genome polymorphic microsatellites occur at intervals of less than 20 kb on average. For example there are about 100 000 scattered sequences with the motif $(CA)_n$ in the mouse genome, with an average spacing of 30 kb. Their frequency precludes detection by probing restriction enzyme digests. Instead, each microsatellite locus is detected separately by a polymerase chain reaction (PCR) initiated by a unique pair of primer sequences located on either side of the locus. Box 4.2 describes PCR and its use in this way. SSLP analysis can be combined with inter-specific and inter-subspecific crosses to achieve very high map densities.

In a mammoth effort, 6580 SSLPs and 797 RFLPs have been mapped onto the mouse genome, with an average spacing of 0.2 centimorgans, equivalent to 400 kb (Dietrich *et al.* 1996). (The centimorgan (cM) is the unit of recombination frequency in normal meiosis. The mouse genome as a whole has a length, in these terms, of about 1500 cM (Copeland *et al.* 1993).)

SSLPs can be used to improve classical genetic methods. For example, in generating a new congenic strain by backcrossing, the offspring may be screened by SSLP analysis at any generation and the best animals selected for the next backcross. In this context the best animals are those showing the highest level of homozygosity for backcross strain markers. In principle, congenicity could be achieved after one generation of backcross if enough animals were produced and a sufficient number of SSLPs investigated. It would be more efficient to screen smaller numbers of animals in successive generations. Obviously once a marker had become homozygous it would be unnecessary to use it to screen a later generation (Markel *et al.* 1997).

CpG islands and expressed sequence tags

CpG islands are blocks of DNA sequence, associated mainly with the 5'-regions of transcription units (genes) and having a high G+C content. Despite, or because of, the high incidence of CpG doublets in these regions, they uncharacteristically fail to become C-methylated. Expressed sequence tags (ESTs) are randomly cloned reverse transcripts of parts of messenger RNA molecules. Neither CpG islands nor ESTs show much intrinsic polymorphism although ESTs in particular are frequently useful as probes to detect linked polymorphisms. The discovery of a CpG island in a DNA sequence is a presumptive indicator of a transcription unit, while the discovery of a sequence that perfectly matches a known EST is definitive evidence of transcription.

Cloned DNA sequences

The contiguous length of DNA that can be cloned as a yeast artificial chromosome (YAC) or bacterial artificial chromosome (BAC) ranges up to megabases. Cosmids can contain more than 100 kb, bacteriophage λ vectors up to 20 kb and multicopy plasmids from <1 to 20 kb. Restriction maps of very large cloned fragments can be generated by employing rare-cutting restriction enzymes such as *Not*I (see above) and special electrophoretic methods (e.g. pulse field gel electrophoresis). Contiguous chromosome regions (contigs) that are spanned by several long overlapping cloned sequences can be defined in this way. These can be analysed more closely by subcloning into cosmid, bacteriophage or plasmid vectors and the use of more frequently cutting restriction enzymes. And of course, other mapping tools such as those described above can be employed in the context of contigs.

References

Copeland, N.G., Jenkins, N.A., Gilbert, D.J., Eppig, J.T., Maltais, L.J., Miller, J.C., Dietrich, W.F., Weaver, A., Lincoln, S.E., Steen, R.G., Stein, L.D., Nadeau, J.H., and Lander, E.S. (1993) A genetic linkage map of the mouse: current applications and future prospects. *Science*, **262**, 57–66.

Dietrich, W.F., Miller, J., Steen, R., Merchant, M.A., Damron-Boles, D., Husain, Z., Dredge, R., Daly, M.J., Ingalls, K.A., O'Connor, T.J., *et al.* (1996) A comprehensive genetic map of the mouse genome. *Nature*, **380**, 149–152.

Markel, P., Shu, P., Ebeling, C., Carlson, G.A., Nagle, D.L., Smutko, J.S., and Moore, K.J. (1997) Theoretical and empirical issues for marker-assisted breeding of congenic mouse strains. *Nature Genetics*, **17**, 280–284.

Nagamine, C.M., Shiroishi, T., Miyashita, N., Tsuchiya, K., Ikeda, H., Takao, N., Wu, X.L., Jin, M.L., Wang, F.S., Kryukov, A.P., *et al.* (1994) Distribution of the molossinus allele of *Sry*, the testis-determining gene, in wild mice. *Molecular Biology & Evolution*, **11**, 864–874.

Saiki, R.K., Scharf, S., Faloona, F., Mullis, K.B., Horn, G.T., Erlich, H.A., and Arnheim, N. (1985) Enzymatic amplification of beta-globin genomic sequences and restriction site analysis for diagnosis of sickle cell anemia. *Science*, **230**, 1350–1354.

Saiki, R.K., Gelfand, D.H., Stoffel, S., Scharf, S.J., Higuchi, R., Horn, G.T., Mullis, K.B., and Erlich, H.A. (1988) Primer-directed enzymatic amplification of DNA with a thermostable DNA polymerase. *Science*, **239**, 487–491.

Further reading

General

Silver, L.M. (1995) *Mouse genetics: concepts and applications*, OUP, Oxford.

Inbred and RI strains

Festing, M.W. (1979) *Inbred strains in biomedical research*, Macmillan, London.

Interspecific crosses

Avner, P., Amar, L., Dandolo, L., and Guenet, J.L. (1988) Genetic analysis of the mouse using interspecific crosses. *Trends in Genetics*, **4**, 18–23.

PCR

Mullis, K., Faloona, F., Scharf, S., Saiki, R., Horn, G., and Erlich, H. (1992) Specific enzymatic amplification of DNA in vitro: the polymerase chain reaction. *Cold Spring Harbor Symposia on Quantitative Biology*, **51 Pt 1**, 236–273.

Erlich, H.A. and Arnheim, N. (1992) Genetic analysis using the polymerase chain reaction. *Annual Review of Genetics*, **26**, 479–506.

Mapping

Copeland, N.G. and Jenkins, N.A. (1991) Development and applications of a molecular genetic linkage map of the mouse genome. *Trends in Genetics*, **7**, 113–118.

FISH

Trask, B.J. (1991) Fluorescence *in situ* hybridization: applications in cytogenetics and gene mapping. *Trends in Genetics*, **7**, 149–154.

CHAPTER FIVE

Sex determination

The X and Y chromosomes

Female mammals carry two X chromosomes while males have one X and one Y chromosome, the X being longer than the Y. The X and Y chromosomes share only a very short region of homology, at the ends of the chromosomes in both human and mouse (Figure 5.1). In the female the two X chromosomes pair normally at meiosis. In the male, the homologous regions of the X and the Y pair and invariably recombine. As a result, the genes in this region are indistinguishable from autosomal genes on the basis of genetic segregation data and the region is consequently said to be pseudoautosomal. The unique part of the X chromosome contains a great many genes that are required in both sexes. Females obviously have two copies and males one copy of these X-linked genes. The expression of many X-linked genes is equalised between the sexes by X-inactivation (Chapter 6). The unique part of the Y chromosome carries genes involved in sex determination and spermatogenesis.

The *Sry* gene and Sxr mice

In considering the *Sry* gene and sex-reversed (Sxr) mice we distinguish between normal and phenotypic males and females. Normal males and females are reproductively fully functional. On the other hand phenotypic males and females possess the genitalia of the sex in question, though they are not always fully developed. They may also exhibit the appropriate secondary sexual characteristics, but are usually not fertile. This comes about because they are genetically equipped to initiate male or female sexual differentiation, but lack other functions required for correct gonadal development or germ cell maturation.

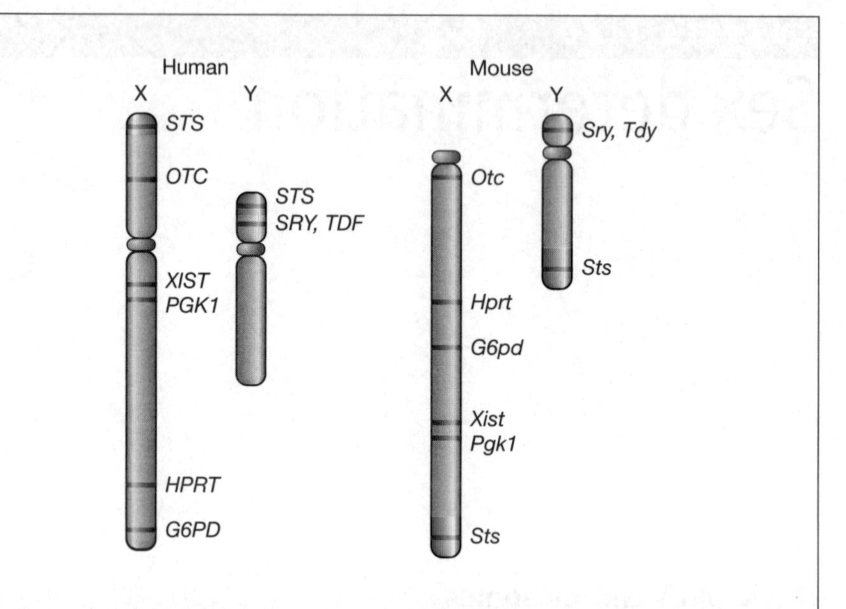

Figure 5.1 Diagram of the X and Y chromosomes of man and the mouse. Pseudoautosomal regions are shown in blue and centromeres as ovals. The positions of the sex-determining genes *SRY/TDF* and *Sry/Tdy*, the X-inactivation centres *XIST* and *Xist* and a few other homologous genes are marked. *STS*, steroid sulphatase; *OTC*, ornithine transcarbamylase; *PGK1*, phosphoglycerate kinase-1; *HPRT*, hypoxanthine-guanine phosphoribosyl transferase; *G6PD*, glucose-6-phosphate dehydrogenase.

Sry

XO individuals (i.e. carrying one X chromosome and no Y) are phenotypic females and XXY individuals are phenotypic males. These rather simple observations led to the conclusion that sex determination in mammals is initiated by a gene or genes carried on the Y chromosome, rather than by the ratio of X chromosomes to autosomes, as in *Drosophila* for example. A hypothetical sex determining region was named *Tdy* on the mouse Y chromosome, and *TDF* (testis determining factor) on the human. The location of *TDF* (the human genetic factor) was pinpointed by studying XX phenotypic males and XY phenotypic females. XX males turn out to carry a piece of the Y on the X chromosome inherited from their fathers, while XY females carry Y chromosome deletions. Both types of chromosomal abnormality arise repeatedly, but not at the same break-points, so that different XX males carry different fragments of the Y chromosome attached to the X and different XY females carry different Y chromosome deletions. If the Y fragments are assumed to carry *TDF* and the Y deletions to lack it, together they define a region of the Y within which *TDF* ought to be located.

Positional cloning involves searching a chromosome region, to which a genetic effect has been mapped, for genes recognised by signature sequences such as open reading frames, promoters and so on, and then testing any genes found against predictions as to evolutionary conservation, tissue-specificity, developmental timing of expression, etc. (see Figure 5.2). The first likely candidate for the role of *TDF* was a gene named *ZFY* which codes for a zinc-finger protein. However, *ZFY* turned out not to have the properties expected, and later another gene in the same chromosomal region called *SRY* was found to correspond to *TDF* (Figure 5.2) (Sinclair *et al.* 1990).

SRY is carried in the expected region of the human Y chromosome, has no X chromosome homologue and codes for a transcription factor that contains a so-called HMG-box DNA binding motif. Mutations in the HMG-box of *SRY*, identified by sequencing the gene isolated from certain XY females, have the same effect as *TDF* deletions (Goodfellow and Lovell-Badge 1993). However, it should be noted that only a small minority of XY females have mutations or deletions in *SRY*. This is not unexpected, since other genes are known to be required for the development of male sex organs and for spermatogenesis (discussed below).

The mouse homologue of *SRY* is called *Sry*. The positions of the sex-determining genes on the human and mouse chromosomes are similar, on the short arm of the Y, which in the mouse is very short indeed, and close to the centromere (see Figure 5.1). Note however, that all of the genes shown on the X chromosomes, as well as the pseudoautosomal regions, occupy different relative positions in the two genomes. To a large extent the chromosomes of humans and mice are made up of syntenic regions, short regions of chromosome in which the order of several or even many genes is the same in the two species. However, the two sets of syntenic regions are scrambled up more or less at random and it is an inescapable conclusion that this is the result of a large number of chromosomal translocations that have occurred on the evolutionary time-scale. Syntenic regions can of course be traced in other mammals as well. Where syntenic regions have

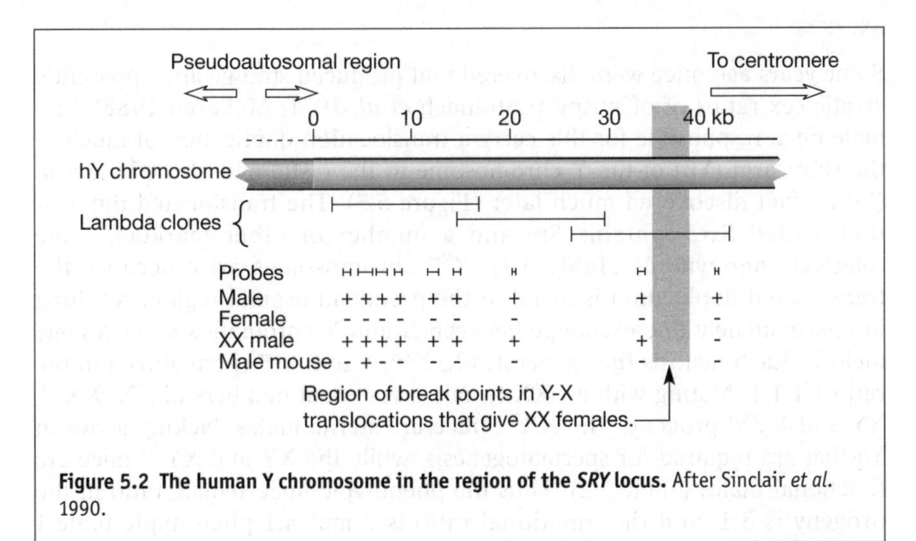

Figure 5.2 The human Y chromosome in the region of the *SRY* locus. After Sinclair *et al.* 1990.

been identified, mapping data from the two species can often be amalgamated and each used to predict as yet undiscovered features of the other.

In addition to the data linking *SRY* mutations to sex reversal, other lines of evidence point to the *SRY/Sry* product as an essential switch mechanism in sex determination. In the mouse a short (11 kb) deletion that includes *Sry* yields a Y chromosome that produces the sex-reversed XY female phenotype. Transgenic manipulation provides further evidence that *Sry* has the sex-determining role. When a 14 kb DNA fragment containing *Sry* but no other genes was introduced by microinjection into 1-cell embryos, 25% of the transgenic XX offspring developed as sterile phenotypic males, that is with the same phenotype as many XX men (Koopman *et al.* 1991). (In the 75% that developed as females the *Sry* gene is assumed not to be expressed, or to be expressed at a level below some threshold necessary for masculinisation. See Chapter 10 for a discussion of variable transgene expression.)

The SRY protein is a member of the HMG superfamily of DNA-binding proteins. SRY proteins are not evolutionarily conserved other than within the DNA-binding domain itself, and nearly all inactivating mutations occur within this DNA-binding domain. SRY protein binds to a sequence within the promoter of the *AMH* gene (see below), although not as strongly as it binds to a related sequence, and to a sequence within the promoter of the *P450 aromatase* gene (Haqq *et al.* 1993). Expression of the *AMH* gene is upregulated in the male, while that of the *P450 aromatase* gene is downregulated. Although there seems to be little doubt that these are targets for the SRY protein, it clearly does not act as a conventional *trans*-acting transcriptional activator or repressor. Its action may relate to the fact that it has been found to force double-stranded DNA to bend through 85° *in vitro* (Giese *et al.* 1992; Ferrari *et al.* 1992).

Sxr mice

Some years ago mice were discovered that produced strange and apparently erratic sex ratios of offspring (Cattanach *et al.* 1971; McLaren 1988). The male mice responsible for this carry a translocation duplication of much of the short arm (Yp) of the Y chromosome to the distal end of the long arm (Yq), a fact discovered much later (Figure 5.3). The translocated duplication, called *Sxr*, contains *Sry* and a number of other markers, some relatively anonymous (Table 5.1). X^{Sxr} chromosomes arise because the translocated duplication is distal to the pseudoautosomal region, resulting in approximately one exchange between X and Y chromatids at each male meiosis. Each meiosis thus generates X, X^{Sxr}, Y and Y^{sxr} spermatozoa in the ratio 1:1:1:1. Mating with an XX female gives equal numbers of XX, XX^{Sxr}, XY and XY^{Sxr} progeny. The XX^{Sxr} mice are sterile males, lacking genes in Yq that are required for spermatogenesis, while the XY and XY^{Sxr} mice are functional males (Table 5.2). Thus the phenotypic male/female ratio in the progeny is 3:1, and the functional ratio is 2 males:1 phenotypic male:1

Table 5.1 Markers on the short arm, Yp, of the mouse Y chromosome.

Zfy1	Zinc finger protein
Smcy	Selected mouse cDNA
Hya	Male-specific antigen
Zfy2	Zinc finger protein
Sry	Sex determining region
Rbm	RNA binding motif
Sx1	Repetitive sequence
Centromere	

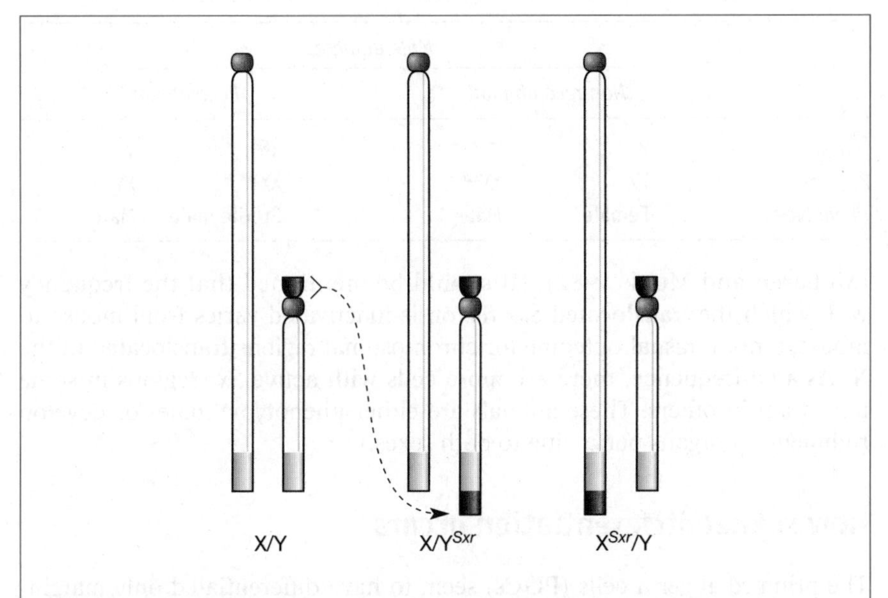

Figure 5.3 The chromosomal translocation carried by Sxr mice. The normal position of the *Sry* gene is on the short arm of the Y chromosome. Translocation to the distal end of the pseudoautosomal region generates the *Sry*-carrying X and Y chromosomes.

female. Half of the functional males are reproductively normal, while the others recapitulate the reproductive performance of their sires (Table 5.3).

Interestingly, the X^{Sxr} chromosome can be transmitted through another generation by taking advantage of a translocation between the X chromosome and chromosome 16 (X(T16H), called Searle's translocation). X-inactivation (see Chapter 6) is normally a random process, one X being inactivated in half of the cells and the other X in the remainder. The effect of Searle's translocation is to suppress the inactivation of the translocated region of the X(T16H) chromosome, while conversely the other X chromosome is inactivated. In mice that carry both an X(T16H) chromosome and an X^{Sxr} chromosome the *Sxr* region is inactivated and they develop into functional females, able to transmit the X^{Sxr} chromosome to their offspring

Table 5.2 Phenotypes of mice with normal chromosomes and chromosomes carrying *Sxr* translocations.

Sex chromosome combination	*Phenotype*
X/X	Normal female
X/Y	Normal male
X^{Sxr}/X	Sterile male
X/Y^{Sxr}	Normal male

Table 5.3 Offspring of a cross between an X/Y^{Sxr} male and an X/X female.

	Male gametes			
	Non-recombinant		*Recombinant*	
Types	X	Y^{Sxr}	X^{Sxr}	Y
Zygotes	XX	XY^{Sxr}	XX^{Sxr}	XY
Phenotypes	Female	Male	Sterile male	Male

(McLaren and Monk 1982). (It should be mentioned that the frequency with which the translocated *Sxr* region is inactivated varies from mouse to mouse, a not unusual outcome for chromosomal regions translocated to the X. As a consequence, there are more cells with active *Sxr* regions in some mice than in others. These animals are either phenotypic males or develop rudimentary organs pertaining to both sexes.)

How sexual differentiation occurs

The primordial germ cells (PGCs) seem to have differentiated only marginally from their ES cell precursors. Following a period in culture and treatment with bFGF, they share with ES cells the property of forming chimeras when introduced into host blastocysts and contributing to the germ line.

The PGCs can be recognised in the gastrulating embryo because they contain unusually large amounts of an alkaline phosphatase and can be differentially stained with a chromogenic substrate. Using this method of identification, they are first recognised in the posterior part of the extra-embryonic mesoderm at about $7\frac{1}{2}$ dpc and then migrate in an anterior direction, multiplying as they go, until they settle down between $10\frac{1}{2}$ and 11 dpc in a region of the developing embryo called the genital ridge. Subsequently, the PGCs develop into gametes while somatic cells in the ridge provide the appropriate environment, depending on the sex of the individual. In particular, they provide supporting cells, follicle cells in the ovary and Sertoli cells in the testis, that communicate through gap junctions with the developing germ cells.

Perhaps surprisingly, the supporting cells rather than the PGCs determine the sex of the animal. The original evidence for this comes from XX↔XY chimeras (Burgoyne *et al.* 1988; McLaren 1991). In such animals the supporting cells may develop in a male or female direction depending on the relative contributions of the two cell types to the chimera. In a predominantly XY chimera a larger number of the supporting cells are XY and this leads them to develop in a male direction. Subsequently, both XX and XY PGCs begin to differentiate down the male path. Due to their requirement for Y chromosome functions, noted previously, the spermatogonia that develop from XX PGCs degenerate; the XY spermatogonia are functional however, and allow a normal fertile male phenotype to develop. In a predominantly XX chimera the supporting cells develop in the female direction, becoming follicle cells, and both XX and XY PGCs develop into immature oocytes. In this case most of the XY oocytes degenerate and disappear but some are ovulated along with the XX oocytes.

The most telling evidence that follicle cells and Sertoli cells arise from the same cell lineage is that, following the degeneration of oocytes under a variety of artificial situations, the follicle cells transdifferentiate into Sertoli-like cells and in some cases produce anti-Mullerian hormone (see below). For example, transdifferentiation of Sertoli cells into follicle cells is reported to occur in some of the T16H/X^{Sxr} males and hermaphrodites described above. If oocytes are not present up to a certain time in ovarian development, the follicle cells disappear altogether, indicating bidirectional communication between an oocyte and its follicle cells. Sertoli cells, on the other hand, form testis cords even in the absence of spermatogonia.

The determining function of the supporting cells seems to be due to the activity of the factor encoded by the *Sry* gene, which becomes transcriptionally active in male ridge cells at about $10\frac{1}{2}$ dpc. Prior to the onset of *Sry* activity the gonad is asexual (termed the indifferent gonad) and, as well as the sexually undifferentiated germ cells and supporting cells, it contains primordia appropriate to both sexes, the female Müllerian ducts and the male Wolffian ducts. In the female there is as we know no *Sry* gene. In consequence the supporting cells become follicle cells, followed by development of the ovary, the further development of the Müllerian ducts and the eventual degeneration of the Wolffian ducts. Differentiation in the female direction is said to be the default pathway, because it occurs in default of *Sry* expression, which may gratify feminists or again may not. In the male, the *Sry* gene becomes active in the supporting cells; these develop into Sertoli cells and produce a diffusible protein, the anti-Müllerian hormone (AMH) or Müllerian inhibitory substance (MIS), and this initiates the degeneration of the Müllerian ducts by programmed cell death (apoptosis). Cells of another type that lie outside the tubules, the Leydig cells, also differentiate in the developing male gonad and begin to secrete testosterone, which stimulates the development of the Wolffian ducts and affects other sexual appendages (Figure 5.4).

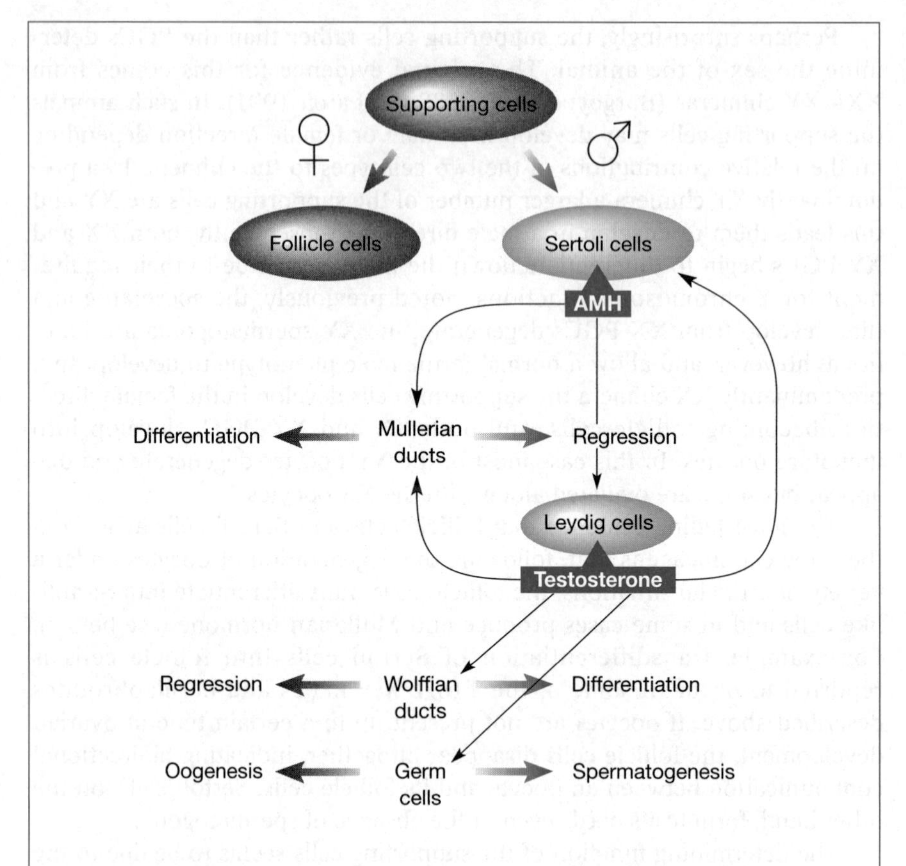

Figure 5.4 Some of the relationships between cell types in sex differentiation. The male cascade is set in motion by the expression of *Sry* in the supporting cells. Large blue arrows indicate the targets for AMH and testosterone. Their effects drive development in the male direction, to the right. In the absence of these effects development proceeds in the female direction, to the left.

AMH is a glycoprotein of the TGF-β superfamily and its receptor, AMH-R, is a TGF-β Type II receptor serine/threonine kinase. *Sry* transcripts are detectable at $10\frac{1}{2}$ dpc, maximal at about $11\frac{3}{4}$ dpc and no longer detectable by $13\frac{1}{2}$ dpc. *Amh* transcripts are first detected at $11\frac{1}{2}$ dpc (Jeske *et al.* 1996). That the expression of AMH depends upon SRY is clear enough, but as yet the mechanism of SRY action is not. It has been suggested that SRY may repress or antagonise an inhibitor of AMH expression. XY mice homozygous for null alleles of *Amh* or *Amh-r* produced by gene targeting are pseudohermaphrodites, with both male and female reproductive organs, consistent with a primary role of AMH in regression of female primordia (Behringer *et al.* 1994; Mishina *et al.* 1996). Human XY individuals homozygous for a null allele of either the *AMH* gene or the *AMH-R* gene show the same effects, a condition called persistent Müllerian duct syndrome.

In the normal female mouse AMH is expressed in the ovary, but only from three days after birth, long after the female sex primordia have ceased to be sensitive to it and are in fact almost fully developed. As anticipated, XX mice carrying a transgene that causes them to express AMH constitutively from an early age have no uterus or oviducts and their ovaries regress (Behringer *et al.* 1990).

Differentiation of the extra-tubular Leydig cells of the testis is also under the ultimate control of SRY, and may be mediated by a signalling molecule called Desert hedgehog which is produced in the Sertoli cells, probably in response to SRY. Males homozygous for a *Dhh* null mutation fail to produce spermatozoa (Bitgood and McMahon 1995). The testosterone secreted by the Leydig cells has an important part to play in masculinisation, as we know from the effects of testosterone and other androgens on female athletes for example. Different mutations in the androgen receptor cause a range of androgen insensitivity syndromes in XY individuals, from a well-proportioned female supermodel phenotype, but lacking functional reproductive organs (testicular feminisation, *Tfm*), to minor impairment of fertility. Several enzymes required for the synthesis of testosterone are produced in the Leydig cells, while 5-α-reductase is required for the synthesis of dihydrotestosterone. Thus several genes involved in the production and utilisation of androgens offer a potential for XY sex reversal when *Sry* itself is unaffected. In addition to the genes already mentioned, others are known to disrupt sex determination, and there are undoubtedly yet others still to be discovered.

A little about cell migration

Two loci in the mouse affect at least three migratory cell systems, the melanocytes which migrate from the neural crest to the skin, the haemopoietic progenitor cells which migrate from the blood islands to the foetal liver to the adult bone marrow, and the PGCs described above. These are the Steel (*Sl*) and Dominant white spotted (*W*) loci. *Sl* codes for the Stem Cell Factor (or Steel factor) which is a mitogen and the ligand for the product of the *W* gene, a transmembrane tyrosine kinase receptor and protooncogene known as *c-kit*. Steel factor is produced along migratory pathways and homing sites in two forms: a soluble form and a form that binds to the extracellular matrix, while the *c-kit* receptor is produced by the migrating cells. As you would expect, *W* mutations are cell-autonomous. This is to say that in *W/W*↔+/+ chimeras the +/+ cells, which express *c-kit*, migrate and proliferate normally while *W/W* cells, which do not express the receptor, fail to do so; Steel factor is provided by cells of both types along the migration route. Conversely *Sl* mutations are not cell-autonomous. In *Sl/Sl*↔+/+ chimeras both the *Sl/Sl* and the +/+ cells express the receptor and migrate more or less normally, the extracellular Steel factor being provided by the +/+ cells along the migration route. In this context we can say that the *Sl/Sl* cells are rescued when incorporated with wild type cells in chimeras, while *W/W* cells are not (Table 5.4).

Table 5.4 Properties of *W* and *Sl* genes and their products.

Locus	*Result from chimeras*	*Product*	*Function*
W	Cell-autonomous	*c-kit*	Tyrosine kinase receptor
Sl	Not cell-autonomous	Steel factor	*c-kit* ligand

References

Behringer, R.R., Cate, R.L., Froelick, G.J., Palmiter, R.D., and Brinster, R.L. (1990) Abnormal sexual development in transgenic mice chronically expressing Müllerian inhibiting substance. *Nature*, **345**, 167–170.

Behringer, R.R., Finegold, M.J., and Cate, R.L. (1994) Müllerian-inhibiting substance function during mammalian development. *Cell*, **79**, 415–425.

Bitgood, M.J. and McMahon, A.P. (1995) Hedgehog and Bmp genes are coexpressed at many diverse sites of cell–cell interaction in the mouse embryo. *Developmental Biology*, **172**, 126–138.

Burgoyne, P.S., Buehr, M., Koopman, P., Rossant, J., and McLaren, A. (1988) Cell-autonomous action of the testis-determining gene: Sertoli cells are exclusively XY in XX-XY chimaeric mouse testes. *Development*, **102**, 443–450.

Cattanach, B.M., Pollard, C.E., and Hawker, S.G. (1971) Sex-reversed mice: XX and XO males. *Cytogenetics*, **10**, 318–337.

Ferrari, S., Harley, V.R., Pontiggia, A., Goodfellow, P.N., Lovell-Badge, R., and Bianchi, M.E. (1992) SRY, like HMG1, recognizes sharp angles in DNA. *EMBO Journal*, **11**, 4497–4506.

Giese, K., Cox, J., and Grosschedl, R. (1992) The HMG domain of lymphoid enhancer factor 1 bends DNA and facilitates assembly of functional nucleoprotein structures. *Cell*, **69**, 185–195.

Goodfellow, P.N. and Lovell-Badge, R. (1993) SRY and sex determination in mammals. *Annual Review of Genetics*, **27**, 71–92.

Haqq, C.M., King, C.Y., Donahoe, P.K., and Weiss, M.A. (1993) SRY recognizes conserved DNA sites in sex-specific promoters. *Proceedings of the National Academy of Sciences of the United States of America*, **90**, 1097–1101.

Jeske, Y.W., Mishina, Y., Cohen, D.R., Behringer, R.R., and Koopman, P. (1996) Analysis of the role of Amh and Fra1 in the Sry regulatory pathway. *Molecular Reproduction & Development*, **44**, 153–158.

Koopman, P., Gubbay, J., Vivian, N., Goodfellow, R., and Lovell Badge, R. (1991) Male development of chromosomally female mice transgenic for Sry. *Nature*, **351**, 117–121.

McLaren, A. (1988) Sex determination in mammals. *Trends in Genetics*, **4**, 153–157.

McLaren, A. (1991) Development of the mammalian gonad: the fate of the supporting cell lineage. *Bioessays*, **13**, 151–156.

McLaren, A. and Monk, M. (1982) Fertile females produced by inactivation of an X chromosome of 'sex-reversed' mice. *Nature*, **300**, 446–448.

Mishina, Y., Rey, R., Finegold, M.J., Matzuk, M.M., Josso, N., Cate, R.L., and Behringer, R.R. (1996) Genetic analysis of the Müllerian-inhibiting substance signal transduction pathway in mammalian sexual differentiation. *Genes & Development*, **10**, 2577–2587.

Sinclair, A.H., Berta, P., Palmer, M.S., Hawkins, J.R., Griffiths, B.L., Smith, M.J., Foster, J.W., Frischauf, A.M., Lovell Badge, R., and Goodfellow, P.N. (1990) A gene from the human sex-determining region encodes a protein with homology to a conserved DNA-binding motif. *Nature*, **346**, 240–244.

Further reading

General

Gilbert, S.F. (1997) *Developmental biology*, Sinauer Associates, Sunderland, Mass.

McLaren, A. (1988) Sex determination in mammals. *Trends in Genetics*, **4**, 153–157.

Sry

Goodfellow, P.N. and Lovell-Badge, R. (1993) SRY and sex determination in mammals. *Annual Review of Genetics*, **27**, 71–92.

Werner, M.H., Huth, J.R., Gronenborn, A.M., and Clore, G.M. (1996) Molecular determinants of mammalian sex. *Trends in Biochemical Sciences*, **21**, 302–308.

Pevny, L.H. and Lovell-Badge, R. (1997) Sox genes find their feet. *Current Opinion in Genetics & Development*, **7**, 338–344.

Sertoli and follicle cells

McLaren, A. (1991) Development of the mammalian gonad: the fate of the supporting cell lineage. *Bioessays*, **13**, 151–156.

Steel and W

Besmer, P. (1991) The kit ligand encoded at the murine Steel locus: a pleiotropic growth and differentiation factor. *Current Opinion in Cell Biology*, **3**, 939–946.

CHAPTER SIX

Uniparental expression, X-inactivation and DNA methylation

This chapter is quite detailed for several reasons. First, because X-chromosome inactivation and gametic imprinting make significant contributions to the biology and genetics of the laboratory mouse, secondly because the use of transgenic technology has significantly advanced our understanding of these phenomena and thirdly because their mechanistic basis is still not completely clear.

- The term uniparental expression refers to the fact that a number of genes are expressed or not depending upon their parent of origin, i.e. upon whether they derive from the oocyte or from the spermatozoon. Such monoallelic expression is often not universal; that is, it may occur only at particular times in development or in particular tissues. It comes about through the marking of one of the two alleles during gametogenesis which persists through the zygote into the somatic cells of the foetus and adult. As a result the expression of the marked and unmarked alleles differs in the offspring; in some cases the active allele is the marked one and in others the unmarked one, and some genes are marked during male and others during female gametogenesis. The nature of the gametic marks and how they are created and maintained is the subject of current investigation.
- X-chromosome inactivation transcriptionally silences many but not all genes on one of the two X chromosomes of somatic cells in females. It is accompanied by condensation of the affected chromosome that persists through interphase. The condensed chromosome associates with the nuclear membrane and is called the Barr body after its discoverer.

DNA methylation

A well-characterised mammalian enzyme called DNA-cytosine-5-methyltransferase (or DNA hemimethylase) is coded for by the *Dnmt* gene. As its name suggests, it attaches a methyl group to carbon atom 5 of cytosine bases in DNA. There are two significant limitations to its action:

- To be methylated the cytosine base must have a guanine base as its 3' neighbour, in the configuration 5'-cytidine-3'-phosphate-5'-guanosine-3'. This is called a CpG doublet, while the methylated form is known as ^{Me}CpG.
- CpG is a mini-palindrome, so that the bases on the complementary DNA strand form an identical CpG doublet when read in the 5' to 3' direction (see Figure 6.1). The enzyme efficiently methylates the cytosine of a CpG doublet when the cytosine of the complementary doublet is already methylated, but is 10- to 30-fold less efficient when both doublets are unmethylated (Yoder *et al.* 1997). This adapts it well for maintaining cytosine methylation at already methylated sites during cell proliferation; in the process of DNA synthesis unmethylated cytosines are incorporated into the newly synthesised strands and these are methylated by the hemimethylase if the base-paired doublets are already methylated.

Completely unmethylated CpG doublets become methylated in the course of development, a process known as *de novo* methylation; also, methylated sites are demethylated. In adult tissues, between 60% and 90% of cytosine residues are methylated (quoted in Kass *et al.* 1997). The hemimethylase shares several conserved motifs with *de novo* methylases from lower eukaryotes but possesses an additional N-terminal domain. ES cells homozygous for an inactivating mutation in the *Dnmt* gene exhibit *de novo* methylating activity, suggesting that a second methyltransferase is responsible for *de novo* methylation (Lei *et al.* 1996).

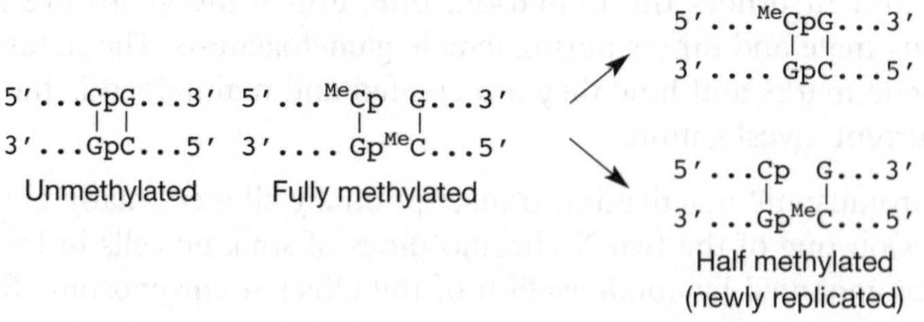

5'...CpG...3'
3'...GpC...5'
Unmethylated

5'...^{Me}Cp G...3'
3'.... Gp^{Me}C...5'
Fully methylated

5'...^{Me}CpG...3'
3'.... GpC...5'

5'...Cp G...3'
3'...Gp^{Me}C...5'
Half methylated
(newly replicated)

Figure 6.1 Mammalian DNA methylation. Relationahip of CpG doublets in the complementary strands of the DNA duplex, showing unmethylated, fully methylated and hemimethylated sites.

Cytosine methylation has an interesting evolutionary consequence. Mammalian DNA suffers a remarkable deficit of CpG doublets relative to the number predicted from its overall base composition (or, equally, the frequency of GpC doublets). The most plausible explanation of this is based on a tendency for cytosine bases to be deaminated. Deamination of 5-methyl-cytosine produces thymine with the result that, if repair does not intervene, an adenine would be incorporated into one of the newly synthesised strands instead of guanine at the next replication. This process would add an extra component, peculiar to CpG doublets, to the underlying equilibrium of base transitions (C ↔ T; G ↔ A) and transversions shifting the equilibrium in favour of the derivative TpG and CpA doublets. In fact, as expected, these doublets are over-represented in mammalian DNA (see Figure 6.2).

Another feature of mammalian DNA related to methylation is the existence of relatively short regions that contain a high proportion of GC base pairs, but with no deficit of CpG doublets relative to expectation (or relative to GpC doublets, which amounts to the same thing). These CpG islands are undermethylated and are frequently located close to or around the 5' ends of transcription units (Bird 1986).

Methylation in or around genes predominantly correlates with transcriptional repression and may have evolved as a device for keeping most genes silent in most tissues. In accordance with this idea, methylation has no effect on gene expression in the early embryo, in primordial germ cells and in embryonic stem cells, and becomes both effective and essential when the major cell lineages begin to diverge at the time of gastrulation (see below). In the developing foetus and after birth methylation is cell lineage- or tissue-specific. It seems likely that methylation that relates to the mark-

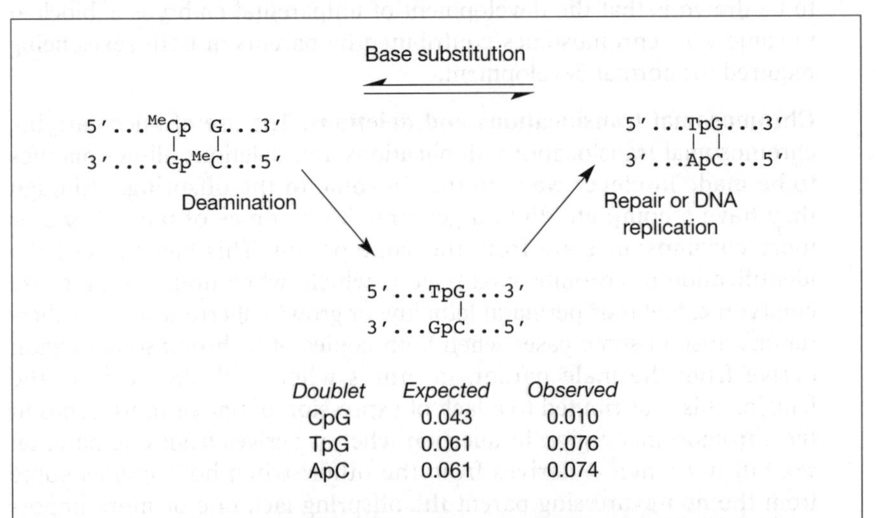

Doublet	Expected	Observed
CpG	0.043	0.010
TpG	0.061	0.076
ApC	0.061	0.074

Figure 6.2 Deamination of methylated CpG sites produces TpG and CpA doublets. The expected frequencies are those calculated from overall base composition assuming random assortment of bases (which roughly equate with observation for other doublets).

ing of genes that are expressed uniparentally differs in important ways from tissue-specific methylation. There is growing evidence that the methylation that we have labelled as tissue-specific both influences and is influenced by chromatin structure (Kass *et al.* 1997). There is reason to believe that methylation and expression relate to each other through an interplay of many components: DNA sequence, methylase, transcription factors (activators and repressors), nucleosomes, histone H1, histone acetylases and deacetylases, methylated DNA binding proteins and others. Not all components need be involved in all instances; of those mentioned the DNA sequence is of course gene-specific while the transcription factors and other DNA binding proteins are in many cases tissue-specific.

Uniparental expression

Three lines of evidence led first to the hypothesis and then to demonstrations that some genes are uniparentally expressed.

- **Pronuclear transplantation and parthenogenesis.** There is a period following fertilisation during which pronuclei can be removed from the oocyte and discarded or transferred between oocytes (Chapter 3). By this means oocytes can be produced that contain one original pronucleus and a second from another, donor oocyte. Both pronuclei can be either paternal or maternal in origin (uniparental), or else one can come from a parent of each sex (biparental) to monitor the adverse effects of the transplantation procedure. Such experiments demonstrate that development is arrested when both nuclei have the same parent of origin (McGrath and Solter 1984; Surani *et al.* 1986). The conclusion to be drawn is that the development of uniparental embryos is blocked in some way, chromosomes contributed by parents of both sexes being required for normal development.
- **Chromosomal translocations and deletions.** The use of mice carrying chromosomal translocations, duplications and deletions allows matings to be made in clever ways so that in some of the offspring, although they have a complete diploid genome, both copies of part of one or more chromosomes are from the same parent. This has allowed the identification of chromosomal regions which, when uniparental, cause embryonic, foetal or perinatal lethality, or growth aberrations. The aberrations arise in some cases when both copies of a chromosomal region derive from the male parent, in others when both derive from the female. This is attributed to a lack of expression of one or more genes in the chromosomal region in question when it derives from one parental sex but not when it derives from the other; when both copies come from the non-expressing parent the offspring lack one or more important functions (Cattanach and Kirk 1985). Again, some quite substantial chromosomal deletions give a normal phenotype in combination with an intact copy of the same chromosome, but only when the intact chro-

mosome derives from the parent of one sex. This is similarly attributed to a failure of expression of a gene or genes in the region of the intact chromosome corresponding to the deletion, when it derives from the parent of the other sex. These conclusions have been fully justified by molecular and biochemical analysis. Regions that exhibit this property have been detected on chromosomes 2, 6, 7 (two regions), 11, 12, 17 (two regions) and 18, and uniparentally expressed genes have been identified in six of these nine regions (Beechey and Cattanach 1996).

- **Gene inactivation.** Genes suspected of being uniparentally expressed have been deliberately inactivated by homologous recombination in ES cells, and in other cases genes inactivated in this way have subsequently been found to be uniparentally expressed. Just as with the chromosomal deletions referred to above, it is found that the intact allele is active when it comes from the parent of one sex but not the other. Here, however, we know which gene is affected and can proceed directly to determine levels of RNA and protein expression and to compare the configuration and other properties of the active and inactive genes. Some of the genes known to be uniparentally expressed are listed in Table 6.1.

Table 6.1 Uniparentally expressed mouse genes. For literature references see Beechey and Cattanach (1996).

Gene	*Description of gene product*	*Chromosome*
	Paternally expressed	
Snrpn	RNA splicing	7 proximal
Znf127	Zinc finger protein	7 proximal
Ins2	Insulin	7 distal
Igf2	Insulin-like growth factor 2	7 distal
U2afbp-rs	Protein involved in RNA splicing	11
Mas	Protooncogene	17
Ins1	Insulin	19
	Maternally expressed	
Mash-2	Trophoblast-specific transcription factor	7 distal
H19	Untranslated RNA	7 distal
p57^{KIP2}	Inhibitor of cyclin-dependent protein kinase	7 distal
Igf2r = Igf2/MPR	Non-signal-transducing receptor for IGF2	17

An important feature of uniparental expression is that the properties of the gene must be reset, where necessary, at each generation. The potential for expression (or lack of it) of the alleles inherited from the two parents is realised in the somatic cells of the offspring (Figure 6.3). However, when a

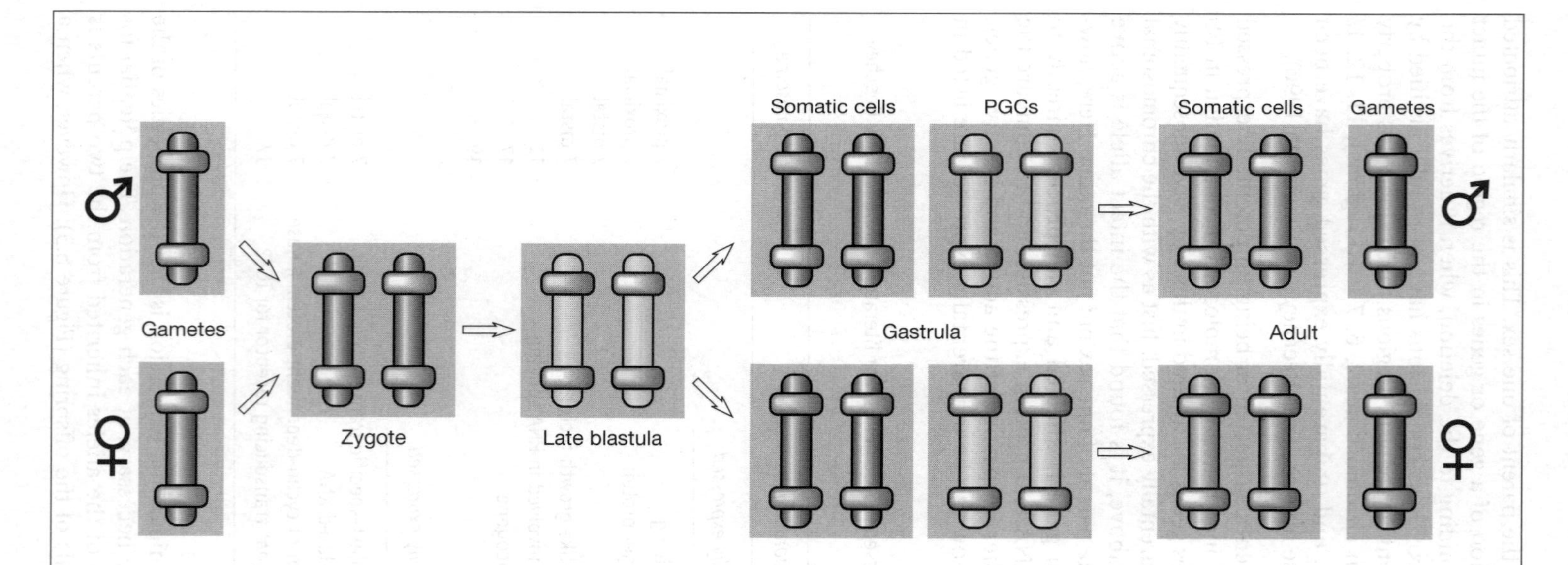

Figure 6.3 The resetting of marked genes during development. The diagram shows a single chromosome (vertical bars); the intensity of the blue colour indicates the overall level of methylation. the chromosome is shown to carry one gene that is marked by methylation in male gametes and another marked in female gametes. These are shown as horizontal bars, blue when marked, white when unmarked. Thus the upper 'gene' is marked in sperm, the lower in oocytes. The overall level of methylation is greatly reduced in the blastula while the methylation state of the marked genes is unchanged. During gastrulation the somatic and germinal lineages diverge; marking is eliminated in the primordial germ cells but not in the somatic lineage, while at this time the general methylation level increases. At a later stage in gametogenesis the uniparentally expressed genes are marked appropriately in the two sexes. As development proceeds, the genes subject to general methylation are selectively demethylated in the different somatic lineages.

gene that is uniparentally expressed passes from a parent of one sex to offspring of the other its potential for expression must be reversed in the germ line of these offspring. Of course, genes may also be reset when passed to offspring of the same sex, but if so they are reset to have the same potential as before. Resetting goes on generation after generation, and one allele of each pair has its potential reversed each time. This necessary potential for reversal limits the range of modifications that we can conceive of as being responsible. It is reasonable to suppose that resetting will occur while the gametes of the two sexes or their precursors are separate, i.e. during gametogenesis, and there is much evidence to support this.

Uniparental gene expression and gametic imprinting

For many years the term imprinting, or genetic or gametic imprinting, has been used to refer to the phenomenon that we have called uniparental gene expression. Recently, it has also been used to refer to the marking of genes for uniparental expression, i.e. to the 'memory' of its parent of origin that the gene carries with it into the next generation. Although one implies the other, we shall see that they are far from being the same. For one thing, genes that are uniparentally expressed are not always marked. For another, in one reasonably well established case of marking, the marked gene is expressed, while in a second case the unmarked gene (from the other parent) is expressed. Again, because of the resetting phenomenon it could be argued that both alleles are marked, in different ways. In this discussion the term imprinting is reserved for the differential marking of a gene during gametogenesis in the two sexes while uniparental expression is called just that.

Imprints have been tentatively identified in at least three genes, all involving methylation. Two main technical difficulties are intrinsic to the search for imprints:

- Maternal and paternal genes (as opposed to their products) need to be distinguished. This can be achieved by reciprocal matings that bring together distinguishable alleles of the gene. Polymorphisms may be available in different strains of laboratory mice, or the gene can be modified by gene targeting. In yet other cases interspecific crosses (e.g. between laboratory mice and M. *castaneus*) allow the two alleles to be distinguished.
- An important tool in the detection of methylated sites employs methylation-sensitive DNA restriction enzymes (Box 6.1). Embryos are of course very small, severely restricting the amount of material available for analysis. Instead, an ingenious PCR technique (Box 6.2) has been employed. A more general but less utilised method exploits the fact that bisulphite deaminates cytosine to form uracil, while 5-methyl-cytosine is unaffected. Following PCR of a stretch of bisulphite treated DNA the altered bases can be detected by determining the nucleotide sequences of cloned PCR products (Olek *et al.* 1996).

BOX 6.1: DETECTING THE METHYLATION STATUS OF RESTRICTION ENZYME RECOGNITION SITES BY SOUTHERN BLOTTING

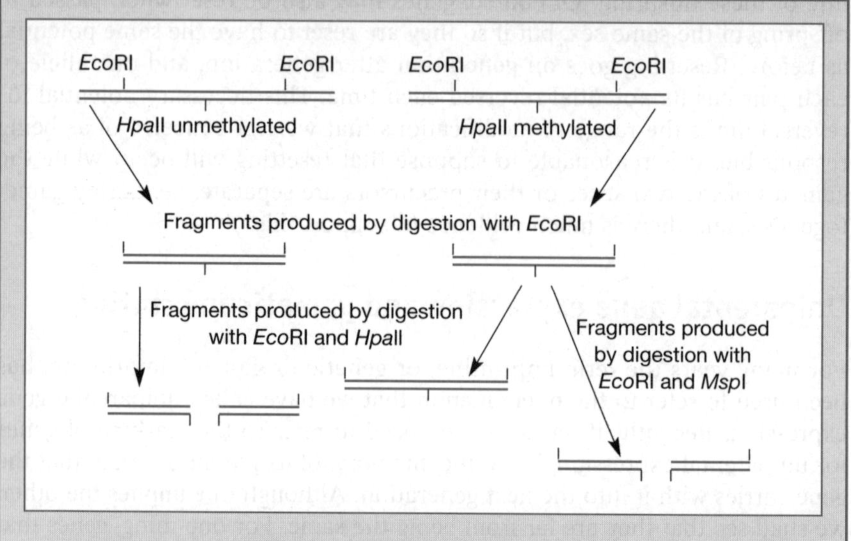

The recognition sequences of quite a few restriction enzymes, for instance *Hpa*II and *Hha*I, contain at least one CpG doublet. Several of these enzymes fail to cleave their restriction sites when the cytosines are methylated. The DNA is digested with a methylation-insensitive restriction enzyme so as to give large fragments that contain recognition sites for a methylation-sensitive enzyme. When the DNA is incubated with the latter enzyme these fragments will be unchanged if the sites are methylated, and further fragmented if they are not. The fragments and sub-fragments are visualised by Southern blotting, using a hybridisation probe specific for the short DNA region under study. The sizes of any sub-fragments produced can reveal which sites are methylated and the relative quantities of fragments and sub-fragments allow the proportion of each site that is methylated to be estimated. A significant limitation of the method is that many CpG doublets do not form part of any known methylation-sensitive restriction site, and cannot be investigated in this way.

In the Figure, the notional DNA duplex is shown in blue and vertical lines show the positions of restriction enzyme recognition sites. The fragment on the left is unmethylated and consequently the methylation-sensitive site (CCGG) is cleaved by *Hpa*II. The fragment on the right is methylated and resists cleavage by *Hpa*II but is cleaved by the methylation-insensitive enzyme *Msp*I, which recognises the same CCGG site.

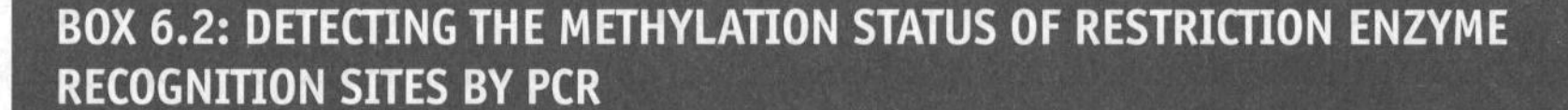

BOX 6.2: DETECTING THE METHYLATION STATUS OF RESTRICTION ENZYME RECOGNITION SITES BY PCR

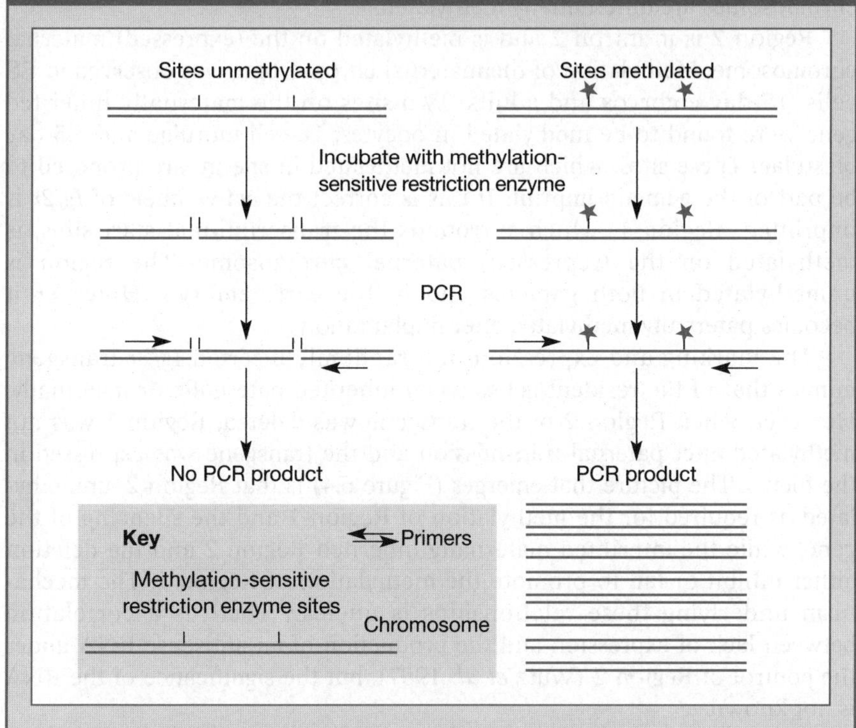

A primer pair is selected that promotes the amplification of a short sequence that includes the methylation site(s) to be investigated. The DNA is incubated with the enzyme and then used as template for PCR with these primers. If the site is methylated a fragment of the predicted size will be amplified. If it is not methylated, incubation with the restriction enzyme cleaves the DNA, the primer target sites are consigned to separated fragments and amplification is not possible.

The Figure illustrates this procedure. The DNA duplex is coloured blue and two methylation-sensitive restriction enzyme sites are indicated. When the unmethylated DNA on the left is incubated with the restriction enzyme the primer target sites are segregated on different fragments and no PCR amplification is possible. When the sites are protected by methylation the region between the primers remains intact and amplification occurs.

The *Igf2r* gene and the *Snrpn* gene

The *Igf2r* gene, located on chromosome 17, codes for a receptor for IGF2 (see Table 6.1) which does not transduce a signal but instead acts to remove IGF2 from the foetal circulation. The paternal allele is not expressed in the foetus, and when a silent paternal allele is combined with a maternal null allele, the foetuses grow more rapidly due to an increased IGF2 level and the pups are larger than normal. The gene contains two CpG islands, Region 1 and Region 2, that were recognised because maternal and paternal chromosomes are differentially methylated.

Region 2 is in intron 2 and is methylated on the (expressed) maternal chromosome. Methylation of the maternal chromosome was observed in ES cells, 15-day embryos and adults. Two sites on the maternally inherited gene were found to be methylated in oocytes, 16-cell morulae and 3.5-day blastulae. These sites, which are not methylated in sperm, are proposed to be part of the gametic imprint. If this is correct the active allele of *Igf2r* is imprinted. Region 1, which surrounds the transcriptional start sites, is methylated on the (repressed) paternal chromosome. The region is unmethylated in both gametes and in the early embryo. However it becomes paternally methylated after implantation.

The marking and expression of a randomly inserted *Igf2r* transgene mimics that of the resident genes when inherited paternally or maternally. However, when Region 2 of the transgene was deleted, Region 1 was not methylated after paternal transmission and the transgene was expressed in the foetus. The picture that emerges (Figure 6.4) is that Region 2, unmethylated, is required for the methylation of Region 1 and the silencing of the gene, while the imprinted maternally inherited Region 2 and the deletion either inhibit or fail to promote the methylation of Region 1. The mechanism underlying these relationships is unclear. There is a correlation between lack of expression and the production of an antisense RNA under the control of Region 2 (Wutz *et al.* 1997), but the significance of the RNA is not known.

The *Snrpn* gene shows a similar although not identical pattern of methylation in relation to expression, but in this case the gene is paternally expressed (Table 6.2). The repressed maternal gene is methylated in a 5' region, and the active paternal gene in a region at the 3' end of the transcription unit rather than within an intron. Possibly a more significant difference from the *Igf2r* gene, however, is that both the maternal and

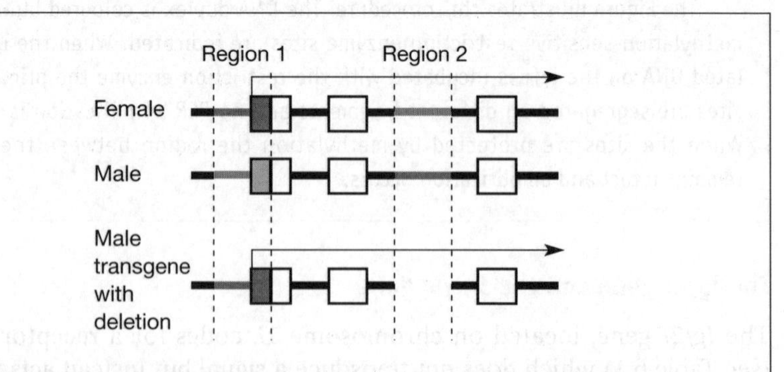

Figure 6.4 The uniparentally expressed *Igf2r* gene. Three exons are shown as boxes. Region 1 and Region 2 are demarcated by broken lines and are coloured blue when methylated, grey when not. The presence of an arrow signifies that the gene is transcribed when configured as shown.

paternal alleles appear to be imprinted, with different methylation patterns, during gametogenesis (Shemer *et al.* 1997).

Table 6.2 Similar properties of the *Igf2r* and *Snrpn* genes. Column 2 lists the parental sex of the active genes, column 3 the parental sex of the imprinted genes, columns 5–6 the positions of the two methylated regions of each gene and the sex in which each is methylated.

Gene	*Uniparental expression*	*Imprinting*	*Methylation region*		
			Name	*Location*	*Parentage*
Igf2r	Maternal	Maternal	Region 1	5'	Paternal
			Region 2	intron 2	Maternal
Snrpn	Paternal	Both	DMR 1	5'	Maternal
			DMR 2	3'	Paternal

The *H19*, *Igf2* and *Ins2* genes

A cluster of genes that are expressed uniparentally lies near the distal end of chromosome 7, including the *H19*, *Igf2* and *Ins2* genes and more distantly the *Mash2* gene (Box 6.3). *Igf2* mRNA is widely distributed in the post-implantation embryo and disappears from most tissues about two weeks after birth, while *Ins2* mRNA is present at a high level in the visceral endoderm of the yolk sac. The *H19* gene is quite unusual, having an RNA product that is spliced and polyadenylated, but no protein product. There is a striking reciprocal relationship between the expression of the *H19* gene on the one hand and of the *Igf2* and *Ins2* genes on the other. Thus:

- when the *H19* gene is expressed, normally when it is maternally transmitted, the alleles of the *Igf2* and *Ins2* genes on the same chromosome are not expressed
- The *Igf2* and *Ins2* alleles on the other, paternally transmitted chromosome are expressed, provided that the *H19* gene on that chromosome is silent, as it normally is

Many of the observations made and conclusions drawn about the interactions in *cis* between these genes are summarised in Box 6.3. The once current idea that the state of the *H19* gene simply determines that of the *Igf2* and *Ins2* genes seems no longer to be tenable. The reciprocal interactions of the *H19* gene and *Igf2*/*Ins2* gene pair, together with the effect of the downstream enhancers on both, suggest that looping of the interphase chromosome may bring together the downstream enhancers, the *H19* promoter and the *Igf2*/*Ins2* enhancer, together with structural and transcriptional proteins appropriate to the state of the various DNA elements and the tissue.

BOX 6.3: INTERPLAY OF DNA ELEMENTS IN THE H19-INS2 REGION OF CHROMOSOME 7

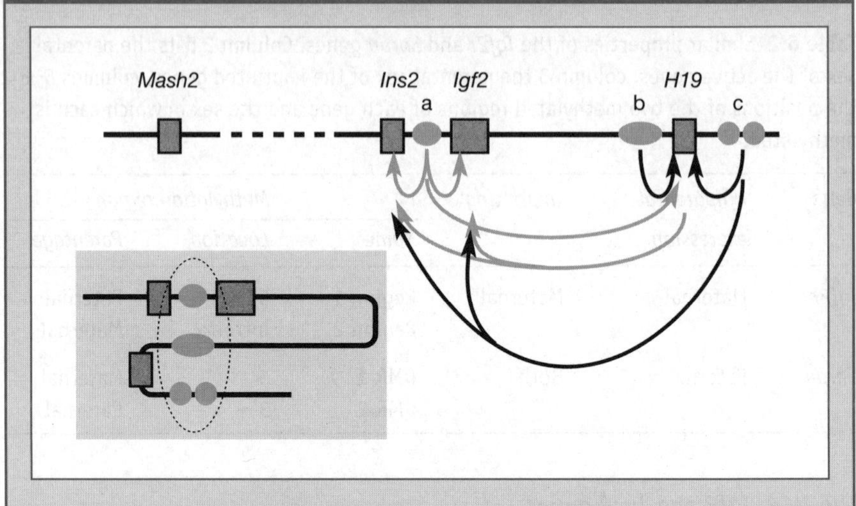

Organisation of the *H19*, *Igf2* and other genes on distal mouse chromosome 7. The centromere is located well beyond the left side of the diagram. Genes are shown as open boxes, and regulatory elements as blue shapes. Positive and negative controls are indicated with black and blue arrows respectively, running from the controlling to the controlled element in each case. About 350 kb of DNA lies between *Mash2* and *Ins2* and another 95 kb between *Ins2* and *H19*. The inset shows how the controlling elements might be brought together by looping of the chromosome, allowing interactions between the *trans*-acting factors bound to them and generating a compound transcription complex. Some of the relevant references are Ainscough *et al.* 1997; Elson and Bartolomei 1997; Forne *et al.* 1997; Hu *et al.* 1997; Moore *et al.* 1997; Ripoche *et al.* 1997; Tremblay *et al.* 1997; Wutz *et al.* 1997.

Experiments with transgenes, one a YAC that carries both the *H19* and *Igf2* genes, others carrying different deletions, and experiments in which different parts of the resident gene were deleted or substituted, have brought out a picture of the interactions between various DNA elements and the *H19* gene product which, however, is far from complete at this time. There are regulatory regions both upstream and downstream of the *H19* gene:

- a differentially methylated region lies between the *Igf2* and *Ins2* genes (a in Figure)
- 2 kb upstream of the *H19* transcription initiation site there is a putative imprint, a 2 kb GC-rich region that becomes differentially methylated in maternal and paternal alleles during gametogenesis (b)
- two endoderm-specific enhancers lie within 7 kb downstream of the 3 kb *H19* transcription unit (c)
- other regions of the two genes become differentially methylated during embryonic development

▶

Some of the more notable observations are as follows:

- there is biallelic expression of the *Igf2* gene when an inactivated *H19* gene is inherited maternally
- the product of the *H19* gene seems to be involved in the transcriptional silencing and internal methylation of the gene, but not in imprinting region b
- the enhancer elements (c) are required for the expression of both the *H19* and *Igf2* genes in *cis* configuration

Deletion of the imprinted enhancer element (a) leads to transcriptional activation not only of *Igf2* but also of *H19*, irrespective of maternal or paternal inheritance.

Evolution of uniparental expression

As far as we know, gametic imprinting by methylation may be the only fundamental signal for uniparental expression. However it seems that when the time comes to implement the gametic imprint, different mechanisms are invoked to achieve uniparental expression of different genes. Equally, there is no reason to suppose that there is a single source of selective pressure favouring uniparental expression (Hurst and McVean 1998). At the same time the number of imprinted genes known to affect growth processes suggests that growth rates have affected the fixation of a number of gene imprints.

The most convincing case to have been put forward relates to the reciprocal expression of the *Igf2* and *Igf2r* genes from maternally and paternally inherited alleles (Moore and Haig 1991). IGF2 is an important growth factor in the foetus, while the MPR receptor does not transduce a signal upon binding IGF2 but acts as a 'sink', reducing its availability. Thus the actions of the proteins are antagonistic in terms of foetal growth. The explanation supposes that in a promiscuous mammal with a large litter size, sometimes with plural paternity, larger pups will be fitter and will consequently confer a selective advantage on the genomes of their fathers; hence the paternal silencing of the *Igf2r* gene. On the other hand the mother, if she survives, will go on to have more litters. If large pups threaten her survival, then smaller pups will confer an advantage on her genome; hence the maternal silencing of the *Igf2* gene. There is direct experimental support for this explanation from the effects of deletions of the *Igf2, H19* and *Igfr* genes, all of which produce the predicted effects on the growth of pups (Willison 1991; Bartolomei *et al.* 1991).

X-chromosome inactivation

The maternal and paternal X chromosomes are designated X^M and X^P, and the active and inactive X chromosomes can be referred to as Xa and Xi.

One X chromosome is inactivated in each female somatic cell and X-inactivation occurs randomly at a multicellular stage of embryogenesis. Consequently every female tissue is a mosaic of two cell types, one of which carries an inactive X^M and the other an inactive X^P.

X-inactivation is non-random in the early stages of development. X^P is inactivated in the morula and early blastocyst while X^M is not. In the ICM of the later blastocyst both X chromosomes become active, while inactivation of X^P persists in the extraembryonic tissues. In the embryo, random inactivation occurs following implantation. Thus at early stages of development there is significant uniparental expression of X-linked genes which persists in the extraembryonic tissues. Whether or not the underlying mechanism resembles the imprinting of uniparentally expressed autosomal genes is not clear.

The solitary male X is inactivated during spermatogenesis. Inactivation in the early female blastocyst may therefore represent the persistence of that inactivation. In the trophoblast however, the *Xist* gene (see below) is required for the inactivation of X^P while spermatogenesis is normal even when the X chromosome carries a deletion of *Xist* (Marahrens *et al.* 1997), indicating that the mechanism of X-inactivation in the trophoblast is different from that operating during spermatogenesis. It appears that X^M and X^P are differentially marked during gametogenesis and that this determines uniparental expression in the extraembryonic tissues.

The *Xic* region

The human X-inactivation centre (*XIC*, or in the mouse *Xce*, the X-controlling element) is a region of the X chromosome that brings about X-inactivation in *cis*. The evidence for this derives from rearrangements and deletions of the X chromosome, translocations between the X and autosomes and from genetic studies. *XIC* and *Xce* lie within a long syntenic region of the human and mouse chromosomes respectively. For the sake of simplicity both will be referred to as *Xic*. The most important observations are as follows:

- X chromosomes carrying deletions of *Xic* fail to be inactivated.
- In reciprocal X–autosome translocations the autosome is partially inactivated if *Xic* has been translocated, while the X continues to be inactivated if it has not.
- There are allelic forms of *Xic* that exhibit different 'strengths' (four in the laboratory mouse). In mice that carry two copies of the same allele equal numbers of the two X chromosomes are somatically inactivated, but when the alleles are different the chromosome carrying the stronger is more frequently inactivated (Cattanach *et al.* 1969; Johnston and Cattanach 1981).
- When randomly integrated into an autosome, a 450 kb YAC derived from the *Xic* region imported the properties of *Xic*. The YAC was intro-

duced into male ES cells where it had no discernible effect as long as they remained undifferentiated. Upon differentiation the single male X chromosome was inactivated in some cells, while in others part of the autosome carrying the YAC was inactivated (Lee *et al.* 1996).

- The behaviour of aneuploid (e.g. XXX/AA) and polyploid embryos supports the interpretation that, for every two haploid sets of autosomes present, one X chromosome will remain active; thus two X chromosomes are inactivated in a tetraploid embryo, two in a triple-X embryo and so on.

Together these observations support the idea that the action of *Xic* can be broken down into two components, X-inactivation and the so-called counting of X chromosomes in relation to autosomes.

The *Xist* gene

The *Xist* gene, which lies within the *Xic* region, is responsible for X-inactivation, though seemingly not for the counting mechanism. Some of the observations relating to *Xist* are listed below.

- *Xist* is the only gene known to be active on the inactive X but not the active X chromosome (this suggested the name Xi-specific transcript). Thus it is uniquely poised to effect X-inactivation in *cis*.
- The product of the gene is a polyadenylated and alternatively spliced RNA, its largest variant being about 15 kb long in the mouse (Brockdorff *et al.* 1992) and 17 kb in humans (Brown *et al.* 1992). No significant protein product is known or likely, and the RNA accumulates in the nucleus.
- The amount of *Xist* RNA present in cells correlates with X-inactivation. Only a very small amount of *Xist* RNA is found in male and female ES cells; this can be visualised by FISH (Chapter 4) as a tiny localised signal. Upon differentiation, the signal disappears from the active X chromosome in both sexes, and becomes very intense on the inactive female X. This strong signal is coextensive with the inactive X and the *Xist* RNA is said to 'paint' the chromosome (Clemson *et al.* 1996). The transition between low-level and high-level *Xist* expression is at least partly due to a very considerable change in the stability of the RNA (Herzing *et al.* 1997; Panning *et al.* 1997), presumably mediated by changes in the activity or localisation of one or more cellular proteins.

The circumstantial evidence that the *Xist* gene is responsible for X-inactivation is fully supported by transgenic experiments. *Xist* is required for inactivation of X^P in the trophoblast and, in *cis*, for X-inactivation in female embryos and differentiated ES cells (Penny *et al.* 1996; Marahrens *et al.*

1997). X-inactivation does not normally occur in differentiated male ES cells, but supernumerary autosomal *Xist* transgenes bring about X-inactivation (Herzing *et al.* 1997). When X^P carries a *Xist* deletion that precludes its inactivation, X^M remains active in the female trophoblast, supporting the view that trophoblast X-inactivation involves imprinting; persistent methylation consistent with imprinting is observed at three *Xist* sites in oocytes and early embryos but not in spermatozoa (Ariel *et al.* 1995; Zuccotti and Monk 1995). In differentiated female ES cells, on the other hand, the evidence suggests that a chromosome carrying a deleted *Xist* gene is targeted for inactivation just as frequently as the normal homologue, but cannot respond to the stimulus (Penny *et al.* 1996). A female mouse inheriting a deleted maternal *Xist* gene develops normally; X^P is found to be universally inactivated in the pups, probably because of selection *in vivo* against those cells in which both X chromosomes remain active (Marahrens *et al.* 1997; Penny *et al.* 1996).

Although the *Xist* gene normally remains active on the inactive X in adults, this may not be necessary to maintain the inactivated X chromosome once inactivation has been established. Clones of human–mouse somatic cell hybrids (Chapter 4) carrying a full complement of mouse chromosomes and a single inactive human X chromosome were subjected to selection against the distal part of the human X. In clones that had lost *XIC* the remainder of the human X remained inactive (Brown and Willard 1994). Possibly *Xist* remains active so as to reinitiate the inactivation of genes that have been *incidentally* reactivated in the adult.

Not all genes on Xi are inactivated. We have seen that *Xist* is expressed exclusively from Xi in most cell types. Unsurprisingly, since they are present in two copies in both sexes, genes in the pseudoautosomal region are not X-inactivated. In addition, in the unique part of the X chromosome a minority of genes that remain active, more in humans than in mice, are interspersed with the inactivated genes. This suggests that there may be signals associated with genes that determine whether or not they can be inactivated. Alternatively, there may be no simple code, and the key may lie in the different proteins with which they interact.

X-inactivation is generally held to equate with dosage-compensation, a method of ensuring that only one copy of each X-linked gene is active in both males and females. This is a little hard to accept in view of the homeostatic potential of mammals. However, the evidence shows that supernumerary chromosomes are highly deleterious; females that inherit an *Xist* deletion from their fathers die in the embryonic stages due apparently to a failure to inactivate either X chromosome in the trophoblast cells (Marahrens *et al.* 1997), and cells in which two copies of part of the X chromosome remain active are selected against in *vivo* (McMahon and Monk 1983). Thus dosage compensation remains the most likely advantage gained from X-inactivation.

Global methylation during development

Global changes in methylation occur during very early development. Following fertilisation and up to the blastocyst stage there is a period during which general demethylation occurs (Kafri *et al.* 1992; Monk *et al.* 1987). Uniparentally expressed genes, however, retain their methylation status (i.e. marked or not as the case may be); additionally, certain sites in the *Xist* gene on the X chromosome which are methylated in oocytes but not in spermatozoa retain their different methylation states (Norris *et al.* 1994; Zuccotti and Monk 1995). However, at this time the transcriptional apparatus seems to be indifferent to the methylation status of the genes (Beard *et al.* 1995) and expression is generally biallelic.

In principle, demethylation during cellular proliferation could be a passive process; in the absence of hemi-methylase activity, methylated sites would first become hemi-methylated and then hemi-methylated sites would become rare. However demethylation does seem to proceed in the absence of DNA replication and there is experimental evidence of the existence of an active demethylation process, possibly catalysed by a ribozyme (Weiss *et al.* 1996).

Following implantation the DNA remains hypomethylated in cells of the extraembryonic lineage, but at the same time the paternal X chromosome, with its persistently methylated and consequently active *Xist* gene, is inactivated in the female trophectoderm (the male trophectoderm, of course, carries no paternal X). In the post-implantation epiblast, there is divergence of the somatic and germinal lineages (Figure 6.3). The somatic lineage now becomes highly methylated (Jaenisch 1997), although the CpG islands and the unmarked alleles of uniparentally expressed genes are spared. In females, X-inactivation is initiated at random in different cells (Jahner *et al.* 1982; Monk *et al.* 1987). The methylation of the somatic cells is followed by selective demethylation, different patterns developing in sublineages as they diverge. These relate at least in part to gene expression or the potential for gene expression at a later time. In contrast, the hypomethylated state developed in the blastocyst persists at first in the primordial germ cells (PGCs). Additionally, the methylation that marks the uniparentally expressed genes is removed and the two chromosomes of each parental pair become equivalent (Shemer *et al.* 1997). During gametogenesis the expression of these genes is in fact biallelic (Szabo and Mann 1995) and both X chromosomes are active in females. Later in gametogenesis the appropriate methylation marks are put in place *de novo* in each sex, both chromosomes of each pair now being identically marked (or not marked as the situation demands). As gametogenesis proceeds, the general level of gametic DNA methylation is increased to high levels in both sexes.

It is evident that both methylation and demethylation are precisely controlled processes both globally and, in at least some cases, down to the level of individual methylation sites. However, the nature of the controlling process or processes is still unclear, even in outline.

Inactivation of the cytosine-5-methyltransferase gene

Inactivation of the *Dnmt* gene (see above) by gene targeting in ES cells has given some important insights into both imprinting and X-inactivation. The cells proliferate normally despite their lack of methylation activity, but die upon differentiation. Similarly, embryos homozygous for a *Dnmt*⁻ allele are initially viable but become grossly abnormal following gastrulation and die at 11 dpc (Lei *et al.* 1996). The expression of genes normally expressed uniparentally is altered in these *Dnmt*⁻ homozygotes. Thus, the normally silent paternal allele of the *H19* gene is activated upon demethylation, while the active paternal and maternal alleles, respectively of the *Igf2* and *Igf2r* genes, are repressed (Li *et al.* 1993). Similarly the *Xist* gene on an active X chromosome is activated by demethylation in somatic cells (Beard *et al.* 1995). While important, these results are not surprising, because it has been known for many years that demethylation, brought about by treating cells with 5-azacytidine or administering the analogue to animals, alters the transcriptional status of methylated genes (Young and Tilghman 1984; Jaenisch *et al.* 1985).

Dnmt⁻ ES cells could be 'rescued' by the replacement of the mutated part of the gene with a *Dnmt* complementary DNA minigene by homologous recombination. General methylation, for example of repetitive DNA sequences associated with constitutive heterochromatin, was restored to normal levels in the rescued ES cells. Similarly, the *Xist* gene was restored to normal and X-inactivation was random upon differentiation *in vivo*. In contrast, uniparental methylation and expression of the imprinted *H19* and *Igf2r* genes was not restored when the cells differentiated (Table 6.3). Similar results were obtained with fibroblast clones recovered from chimeras that contained the rescued ES cells. However a different picture emerged in embryos carrying an ES cell haploid genome derived by mating the chimeras. The methylation and expression of uniparentally expressed genes was normal in these embryos, implying that the alleles present in the rescued cells had been reset to their normal states during gametogenesis in the chimera (Tucker *et al.* 1996a).

Probably the most useful advance at this time would be the elucidation of the mechanism that brings about *de novo* methylation and the inactivation of *de novo* methylation independently of the *Dnmt* gene, if that proves to be possible.

RNA

A striking feature of the *Xist* gene is the very low sequence conservation between mouse and man. The only well-conserved region is an array of short direct repeat sequences located towards the 5' ends of the molecules, nine in the *XIST* gene, eight in the mouse, which have been shown to associate with two proteins approximately 40 kDa in size (Brown and Baldry 1996). Both *H19* and *Xist* are implicated in the inactivation of other loci in *cis* configuration, and the product of each is an RNA that is not a template for protein synthesis and remains in the nucleus in proximity to the gene. Deletion of either of these genes by homologous recombination results in biparental expression of the genes under their control, yet the size of the longest *Xist* product is 17 kb, while the length of the *H19* product is a mere 2.7 kb and there are no marked sequence similarities between the two. Neither RNA contains a convincing long reading frame, and *Xist* RNA is not found in polyribosomes, indicating that the RNAs are not translated. There is no indication that either RNA base-pairs with target RNA or DNA sequences, or indeed that their mechanisms of action are the same. Presumably they act in concert with proteins to prevent transcription of the target genes either directly or indirectly by altering chromatin structure. Some properties of uniparentally expressed genes and of *Xist* are listed in Table 6.3.

Table 6.3 Characteristics of four uniparentally expressed genes.

	H19	*Igf2r*	*Xist*	*Igf2*
Uniparental expression	Maternal	Maternal	Paternal[1]	Paternal
Methylation (other than imprinting)	Paternal	Paternal	Maternal	Maternal
Imprinting methylation	Paternal	Maternal	Maternal	None
Dnmt rescue[2]	No	No	Yes	Partial
Dnmt rescue after germ line transmission[3]	Yes	Yes	Yes	Yes
RNA	Non-coding	Coding	Non-coding	Coding

[1] In the female trophoblast.
[2,3] Refers to the rescue by repair of a *Dnmt* allele in homozygous *Dnmt*⁻ ES cells by homologous recombination.
[2] The cells differentiated either *in vitro* or *in vivo* after introduction into blastocysts. [3] Chimeric mice that developed from the blastocyst injections were mated and their offspring analysed. See Tucker *et al.* 1996a; Tucker *et al.* 1996b.

References

Ainscough, J.F., Koide, T., Tada, M., Barton, S., and Surani, M.A. (1997) Imprinting of Igf2 and H19 from a 130 kb YAC transgene. *Development,* **124**, 3621–3632.

Ariel, M., Robinson, E., McCarrey, J.R., and Cedar, H. (1995) Gamete-specific methylation correlates with imprinting of the murine Xist gene. *Nature Genetics*, **9**, 312–315.

Bartolomei, M.S., Zemel, S., and Tilghman, S.M. (1991) Parental imprinting of the mouse H19 gene. *Nature,* **351**, 153–155.

Beard, C., Li, E., and Jaenisch, R. (1995) Loss of methylation activates Xist in somatic but not in embryonic cells. *Genes & Development,* **9**, 2325–2334.

Beechey, C.V. and Cattanach, B.M. (1996) Genetic imprinting map. *Mouse Genome,* **94**, 96–99.

Bird, A.P. (1986) CpG-rich islands and the function of DNA methylation. *Nature,* **321**, 209–213.

Brockdorff, N., Ashworth, A., Kay, G.F., McCabe, V.M., Norris, D.P., Cooper, P.J., Swift, S., and Rastan, S. (1992) The product of the mouse Xist gene is a 15 kb inactive X-specific transcript containing no conserved ORF and located in the nucleus. *Cell,* **71**, 515–526.

Brown, C.J., Hendrich, B.D., Rupert, J.L., Lafreniere, R.G., Xing, Y., Lawrence, J., and Willard, H.F. (1992) The human XIST gene: analysis of a 17 kb inactive X-specific RNA that contains conserved repeats and is highly localized within the nucleus. *Cell,* **71**, 527–542.

Brown, C.J. and Baldry, S.E. (1996) Evidence that heteronuclear proteins interact with XIST RNA in vitro. *Somatic Cell & Molecular Genetics,* **22**, 403–417.

Brown, C.J. and Willard, H.F. (1994) The human X-inactivation centre is not required for maintenance of X-chromosome inactivation. *Nature,* **368**, 154–156.

Cattanach, B.M., Pollard, C.E., and Perez, J.N. (1969) Controlling elements in the mouse X-chromosome. I. Interaction with the X-linked genes. *Genetical Research,* **14**, 223–235.

Cattanach, B.M. and Kirk, M. (1985) Differential activity of maternally and paternally derived chromosome regions in mice. *Nature,* **315**, 496–498.

Clemson, C.M., McNeil, J.A., Willard, H.F., and Lawrence, J.B. (1996) XIST RNA paints the inactive X chromosome at interphase: evidence for a novel RNA involved in nuclear/chromosome structure. *Journal of Cell Biology,* **132**, 259–275.

Elson, D.A. and Bartolomei, M.S. (1997) A 5' differentially methylated sequence and the 3'-flanking region are necessary for H19 transgene imprinting. *Molecular & Cellular Biology,* **17**, 309–317.

Forne, T., Oswald, J., Dean, W., Saam, J.R., Bailleul, B., Dandolo, L., Tilghman, S.M., Walter, J., and Reik, W. (1997) Loss of the maternal H19 gene induces changes in Igf2 methylation in both cis and trans. *Proceedings of the National Academy of Sciences of the United States of America,* **94**, 10243–10248.

Herzing, L.B., Romer, J.T., Horn, J.M., and Ashworth, A. (1997) Xist has properties of the X-chromosome inactivation centre. *Nature,* **386**, 272–275.

Hu, J.F., Vu, T.H., and Hoffman, A.R. (1997) Genomic deletion of an imprint maintenance element abolishes imprinting of both insulin-like growth factor II and H19. *Journal of Biological Chemistry,* **272**, 20715–20720.

Hurst, L.D. and McVean, G.T. (1998) Growth effects of uniparental disomies and the conflict theory of genomic imprinting. *Trends in Genetics,* **13**, 436–443.

Jaenisch, R., Schnieke, A., and Harbers, K. (1985) Treatment of mice with 5-azacytidine efficiently activates silent retroviral genomes in different tissues. *Proceedings of the National Academy of Sciences of the United States of America,* **82**, 1451–1455.

Jaenisch, R. (1997) DNA methylation and imprinting: why bother? *Trends in Genetics,* **13**, 323–329.

Jahner, D., Stuhlmann, H., Stewart, C.L., Harbers, K., Lohler, J., Simon, I., and Jaenisch, R. (1982) De novo methylation and expression of retroviral genomes during mouse embryogenesis. *Nature,* **298**, 623–628.

Johnston, P.G. and Cattanach, B.M. (1981) Controlling elements in the mouse. IV. Evidence of non-random X-inactivation. *Genetical Research,* **37**, 151–160.

Kafri, T., Ariel, M., Brandeis, M., Shemer, R., Urven, L., McCarrey, J., Cedar, H., and Razin, A. (1992) Developmental pattern of gene-specific DNA methylation in the mouse embryo and germ line. *Genes & Development,* **6**, 705–714.

Kass, S.U., Pruss, D., and Wolffe, A.P. (1997) How does DNA methylation repress transcription? *Trends in Genetics,* **13**, 444–449.

Lee, J.T., Strauss, W.M., Dausman, J.A., and Jaenisch, R. (1996) A 450 kb transgene displays properties of the mammalian X-inactivation center. *Cell,* **86**, 83–94.

Lei, H., Oh, S.P., Okano, M., Juttermann, R., Goss, K.A., Jaenisch, R., and Li, E. (1996) De novo DNA cytosine methyltransferase activities in mouse embryonic stem cells. *Development,* **122**, 3195–3205.

Li, E., Beard, C., and Jaenisch, R. (1993) Role for DNA methylation in genomic imprinting. *Nature,* **366**, 362–365.

Marahrens, Y., Panning, B., Dausman, J., Strauss, W., and Jaenisch, R. (1997) Xist-deficient mice are defective in dosage compensation but not spermatogenesis. *Genes & Development,* **11**, 156–166.

McGrath, J. and Solter, D. (1984) Completion of mouse embryogenesis requires both the maternal and paternal genomes. *Cell,* **37**, 179–183.

McMahon, A. and Monk, M. (1983) X-chromosome activity in female mouse embryos heterozygous for Pgk-1 and Searle's translocation, T(X; 16) 16H. *Genetical Research,* **41**, 69–83.

Monk, M., Boubelik, M., and Lehnert, S. (1987) Temporal and regional changes in DNA methylation in the embryonic, extraembryonic and germ cell lineages during mouse embryo development. *Development,* **99**, 371–382.

Moore, T., Constancia, M., Zubair, M., Bailleul, B., Feil, R., Sasaki, H., and Reik, W. (1997) Multiple imprinted sense and antisense transcripts, differential methylation and tandem repeats in a putative imprinting control region upstream of mouse Igf2. *Proceedings of the National Academy of Sciences of the United States of America,* **94**, 12509–12514.

Moore, T. and Haig, D. (1991) Genomic imprinting in mammalian development: a parental tug-of-war. *Trends in Genetics,* **7**, 45–49.

Norris, D.P., Patel, D., Kay, G.F., Penny, G.D., Brockdorff, N., Sheardown, S.A., and Rastan, S. (1994) Evidence that random and imprinted Xist expression is controlled by preemptive methylation. *Cell,* **77**, 41–51.

Olek, A., Oswald, J., and Walter, J. (1996) A modified and improved method for bisulphite based cytosine methylation analysis. *Nucleic Acids Research,* **24**, 5064–5066.

Panning, B., Dausman, J., and Jaenisch, R. (1997) X chromosome inactivation is mediated by Xist RNA stabilization. *Cell,* **90**, 907–916.

Penny, G.D., Kay, G.F., Sheardown, S.A., Rastan, S., and Brockdorff, N. (1996) Requirement for Xist in X chromosome inactivation. *Nature,* **379**, 131–137.

Ripoche, M.A., Kress, C., Poirier, F., and Dandolo, L. (1997) Deletion of the H19 transcription unit reveals the existence of a putative imprinting control element. *Genes & Development,* **11**, 1596–1604.

Shemer, R., Birger, Y., Riggs, A.D., and Razin, A. (1997) Structure of the imprinted mouse Snrpn gene and establishment of its parental-specific methylation pattern. *Proceedings of the National Academy of Sciences of the United States of America,* **94**, 10267–10272.

Surani, M.A., Barton, S.C. and Norris, M.L. (1986) Nuclear transplantation in the mouse: heritable differences between parental genomes after activation of the embryonic genome. *Cell,* **45**, 127–136.

Szabo, P.E. and Mann, J.R. (1995) Biallelic expression of imprinted genes in the mouse germ line: implications for erasure, establishment, and mechanisms of genomic imprinting. *Genes & Development,* **9**, 1857–1868.

Tremblay, K.D., Duran, K.L., and Bartolomei, M.S. (1997) A 5' 2-kilobase-pair region of the imprinted mouse H19 gene exhibits exclusive paternal methylation throughout development. *Molecular & Cellular Biology,* **17**, 4322–4329.

Tucker, K.L., Beard, C., Dausman, J., Jackson-Grusby, L., Laird, P.W., Lei, H., Li, E., and Jaenisch, R. (1996a) Germ-line passage is required for establishment of methylation and expression patterns of imprinted but not of nonimprinted genes. *Genes & Development,* **10**, 1008–1020.

Tucker, K.L., Talbot, D., Lee, M.A., Leonhardt, H., and Jaenisch, R. (1996b) Complementation of methylation deficiency in embryonic stem cells by a DNA methyltransferase minigene. *Proceedings of the National Academy of Sciences of the United States of America,* **93**, 12920–12925.

Weiss, A., Keshet, I., Razin, A., and Cedar, H. (1996) DNA demethylation *in vitro*: involvement of RNA. *Cell,* **86**, 709–718.

Willison, K. (1991) Opposite imprinting of the mouse Igf2 and Igf2r genes. *Trends in Genetics,* **7**, 107–109.

Wutz, A., Smrzka, O.W., Schweifer, N., Schellander, K., Wagner, E.F., and Barlow, D.P. (1997) Imprinted expression of the Igf2r gene depends on an intronic CpG island. *Nature,* **389**, 745–749.

Yoder, J.A., Soman, N.S., Verdine, G.L., and Bestor, T.H. (1997) DNA (cytosine-5)-methyltransferases in mouse cells and tissues. Studies with a mechanism-based probe. *Journal of Molecular Biology,* **270**, 385–395.

Young, P.R. and Tilghman, S.M. (1984) Induction of alpha-fetoprotein synthesis in differentiating F9 teratocarcinoma cells is accompanied by a genome-wide loss of DNA methylation. *Molecular & Cellular Biology,* **4**, 898–907.

Zuccotti, M. and Monk, M. (1995) Methylation of the mouse Xist gene in sperm and eggs correlates with imprinted Xist expression and paternal X-inactivation. *Nature Genetics,* **9**, 316–320.

Further reading

Methylation

Eden, S. and Cedar, H. (1994) Role of DNA methylation in the regulation of transcription. *Current Opinion in Genetics & Development,* **4**, 255–259.

Kass, S.U., Pruss, D., and Wolffe, A.P. (1997) How does DNA methylation repress transcription? *Trends in Genetics,* **13**, 444–449.

Bestor, T.H. and Tycko, B. (1996) Creation of genomic methylation patterns. *Nature Genetics,* **12**, 363–367.

Imprinting

Barlow, D.P. (1994) Imprinting: a gamete's point of view. *Trends in Genetics,* **10**, 194–199.

Jaenisch, R. (1997) DNA methylation and imprinting: why bother? *Trends in Genetics,* **13**, 323–329.

Reik, W. and Maher, E.R. (1997) Imprinting in clusters: lessons from Beckwith-Wiedemann syndrome. *Trends in Genetics,* **13**, 330–334.

Hurst, L.D. and McVean, G.T. (1998) Growth effects of uniparental disomies and the conflict theory of genomic imprinting. *Trends in Genetics,* **13**, 436–443.

Surani, M.A. (1988) Imprinting and the initiation of gene silencing in the germ line. *Cell,* **93**, 309–312.

X-inactivation

Lyon, M.F. (1993) Epigenetic inheritance in mammals. *Trends in Genetics,* **9**, 123–128.

Migeon, B.R. (1994) X-chromosome inactivation: molecular mechanisms and genetic consequences. *Trends in Genetics,* **10**, 230–235.

Lee, J.T. and Jaenisch, R. (1997) The (epi)genetic control of mammalian X-chromosome inactivation. *Current Opinion in Genetics & Development,* **7**, 274–280.

Kuroda, M.I. and Meller, V.H. (1997) Transient *Xist*-ence. *Cell,* **91**, 9–11.

PART THREE

Transgenic mammals

CHAPTER SEVEN

Making transgenic mammals

Two methods of generating transgenic mammals are well established, the direct microinjection of foreign DNA into embryo pronuclei and the introduction of DNA into the germ line *via* genetically manipulated embryonic stem cells. The first can be applied to any mammal, while the second is restricted to species from which ES cells that colonise the germ line can be established, so far only the mouse. A third route, recently described, is by the replacement of the nucleus of an activated oocyte with the nucleus of a genetically manipulated somatic cell. This route, so far successful in sheep and cows, may also prove eventually to be a general one.

Ingenious methods have been developed for using modified retroviruses as vectors to carry genes into embryos, taking advantage of the integration of retroviral proviruses into the nuclear genome. However, the requirement to package the vector in the virus particle severely limits the size of gene that can be imported into cells in this way. Also, the provirus attracts methylation which, possibly in conjunction with other mechanisms (Laker *et al.* 1998), disables its expression when it passes through the germ line (Jahner *et al.* 1982). Together these disadvantages, at least one of which might have been anticipated, rendered the retrovirus route to transgenic mammals unproductive. More recently it has become apparent that the properties of retrovirus vectors are well suited to gene-trapping projects (see Chapter 8).

To this day the majority of transgenic mammals have been generated by direct microinjection of DNA into pronuclei of 1-cell embryos.

Vectors

In relation to transgenic animals, vectors fulfil a number of functions in addition to mapping (as described in Chapter 4) namely:

- the isolation of genes
- the manipulation of DNA and assembly of DNA fragments into 'constructs' suitable for use in transgenic work
- the propagation of isolated genes and assembled DNA constructs

The most useful vectors for the purpose of gene isolation are those that lend themselves to the production of 'libraries', consisting of overlapping fragments of genomic DNA, ideally encompassing the entire genome several times over. For instance, a bacteriophage λ genomic library of 10^6 viruses, each containing on average 20 kb of DNA, represents 6–7 copies of the mouse genome and the probability that every gene is represented is very high. Another type of library contains complete or partial DNA copies of all the mRNA molecules present in a tissue. Each individual vector genome that has a DNA fragment incorporated into it must be capable of clonal proliferation. Most vectors multiply in concert with a bacterial host which forms colonies spontaneously when spread over an agar surface. Virus vectors destroy their bacterial hosts and form colonies called plaques, cleared of bacteria, when plated on a lawn of cells growing, again, on an agar surface. Various devices are used to ensure that most of the colonies or plaques contain a vector that has incorporated a mammalian DNA fragment. The objective of library methods is of course to identify and isolate the colonies, or clones, that carry a particular DNA sequence. A number of alternative screening methods are available for this purpose, based on nucleic acid hybridisation or reaction with specific antibodies. Many texts deal with this subject area.

The assembly of DNA fragments into a construct suitable for transgenic work is very often a multi-step process, successive steps in assembly each being followed by the isolation and propagation of a vector carrying a new combination of fragments. A vast array of combinatorial techniques is available, comprising specialised vectors and bacterial hosts, DNA restriction and modification enzymes and both off-the-shelf and bespoke synthetic DNAs. In this area much useful information can be found not only in practical manuals but also in the catalogues of commercial suppliers.

Propagation of the final construct is implicit in the assembly process. Separation of the vector DNA from that of the host is usually necessary once the construct is complete, and also at intermediate stages in its assembly, and vectors are designed with this in mind. In the special case of transgenes it is often thought advisable to follow this step by separating the transgene from the vector, which may interfere with transgene expression in some cases (Townes *et al.* 1985; Clark *et al.* 1997). This is usually achieved by cleavage with one or more DNA restriction enzymes, ideally chosen during the design phase of the construct so as not to cleave the transgene internally. The physical separation of the transgene from the vector fragment is generally accomplished by agarose gel electrophoresis.

Different classes of vector are appropriate for the manipulation of DNA fragments in different size ranges (Table 7.1).

Table 7.1 Recombinant DNA vectors.

Vector	*Origin*	*Insert size range*[1]
Multicopy plasmids	Various multicopy plasmids	Up to 20+ kb
Lambda vectors	Bacteriophage λ	Up to 30 kb
Cosmid	Bacteriophage λ	Up to 40 kb
P1 artificial chromosome (PAC)	Bacteriophage P1	80–90 kb
Bacterial artificial chromosome (BAC)	Large bacterial plasmid (F factor)	100–300 kb
Yeast artificial chromosome (YAC)	Yeast chromosome	100–1000+ kb

[1] The upper limits of size ranges marked + are indefinite; however, greater size invariably involves greater difficulty.

Pronuclear microinjection (Figure 7.1)

Prospective embryo donors are treated with gonadotrophins to induce superovulation and mated with normal males. Female mice, for example, are treated with gonadotrophin from pregnant mares' serum (PMSG) and then, 48 hours later, with human chorionic gonadotrophin (HCG). Oocytes are recovered next day by flushing medium through the oviducts, treated

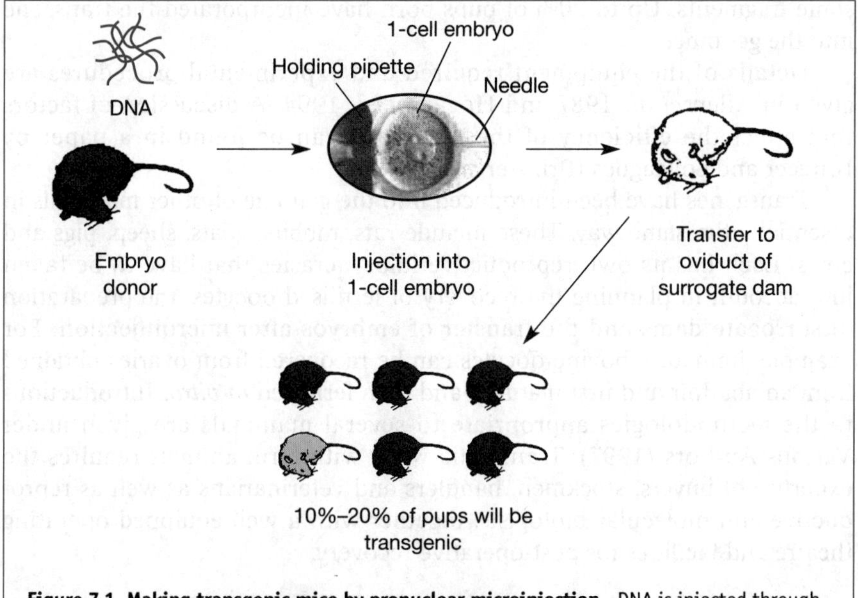

Figure 7.1. Making transgenic mice by pronuclear microinjection. DNA is injected through a fine glass needle into the large male pronucleus of a 1-cell embryo while it is held to a blunt-nosed pipette by suction. Either soon after injection or next day the embryos are introduced by surgery into the oviducts of surrogate dams. Up to 20% of the pups that are successfully delivered are expected to have integrated the transgene.

with hyaluronidase to remove adherent follicle cells and washed by transferring them between microdrops. Unfertilised oocytes are then discarded. Using an inverted microscope with phase contrast optics, or preferably interference optics, and equipped with micromanipulators, DNA is introduced into the oocytes, one by one, by microinjection.

The DNA is generally well characterised and usually what is being introduced is a number of identical linear molecules. Between 50 and 1000 molecules in a microlitre volume are introduced directly into one of the two pronuclei, usually the swollen male pronucleus. Mouse embryos are either introduced immediately into a recipient dam via the oviduct or first allowed to develop to the 2–4 cell stage. The recipients are prepared by mating because, in the mouse, mating is necessary to make a dam physiologically receptive. The mates have previously been vasectomised so that the oocytes produced by the recipient dams will remain unfertilised and incapable of development; consequently only the newly introduced embryos develop.

This process is remarkably efficient. Up to 60% of embryos survive injection and up to 30% of the embryos transferred to the oviduct survive to birth. As a rule, the overall survival through to birth of oocytes injected with DNA is around 15% (Brinster *et al.* 1985; Page *et al.* 1995) but is lower than the survival of mock-injected oocytes (Canseco *et al.* 1994; Page *et al.* 1995), presumably due to a lethal effect of the DNA, possibly the insertional inactivation of haplo-insufficient genes (although such genes are rare) or local disruption of chromosome structure or the loss of chromosome fragments. Up to 20% of pups born have incorporated the transgene into the genome.

Details of the equipment required and experimental procedures are given in Allen *et al.* 1987 and Hogan *et al.* 1994. A discussion of factors that affect the efficiency of the technique can be found in a paper by Brinster and colleagues (Brinster *et al.* 1985).

Transgenes have been introduced into the genome of other mammals in essentially the same way. These include rats, rabbits, goats, sheep, pigs and cows. Each has its own reproductive idiosyncrasies that have to be taken into account in planning the recovery of fertilised oocytes, the preparation of surrogate dams and the transfer of embryos after microinjection. For example, immature bovine oocytes can be recovered from ovaries obtained from an abattoir and first matured and then fertilised *in vitro.* Introductions to the methodologies appropriate to several mammals are given under Various Authors (1997). Transgenic work with farm animals requires the expertise of buyers, stockmen, handlers and veterinarians as well as reproductive and molecular biologists, together with a well equipped operating theatre and facilities for post-operative recovery.

Stocks and strains

In the choice of mouse strains for the donation of oocytes, in general more emphasis has been placed on ease of production and manipulation than on the expression of transgenes. Thus the mice that are mated to produce the 1-cell embryos are most frequently F_1 offspring of a cross between inbred lines such as C57BL/6 and CBA/2. The parents, with hybrid vigour, mate fairly avidly and successfully and the oocytes are nourished in an F_1 ovary. However the most important animals in the operation, the transgenic offspring, are genetically F_2 animals, and consequently vary enormously in genotype. Recovery from this situation is possible through backcrossing but this is very time-consuming. If F_1 dams must be used it would certainly be better to mate them to one of the parental lines, e.g. C57BL/6. The founder animals would then have 75% C57BL/6 chromosomes and there would be a 3:1 probability that the transgene would integrate into a C57BL/6 chromosome. The best practice is to employ pure inbred 1-cell embryos, using the principal standard C57BL/6 line if there is no reason not to do so, accepting some inconvenience and inefficiency at the microinjection stage but avoiding potential genetic background problems. The question of genetic background is taken up in Chapter 10.

Mice of the FVB/N strain are reproductively vigorous and their oocyte pronuclei are exceptionally large, facilitating microinjection (Taketo *et al.* 1991). FVB/N has the virtue of being an inbred strain, but without the wealth of characterisation available for the C57BL/6 strain.

Interestingly, farm animals present fewer problems, most being moderately inbred and, more importantly, quite homogeneous morphologically and physiologically within a breed.

Genomic modification of embryonic stem cells (Figure 7.2)

Isolation of embryonic stem cells

Embryonic stem cells (ES cells) were first derived from explanted mouse blastocysts (Evans and Kaufman 1981) and blastocyst inner cell masses isolated by immunosurgery (Martin 1981). Some opinion holds that delayed blastocysts are a more effective source of ES cells. Delay is induced by ovariectomising the dam at 2.5 dpc and then administering a progesterone analogue. This prevents the uterus from becoming receptive and the hatched blastocysts do not implant. Blastocysts were originally explanted on top of feeder cells in multi-well tissue culture plates containing conditioned medium, i.e. medium in which chosen cells have previously been cultivated and containing mitogens secreted by those cells. Recently, it has become routine to supplement the medium with LIF (Chapter 3) as well as serum, and in fact the feeder layer can be dispensed with. The blastocysts spread out (the undelayed blastocysts first hatch) and various cell types proliferate, including inner cell mass (ICM) cells. The latter form clumps that are picked out, gently disaggregated and re-plated, either on a feeder layer or in LIF-supplemented medium or both. In skilled hands about 10% of strain 129/Sv embryos yield an ES cell clone. These rather taxing proce-

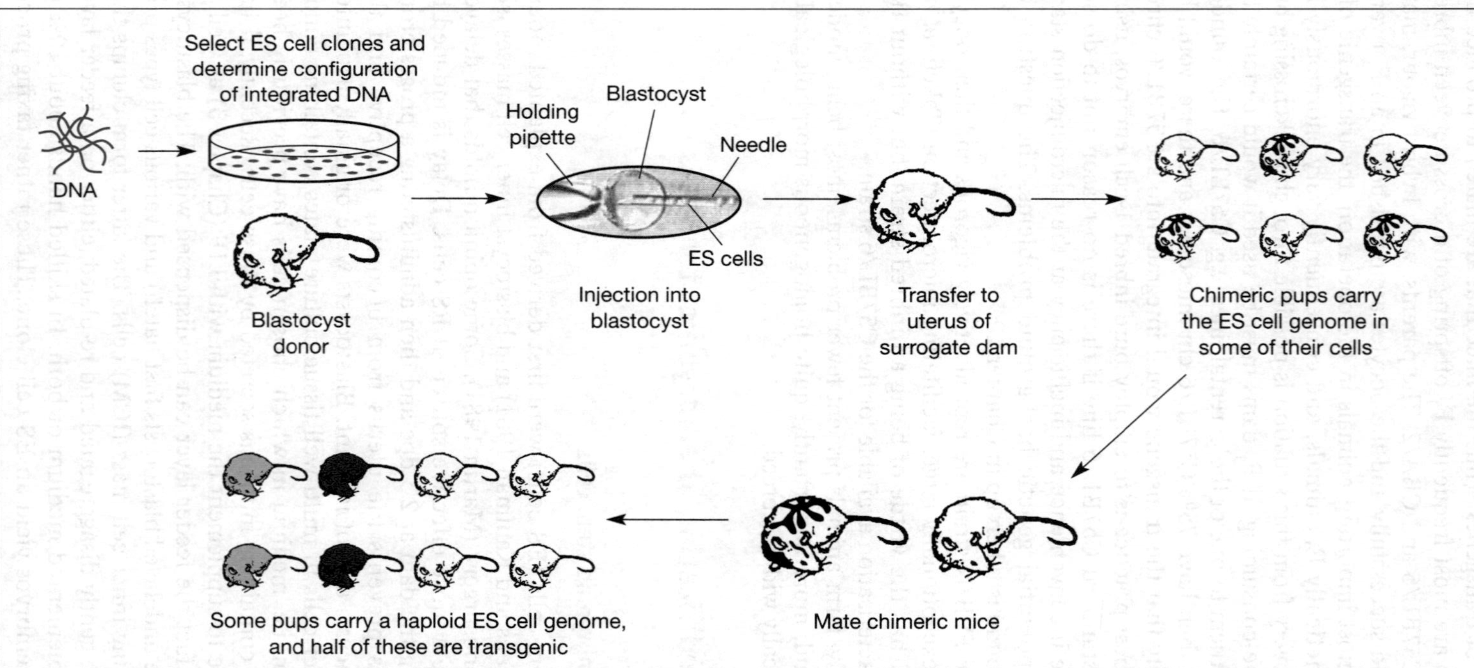

Figure 7.2 Making transgenic mice by the ES cell route. DNA is introduced into ES cells by electroporation and clones that have integrated the transgene are isolated by growth in selective medium. Following verification, cells from one or more of these clones are introduced into blastocysts through a glass needle and the blastocysts are transferred into the uterus of surrogate dams. Those blastocysts that are successfully colonised by ES cells are chimeric. When the chimeras are mated, some of their offspring carry a haploid ES cell genome, and half of these (blue) will carry the transgene.

dures are described in detail in Robertson 1987; Abbondanzo *et al.* 1993; Brook and Gardner 1997. Fortunately, it is usually possible to obtain established ES cells from a practising laboratory.

There is general agreement that ES cells are most easily derived from the inbred 129/Sv mouse strain or one of the closely related inbred strains derived from it, less easily from inbred C57BL mice and with great difficulty or not at all from other mouse strains. A recent report suggests that one way around the problem may be to employ transgenic blastocysts (McWhir *et al.* 1996). The *Oct3/4* promoter is active in mouse ES cells, but becomes inactive when they differentiate. When the expression of the *neo* gene is under the control of this promoter, differentiated cells are selectively killed by the neomycin analogue G418 while undifferentiated cells are spared. Using this method ES cells were successfully isolated from hybrid mice with a predominantly CBA genome (although germ-line colonisation was demonstrated in only one line). The immunosurgery approach used in one of the intitial reports of germ-line competent ES cells (Martin 1981) was recently refined by using manual microsurgery to free epiblast from primitive endoderm (Brook and Gardner 1997). Fifty percent of microdissected epiblasts from 5th day blastocysts developed ES cell lines, while primitive endoderm and trophectoderm from the same strain 129 embryos did not. When the technique was applied to delayed blastocysts, ES cells were derived from 100% of strain 129 epiblasts and from 50% of epiblasts from the relatively intractable CBA strain.

Isolation of ES-like cells from a variety of species has been reported, for example rabbit (Schoonjans *et al.* 1996), rhesus monkey (Thomson *et al.* 1995), marmoset (Thompson *et al.* 1996) and cow (Cibeli *et al.* 1998a), and the rabbit and cow cells contributed to chimeras. Cibelli *et al.* (1998a) also reported the isolation of transgenic ES-like cells from blastocysts developed from oocytes to which diploid nuclei from transgenic fibroblasts had been transferred (nuclear transfer is described in the next section). However, as yet there are no reports of germ-line colonisation and transmission from ES cells other than those derived from a few mouse strains.

Genomic modification

Germ line transmission of mouse ES cell genomes was first demonstrated in 1984 (Bradley *et al.* 1984) and directed germ line modification in 1986 (Gossler *et al.* 1986). The fundamental importance of these developments cannot be overemphasised.

- Previously, all known genetic changes in mammals were due to random mutational events, whether spontaneous or induced by mutagenic agents. Although many of these were interesting and useful, the task of linking mutation to function and identifying and isolating the gene responsible was very formidable. Germ line transmission of ES cell genomes makes it possible to work in the opposite direction, so to speak, by introducing mutations into chosen genes.

- Previously, different changes in the same gene might or might not turn up, depending on chance. Now the same gene can be altered in different ways.
- Previously, only lethal mutations and those that produced readily observed morphological effects could be identified. Now any gene can be inactivated or altered. Perhaps the most remarkable general result to emerge from the new technology is that many genes assumed to be essential have been found instead to be dispensable.

Although the objective of genomic modification of ES cells is usually to alter the structure of a chosen resident gene by homologous recombination (Chapter 9), a very similar approach can be employed to bring about the random integration of foreign DNA into the chromosomes (Chapter 8). Random integration occurs at a much higher frequency than homologous recombination.

Methods became available about 25 years ago for introducing DNA into cell types such as fibroblast cell lines (Willecke and Ruddle 1975). Cloned DNA can be introduced into ES cells by treating them with a particulate co-precipitate of DNA with calcium phosphate or by lipofection, but it is usually introduced by electroporation (Box 7.1).

There is a very great difference between 1-cell embryo injection and cell electroporation in the frequency with which the foreign DNA becomes integrated. Up to 3% of microinjected embryos develop into transgenic pups but only 1 in 10^4 to 10^5 electroporated cells stably integrates the foreign

BOX 7.1: CALCIUM PHOSPHATE CO-PRECIPITATION, ELECTROPORATION AND LIPOFECTION

DNA can be introduced with quite high efficiency into cells in a mass culture by calcium phosphate co-precipitation (cells in monolayer culture) or electroporation (cells in suspension). Efficiency of uptake is cell type-specific and perhaps 10–30% of the cells take up DNA when either of these procedures is optimised.

To make a co-precipitate calcium chloride is added to a solution containing DNA and sodium phosphate; insoluble calcium phosphate forms a fine precipitate which incorporates the DNA. Cells take up the precipitate by pinocytosis or phagocytosis.

Electroporation has overtaken calcium phosphate co-precipitation as the most commonly employed method of introducing DNA into animal cells. Any macromolecule can be introduced into cells by this method, which entails the transient disruption of the cell membrane by an electrical potential. A number of different protocols are available. In general, higher voltages are associated with shorter exposure times.

In lipofection the DNA makes electrovalent complexes with positively charged groups external to very small spheroidal lipid vesicles. These are designed to fuse with cell membranes, and in the course of this interaction the DNA is introduced into the cells. This method can be very efficient, it involves simple procedures (if commercially available reagents are employed) and it can be used either for transient expression studies or for generating stable transfectants.

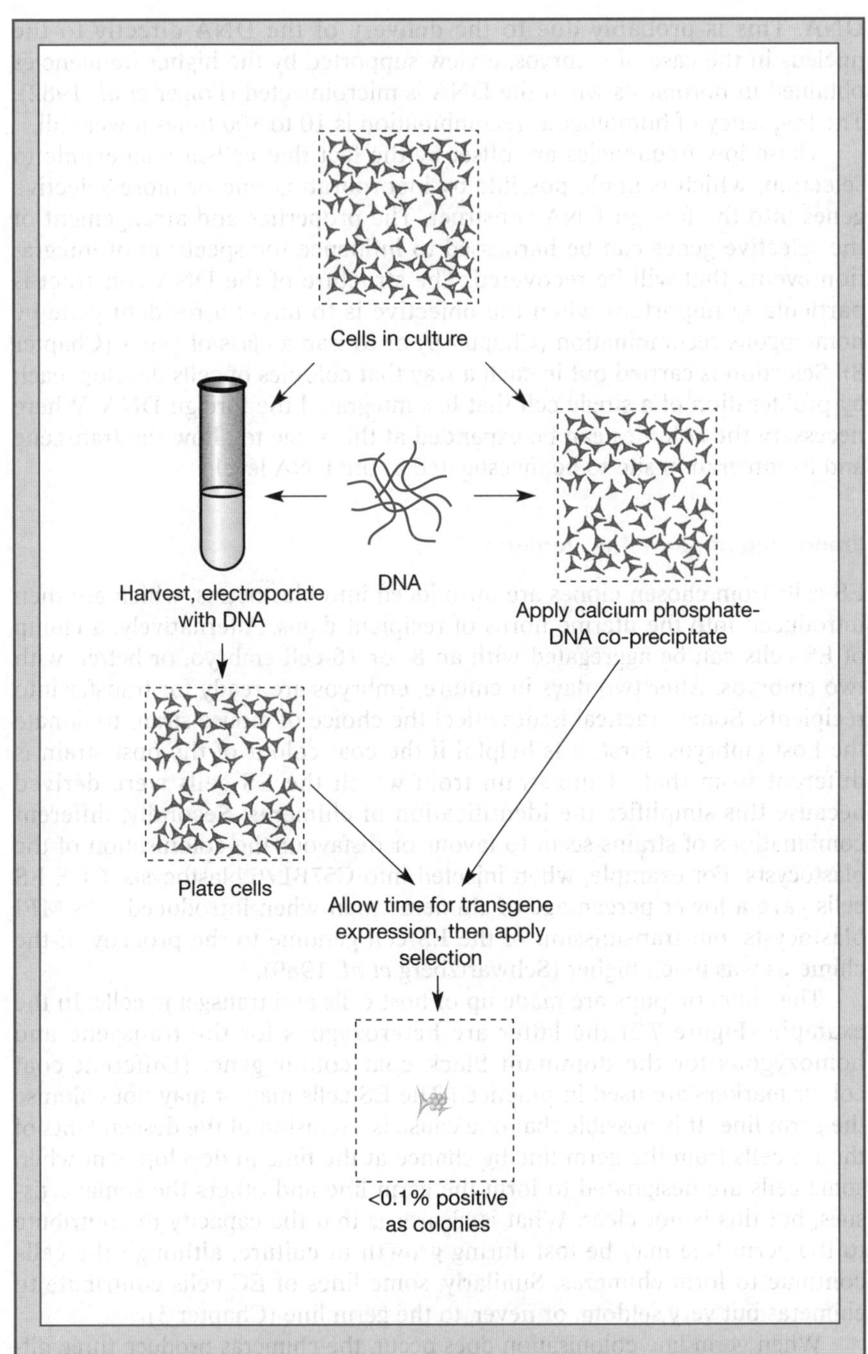

Electroporation and calcium phosphate co-precipitation methods of introducing DNA into cultured cells. The cells are shown as though adhering to the surface of a small section of a culture dish after growth to a point at which they are still not covering the surface of the dish. The great majority of cells that fail to take up and express the foreign DNA are killed by selection while the surviving cells (blue) proliferate to form colonies.

DNA. This is probably due to the delivery of the DNA directly to the nucleus in the case of embryos, a view supported by the higher frequencies obtained in fibroblasts when the DNA is microinjected (Folger *et al.* 1982). The frequency of homologous recombination is 10 to 100 times lower still.

These low frequencies are offset by the fact that cells are amenable to selection, which is made possible by incorporating one or more selective genes into the foreign DNA construct. The properties and arrangement of the selective genes can be harnessed to influence the spectrum of integration events that will be recovered. The structure of the DNA construct is particularly important when the objective is to target a resident gene by homologous recombination (Chapter 9) or to trap a class of genes (Chapter 8). Selection is carried out in such a way that colonies of cells develop, each by proliferation of a single cell that has integrated the foreign DNA. Where necessary the colonies can be expanded at this stage to allow the transgene and its integration site to be investigated at the DNA level.

Production of germ line chimeras

ES cells from chosen clones are introduced into blastocysts, which are then introduced into the uterine horns of recipient dams. Alternatively, a clump of ES cells can be aggregated with an 8- or 16-cell embryo, or better, with two embryos. After two days in culture, embryos are ready for transfer into recipients. Some practical issues affect the choice of mouse strain to donate the host embryos. First, it is helpful if the coat colour of the host strain is different from that of the strain from which the ES cells were derived because this simplifies the identification of chimeras. Secondly, different combinations of strains seem to favour or disfavour the colonisation of the blastocysts. For example, when injected into C57BL/6 blastocysts, CCE ES cells gave a lower percentage of chimeras than when introduced into MFI blastocysts, but transmission of the ES cell genome to the progeny of the chimeras was much higher (Schwartzberg *et al.* 1989).

The chimeric pups are made up of host cells and transgenic cells. In the example (Figure 7.2) the latter are heterozygous for the transgene and homozygous for the dominant black coat colour gene. (Different coat colour markers are used in practice.) The ES cells may or may not colonise the germ line. It is possible that one cause is exclusion of the descendants of the ES cells from the germ line by chance at the time in development when some cells are designated to form the germ line and others the somatic tissues, but this is not clear. What is clearer is that the capacity to contribute to the germ line may be lost during growth in culture, although the cells continue to form chimeras. Similarly, some lines of EC cells contribute to chimeras but very seldom, or never, to the germ line (Chapter 3).

When germ line colonisation does occur, the chimeras produce three different types of gamete, assuming targeting of one allele of an autosomal gene or transgene insertion at a single site: (i) those derived from the host blastocyst, (ii) those with an ES cell genome and carrying the modified gene or transgene and (iii) those with an ES cell genome but not carrying the genomic modification. When the chimeras are mated to normal mice the pups to which the type (ii) gametes contribute are heterozygous for the modification. Homozygotes can be derived by mating mice of this generation together.

Blastocysts are taken at random for ES cell transfer. When male ES cells are transferred into a female blastocyst the germinal ridge will sometimes be male (see Chapter 3). The germ cells with a female chromosome complement fail to develop beyond the spermatogonium stage, and all the mature spermatozoa are ES cell-derived. These mice are somatic chimeras, but if fertile they produce only offspring with one set of ES cell chromosomes, half of which will carry the modified gene.

The chimera stage takes up a minimum of nine weeks (three weeks gestation and six weeks for the chimeric pups to reach sexual maturity) at the end of which time there may be no gametes with an ES cell genome. In some cases this delay can be avoided. ES cells can be aggregated with tetraploid embryos produced by electrofusion at the 2-cell stage or by treatment with cytochalasin B. As development of the aggregate proceeds, the trophoblast comes to consist entirely of tetraploid cells, while the embryo proper derives mainly from the diploid ES cells. In the first applications of this method (Nagy *et al.* 1990; Nagy *et al.* 1993) the great majority of the pups died soon after birth. However, the method could be used to determine the effect of a dominant transgene on pre-natal development. More recently the method has been used successfully to deliver ES cell-derived live animals (Ueda *et al.* 1995; Wang *et al.* 1997), but with poor efficiency.

Reporter genes

Some applications call for the expression of a reporter gene under the control of a cellular promoter, usually to explore the cellular specificity and developmental profile of the promoter. In such cases a reporter gene that serves both this purpose and that of selection in culture can sometimes be employed. An example is the β-geo reporter which specifies a protein that has both neomycin phosphotransferase and β-galactosidase activity, the former allowing selection of expressing cells in culture and the latter allowing visualisation of expressing cells in cytological preparations. However this method can be used only if the promoter under study is active in ES cells (see Chapters 8 and 9). Frequently the selective marker gene requires to be under the control of a separate promoter, and this raises the possibility that the activity profile of the gene under study will be influenced by reporter proximity of the two genes (see, for example, Townes *et al.* 1985; Clark *et al.* 1997). A solution to this problem was developed recently. By the use of a site-specific recombination system (see Chapter 9) the selective marker gene can be excised after it has fulfilled its function in selection and before transgenic mice are generated (Dale and Ow 1991).

In vitro differentiation

The tendency of ES cells to differentiate can be turned to advantage. Differentiation can be made more synchronous by cultivating the cells in suspension culture (Martin and Evans 1975), where they aggregate to form 'embryoid bodies' with an outer layer of extraembryonic endoderm cells surrounding mesoderm and ectoderm (Doetschman *et al.* 1985). If suspension culture is continued, about half of the embryoid bodies develop further into 'cystic embryoid bodies' which express genes characteristic of the embryonic visceral yolk sac, such as the α-fetoprotein, transthyretin and hepatocyte nuclear factor genes (Abe *et al.* 1996). If, instead, the embryoid bodies are plated out, for example on gelatine-coated culture wells, many differentiated cell types appear, but in an ill-organised way. Different treatments can be used to bias the products towards one or another cell type (see for example Faust *et al.* 1994, Vittet *et al.* 1996, Dani *et al.* 1997).

The haematopoietic lineage has been most intensively studied. The cystic embryoid bodies develop structures like blood islands, and when embryoid bodies are plated blood island-like structures grow out (Bautch *et al.* 1996). These contain haematopoietic precursor and endothelial cells and cells carrying B-cell marker combinations are also found (Potocnik *et al.* 1994). Judged by the pattern of globin genes successively expressed, the cells replicate the *in vivo* switch from embryonic to foetal globins (Lindenbaum and Grosveld 1990). Differentiating ES cells can repopulate the erythroid, myeloid and lymphoid lineages of lethally-irradiated mice (Hole *et al.* 1996) showing that among their number they contain pluripotent haematopoietic stem cells.

Differentiation *in vitro* can be put to good use in transgenic work. For example, effects of inactivating a gene essential in a developmental pathway may be analysed *in vitro* (e.g. Warren *et al.* 1994). Clones of cells carrying lineage-specific genes that have been 'trapped' by random integration (Chapter 8) can be detected by visualising the reporter activity of the integrated DNA, and expression during *in vitro* differentiation can be used as a screen for a particular class of trapped genes (e.g. Forrester *et al.* 1996).

Stocks and strains

ES cell lines derived from inbred strain 129/Sv or one of the closely related inbred strains such as 129/Ev and 129/Ola have been employed in most transgenic work. Unfortunately these are not among the best characterised mouse strains. Now that ES cells can be derived from the most standard reference strain, C57BL/6, it is to be hoped that they will come to be more frequently employed.

Transfer of diploid somatic nuclei (Figure 7.3)

In some ways the generation of transgenic animals by the transfer of diploid nuclei parallels the ES cell route, but there are also important differences between the two methods. The most important similarity is that the starting point of both procedures is the introduction of foreign DNA into cultured cells, permitting selection to be applied. Thus both random DNA integration and gene substitution are attainable. Of the important differences, one is favourable and the other unfavourable. The favourable difference is that a

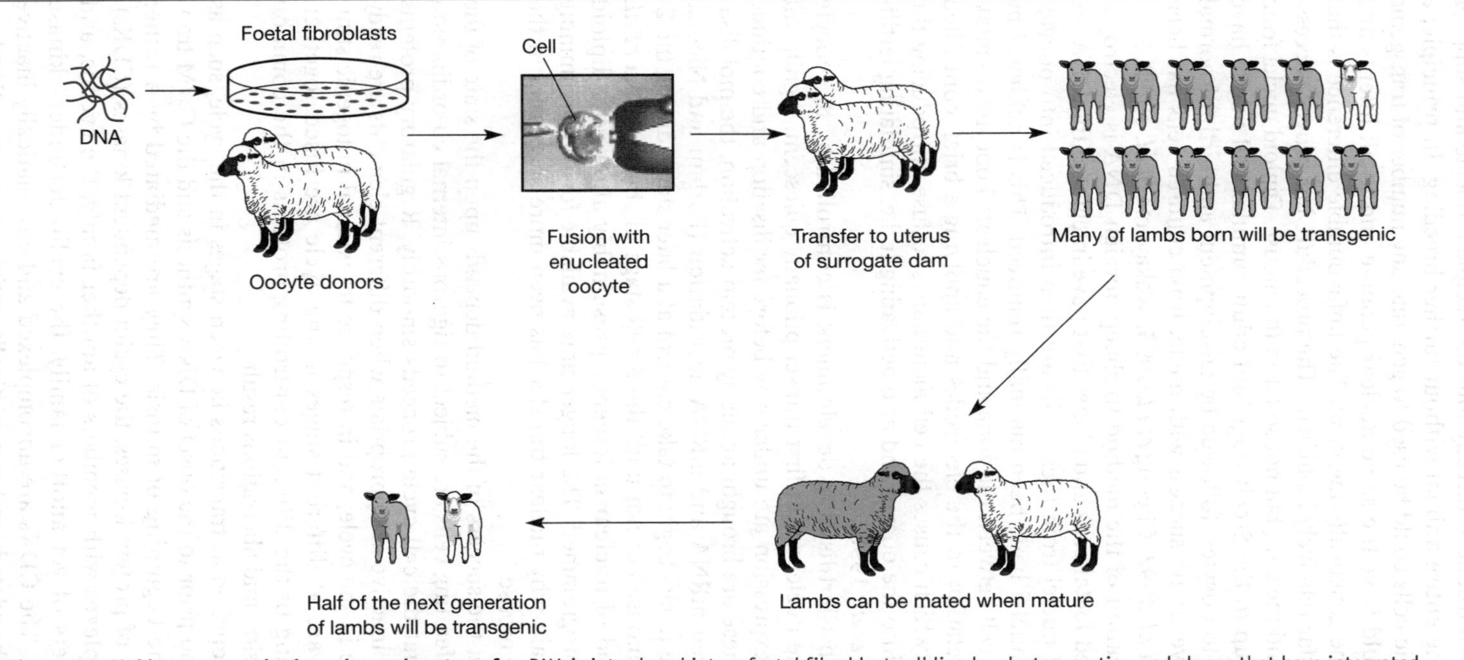

Figure 7.3 Making transgenic sheep by nuclear transfer. DNA is introduced into a foetal fibroblast cell line by electroporation and clones that have integrated the transgene are isolated by growth in selective medium. Oocytes are enucleated by micromanipulation and each is brought together with a transgenic fibroblast in the G_1 phase of the cell cycle. The oocyte and the cell are fused by applying a brief electrical potential, bringing the diploid fibroblast nucleus into the oocyte. (Note: in the picture the cell shown being fused with the oocyte is a blastomere rather than a fibroblast; a fibroblast is much smaller.) Following a period of development *in vitro* the surviving embryos are introduced into surrogate dams. Most of the live-born lambs will carry the same diploid chromosome complement as the cultured fibroblasts and will be transgenic, and of course genetically identical. The stock can be expanded by mating these animals when mature. Assuming a single site of integration of the transgene into the fibroblast chromosomes, half of the lambs born in this generation will be transgenic.

diploid nucleus is introduced directly into the recipient oocyte and supplies the genome of the entire animal without further breeding. In principle, a clone of transgenic cells could be used to generate any number of transgenic animals that would have the same nuclear genome and to all intents and purposes would be genetically identical. The unfavourable difference is that the method is so far relatively inefficient. The procedure has proved successful in the sheep and the cow, but not so far in the mouse. Diploid nuclei from mouse embryos up to the 8-cell stage and adult cumulus cell nuclei have been transferred to oocytes, followed by development and birth of normal mice. So far, however, no success with nuclei from cultured cells has been reported (Kono *et al.* 1992; Cheong *et al.* 1993; Wakayama *et al.* 1998).

In the application of the method to sheep, foreign DNA is electroporated into cultured foetal cells and clones that have integrated the DNA are selected. Diploid nuclei from these cells are then introduced into oocytes from which the nuclei have been manually removed. This is achieved by placing a diploid cell between the zona and the enucleated oocyte, orientating the cells in relation to the electrodes and applying a brief, controlled electric potential which causes the cell membranes to fuse, introducing the somatic nucleus into the oocyte and also activating it, i.e. simulating fertilisation (Campbell *et al.* 1996).

Immediate post-fertilisation development is controlled by the oocyte cytoplasm and the nuclei are at first transcriptionally quiescent. Significant changes in gene expression are under way before fertilisation and continue afterwards, but these are brought about by protein activation, the mobilisation of sequestered mRNA and mRNA degradation (Forlani and Nicolas 1997). The zygote nuclei begin to take control at a later stage, during the 2-cell stage in the mouse but not until the 8-cell stage in sheep (Fulka *et al.* 1996). The period of nuclear quiescence presumably allows the diploid nucleus to be reprogrammed. The longer time available for reprogramming may help to explain why nuclear transfer has been more successful in the sheep than in the mouse.

It is clear that the success of the method depends upon the state of the diploid cell. Proliferating cells are subject to rigorous internal co-ordination which ensures that the cell cycle proceeds smoothly. Regulatory proteins become active or inactive as appropriate when different elements are ready for the next stage in the cycle, and in response to external mitogenic stimuli. If cells that are at different stages in the cycle are fused together, causing the mixing of the different controlling proteins in the common cytoplasm, confusion and aberrations result.

Controls are applied at transitions between stages in the cycle, such as the G_1/S transition prior to the onset of DNA synthesis and the G_2/M transition prior to the beginning of mitosis. They are mediated by a rather bewildering array of protein kinases, the cyclin-dependent kinases (CDKs). These form complexes with members of another family, the cyclins, and also with members of yet another family, the cyclin-dependent kinase inhibitors (CKIs). The CDKs are uncomplexed and enzymatically inactive when hypophosphorylated, and enzymatically active when correctly phosphorylated and complexed with a cyclin. Other proteins such as pRb, the retinoblastoma protein, are phosphorylated by CDK/cyclin complexes. Cultured cells limited by mitogen deprivation come to rest in G_{10}, with a complete diploid set of chromosomes. At this stage very little active CDK/cyclin is present, the inhibitor $p27^{KIP1}$ is active and so is hypophosphorylated pRb, which binds and inactivates proteins such as the transcription factor E2F.

The meiotic divisions of the oocyte are characterised by high levels of maturation promoting factor (MPF) which is a CDK/cyclin complex of Cdc2 and cyclin B. Following ovulation, while the oocyte awaits the stimulus of fertilisation (or activation) before initiating its second meiotic division, the level of MPF is high. Under the direction of MPF the membranes of newly introduced nuclei break down and interphase chromosomes become condensed. S-phase nuclei suffer severe chromosomal damage in these circumstances (Rao *et al.* 1977). Different methods have been used to overcome the problems posed by MPF, such as pinching off and removing the metaphase plate to which MPF seems to be bound (Willadsen 1983) or allowing the level of MPF to subside naturally after activation (McGrath and Solter 1983).

Much the most successful method to date has been to activate a sheep oocyte and simultaneously fuse it with a somatic cell arrested by mitogen deprivation in the G_1 phase of the cycle (Campbell *et al.* 1996; Wilmut *et al.* 1997; Schnieke *et al.* 1997). It seems that the nucleus of a quiescent somatic cell is in an appropriate state to respond to the oocyte regulatory proteins. Development of the fused oocytes was allowed to proceed to the morula or blastocyst stage over six days, and viable embryos were finally transferred to uteri of suitably prepared surrogate dams where some of them implanted and a few developed to term. Very recently, calves were cloned by a similar procedure (Cibelli *et al.* 1998b). As mentioned above nuclei from mouse cumulus cells, which are arrested in the G_0 (post-mitotic) phase, were successfully transferred to mouse oocytes, although transfer of nuclei from other differentiated cell types failed (Wakayama *et al.* 1998).

While animals produced in this way carry the same diploid chromosome complement as the cell line from which they were derived, and presumably the cell line donor animal, it is worth noting that there are likely to be differences between the donor and the clones.

- It is well known that mitochondria are inherited through the female line, from the oocyte, with no contribution from the spermatozoon that fertilises it. In the nuclear transfer protocol the oocyte and the diploid cell are fused and there will consequently be a contribution of mitochondria from the diploid cell to the oocyte, albeit a small one. It is most unlikely that this will have any effect on the phenotype.
- Cell culture tends to promote genetic change in the cell population; cells that mutate to an increased rate of proliferation will tend to displace the unchanged cells in the course of time. The chance that this will happen can be reduced by minimising the time of growth in culture.

- Chromosomal telomeres become shorter during the proliferation of somatic cells both *in vivo* and in culture, and seem to contribute to cell senescence. The clones will have shorter telomeres than the donor animal. However, cultured fibroblasts derived from a bovine foetus produced by transfer of a nucleus from an immediately pre-senescent cell achieved the same number of divisions as fibroblasts from a normal foetus (Cibelli *et al.* 1998b).
- The pattern of chromosomal imprinting may change during cell culture, as it does in ES cells (Dean *et al.* 1998). Biallelic expression of the X chromosome in the female trophoblast would be lethal, for example, and biallelic expression or repression of imprinted autosomal genes would also be problematical. If this is a problem, it is unlikely to affect the offspring of the surviving first generation clones.

These questions provide fertile ground for investigation, and answers are likely to emerge in the near future.

Retroviral vectors

Retroviruses have featured prominently in the development of transgenic mammals. The first modifications of the mouse genome by DNA integration and among the first to be made by the ES cell route were produced by retrovirus infection of embryos (Jaenisch *et al.* 1981) and ES cells (Kuehn *et al.* 1987) respectively. Retrovirus infection was also the most successful route to the random insertional inactivation of genes (Gridley *et al.* 1987). This last route has recently been refined with the development of specialised retroviral vectors for gene trapping (Chapter 8).

Retrovirus replication

Retrovirology is a vast subject with a wealth of detailed information available on a large number of viruses, many of which have special properties. However, a broad understanding of the basic viral replication cycle is sufficient for present purposes. Virus particles are budded off from infected cells bounded by cell membrane which has been modified by the insertion into it of many copies of the viral ENV (envelope) protein. Each particle contains a capsid, made up of polypeptides derived by proteolytic digestion of a precursor polyprotein called the GAG protein. The capsid encloses a number of important viral and cellular products. These include two copies of the viral RNA genome, copies of a cellular tRNA that has been subverted to viral purposes and products of the viral POL gene with reverse transcriptase and hybridase (ribonuclease H) activities. The viral genome is a mRNA-like molecule with a 5' cap structure and a 3' poly-A tail. At both ends it has identical sequences of 100–300 nucleotides in the same 5' to 3' orientation. These are the R regions of the viral genome (Figure 7.4A).

Entry into a new host cell is through interaction between the envelope proteins and cellular receptors. Inside the cell, using the viral genomic RNA as a template and the cellular tRNA which is partially base-paired with it as a primer, the reverse transcriptase synthesises a single-stranded copy DNA complementary to the viral genome. Digestion of the RNA template by the hybridase exposes the DNA complementary to the 5' R region, allowing it to base-pair with the 3' R region of the same or another viral RNA genome. This allows the reverse transcriptase to continue to extend the copy DNA, using as template the RNA to which it has just become base-paired. Soon, synthesis of the second DNA strand, complementary to the copy DNA, is initiated. With the participation of cellular repair enzymes covalent double-stranded circular DNA and DNA concatemers are formed.

The most important thing about these structures is that they bring together sequences from the 5' and 3' ends of the RNA, called U5 and U3, that neighbour the R regions. In the orientation corresponding to the 5' to 3' orientation of the genomic RNA these are in the order U3-R-U5. In the RNA molecule the inner boundaries of U3 and U5 are defined by copies of a short inverted repeat sequence, which now form the outer boundaries of the conjoint U3-R-U5 and are required for the precise integration of the viral DNA into the chromosome. The DNA integrates into the genome under the action of another viral enzyme activity, the integrase, with little site preference (Sandmeyer *et al.* 1990; Withers-Ward *et al.* 1994). The ends of the viral sequence are two copies of the U3-R-U5 region, now called the LTR (Figure 7.4B), and it is bounded by 4–6 nucleotide direct repeats of chromosomal DNA.

The integrated DNA is called a provirus, and sooner or later it is likely to be transcribed like a normal cellular gene. *In vivo* the transcription of the viral RNA is often tissue-specific; the cultured cells on which the viruses are propagated are of course chosen to be transcriptionally permissive. The U3 region contains the viral promoter and transcription begins at the 5' end of the 5' R region and ends under the influence of a polyadenylation signal at the 3' end of the 3' R region, producing the directly repeated termini of the genomic RNA (Figure 7.4B). The genomic RNA is also an mRNA from which a polyprotein called GAG-POL is synthesised from the *gag* and *pol* gene sequences. A smaller RNA, generated by RNA splicing, codes for the ENV protein. ENV is a cell membrane glycoprotein with a transmembrane domain and GAG-POL becomes anchored to the membrane *via* a covalently attached fatty acid. Both proteins are processed at the membrane by a virus-encoded protease activity, and virus particles self-assemble and bud off. The incorporation of the RNA genome in the particles is dependent on a packaging signal (ψ) located close to U5.

Vectors

For use in the insertional inactivation of genes retroviruses required no modification. In contrast they have been highly modified, in hundreds of laboratories, for use as gene vectors or in gene trapping. The principles first adopted (Cepko *et al.* 1984) have remained unchanged, although many

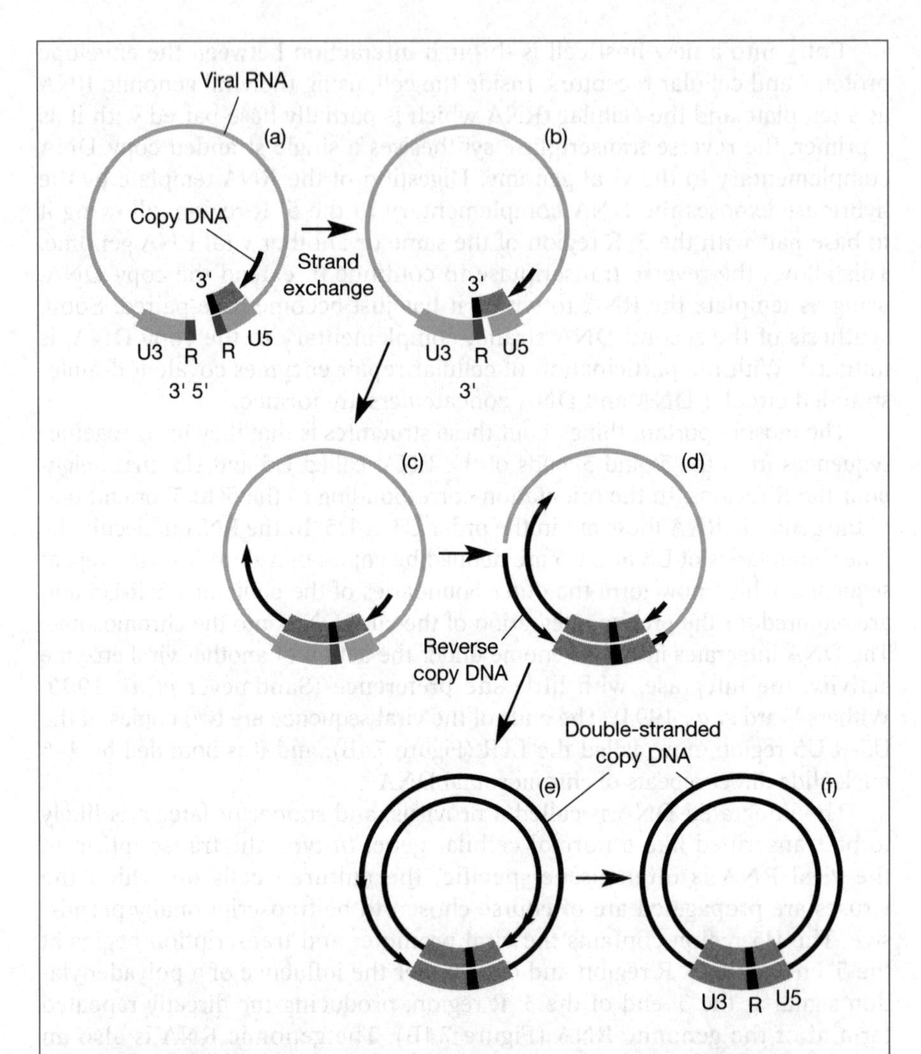

Figure 7.4A Diagram to show how U3, R and U5 regions are brought together during the synthesis of copy DNA. The poly-A sequence at the 3′ end of the viral RNA is not shown. Digestion of the RNA from the first-formed duplex region (a) allows the R region of the copy DNA to pair with another RNA R region, in this case from the 3′ end of the same viral genome (b) so that synthesis of the copy DNA can proceed (c). Reverse copy DNA synthesis begins (d) and continues (e) and finally covalent double stranded molecules are produced by DNA-DNA ligation (f). U3, R and U5 together make up the LTR. Single-strand breaks at the boundaries of the LTR initiate integration into the chromosome. A much shorter staggered break is made in the chromosome. Replicative repair of these staggered breaks brings about the duplication of the LTR and the short chromosomal sequence.

minor changes and some improvements have been introduced over the years. An example of current practice is illustrated in Figure 7.4B. The starting point is provirus, molecularly cloned from genomic DNA together with a small amount of chromosomal DNA at each end. The example shows the replacement of the *gag, pol* and *env* genes with a reporter gene that has its

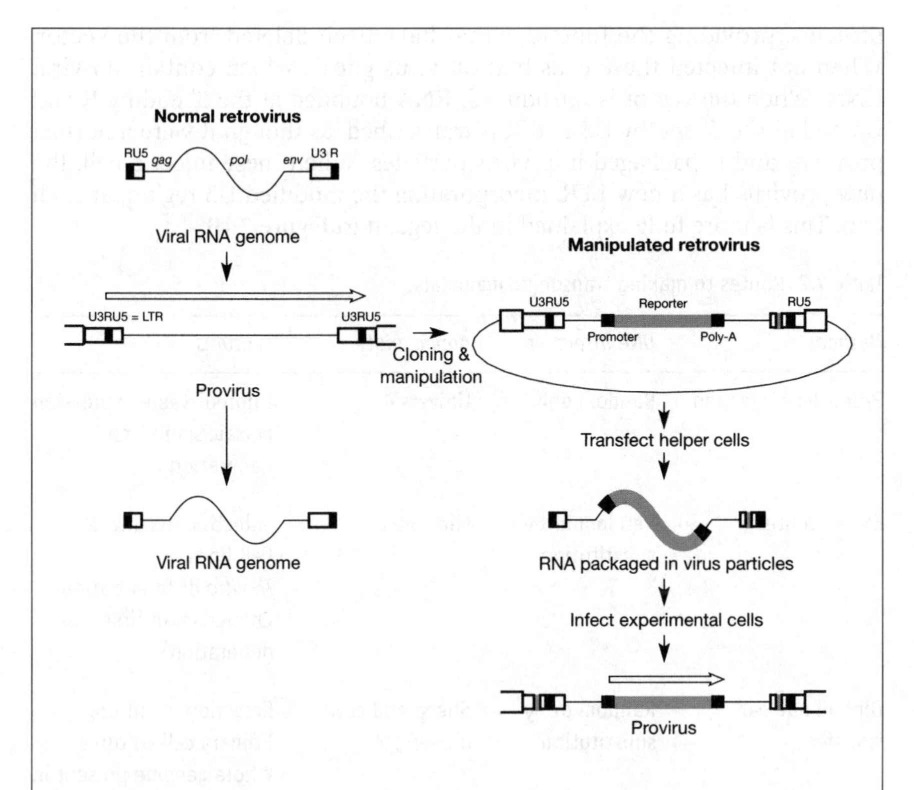

Figure 7.4B Modification of a retrovirus. Aspects of the replication cycle of retroviruses are shown on the left. The RNA genome, top, is packaged in the viral envelope. Upon entry into a cell a copy DNA is made (see Figure 7.4A). This integrates neatly and more or less randomly into one or other of the chromosomes as a provirus. Transcription of the provirus produces more viral RNA which is packaged into newly synthesised envelopes. R, the directly repeated nucleotide sequence; U5 and U3, the parts of the RNA genome adjacent to the 5′ and 3′ copies of R, respectively; U3, R and U5 together form the long terminal repeats (LTRs) at the ends of the provirus (see Figure 7.4A); *gag*, *pol* and *env*, the viral genes. Vector modification is illustrated on the right. Two changes to the structure of the provirus are shown. First, the *gag*, *pol* and *env* genes have been substituted by a reporter sequence. Secondly, the 3′ LTR has been modified, taking advantage of the fact that the promoter for provirus transcription is within the 5′ U3 region which is not itself transcribed. In the formation of the *next* proviral stage the U3 regions of both LTRs are copies of the 3′ U3 region of the present proviral stage. When U3 is modified as shown, inactivating its promoter function but leaving intact the short sequence that participates in integration, the viral RNA will still be transcribed from the vector. This RNA can be packaged in a helper cell line, infect a new cell, be reverse transcribed and become integrated, but there will be no transcription from the new provirus because of the inactivating change in U3.

own promoter and poly-A site. In addition the 3' U3 region has been modified to inactivate its potential promoter but leaving intact the internal integration sequence. The remainder is a multicopy plasmid molecular cloning vector. Special helper cell lines have been produced, stably transfected with genes that constitutively synthesise the GAG-POL and ENV proteins, providing the functions that have been deleted from the vector. When not infected these cells bud off virus ghosts which contain no viral RNA. When the vector is introduced, RNA bounded at the 5' end by R and U5 and at the 3' end by U3 and R is transcribed, as though it were a normal provirus, and is packaged into virus particles. When these infect a cell, the new provirus has a new LTR, incorporating the modified U3 region, at each end. This is more fully explained in the legend to Figure 7.4B.

Table 7.2 Routes to making transgenic mammals.

Method	*DNA insertion*	*Applicability*	*Features*
Pronuclear injection	Random only	Universal	Unpredictable expression Mosaicism in first generation
ES cell route	Random or by substitution	Mice only	Selection in culture Cell lines *In vitro* differentiation Chimerism in first generation[1]
Diploid nuclear transfer	Random or by substitution	Sheep and cows; universal?	Selection in culture Primary cell cultures Whole genome present in first generation
Retroviral vectors	Random only	Universal	Limited gene size Good gene traps

[1] With some exceptions (see text).

References

Abbondanzo, S.J., Gadi, I., and Stewart, C.L. (1993) Derivation of embryonic stem cell lines. *Methods in Enzymology,* **225**, 803–823.

Abe, K., Niwa, H., Iwase, K., Takiguchi, M., Mori, M., Abe, S.I., and Yamamura, K.I. (1996) Endoderm-specific gene expression in embryonic stem cells differentiated to embryoid bodies. *Experimental Cell Research,* **229**, 27–34.

Allen, N.D., Barton, S.C., Surani, M.A., and Reik, W. (1987) Production of transgenic mice. In *Mammalian development,* M. Monk (ed), IRL Press, Oxford, pp. 217–233.

Bautch, V.L., Stanford, W.L., Rapoport, R., Russell, S., Byrum, R.S., and Futch, T.A. (1996) Blood island formation in attached cultures of murine embryonic stem cells. *Developmental Dynamics,* **205**, 1–12.

Bradley, A., Evans, M., Kaufman, M.H., and Robertson, E. (1984) Formation of germ-line chimaeras from embryo-derived teratocarcinoma cell lines. *Nature,* **309**, 255–256.

Brinster, R.L., Chen, H.Y., Trumbauer, M.E., Yagle, M.K., and Palmiter, R.D. (1985) Factors affecting the efficiency of introducing foreign DNA into mice by microinjecting eggs. *Proceedings of the National Academy of Sciences of the United States of America,* **82**, 4438–4442.

Brook, F.A. and Gardner, R.L. (1997) The origin and efficient derivation of embryonic stem cells in the mouse. *Proceedings of the National Academy of Sciences of the United States of America,* **94**, 5709–5712.

Campbell, K.H., McWhir, J., Ritchie, W.A., and Wilmut, I. (1996) Sheep cloned by nuclear transfer from a cultured cell line. *Nature,* **380**, 64–66.

Canseco, R.S., Sparks, A.E., Page, R.L., Russell, C.G., Johnson, J.L., Velander, W.H., Pearson, R.E., Drohan, W.N., and Gwazdauskas, F.C. (1994) Gene transfer efficiency during gestation and the influence of co-transfer of non-manipulated embryos on production of transgenic mice. *Transgenic Research,* **3**, 20–25.

Cepko, C.L., Roberts, B.E., and Mulligan, R.C. (1984) Construction and applications of a highly transmissible murine retrovirus shuttle vector. *Cell,* **37**, 1053–1062.

Cheong, H.T., Takahashi, Y., and Kanagawa, H. (1993) Birth of mice after transplantation of early cell-cycle-stage embryonic nuclei into enucleated oocytes. *Biology of Reproduction,* **48**, 958–963.

Cibelli, J.B., Stice, S.L., Golueke, J., Kane, J.J., Jerry, J., Blackwell, C., Ponce de Leon, F.A. and Robl, J.M. (1998a) Transgenic bovine chimeric offspring produced from somatic cell-derived stem-like cells. *Nature Biotechnology,* **16**, 642–646.

Cibelli, J.B., Stice, S.L., Golueke, J., Kane, J.J., Jerry, J., Blackwell, C., Ponce de Leon, F.A. and Robl, J.M. (1998b) Cloned transgenic calves produced from non-quiescent foetal fibroblasts. *Science,* **280**, 1256–1258.

Clark, A.J., Harold, G., and Yull, F.E. (1997) Mammalian cDNA and prokaryotic reporter sequences silence adjacent transgenes in transgenic mice. *Nucleic Acids Research,* **25**, 1009–1014.

Dale, E.C. and Ow, D.W. (1991) Gene transfer with subsequent removal of the selection gene from the host genome. *Proceedings of the National Academy of Sciences of the United States of America,* **88**, 10558–10562.

Dani, C., Smith, A.G., Dessolin, S., Leroy, P., Staccini, L., Villageois, P., Darimont, C., and Ailhaud, G. (1997) Differentiation of embryonic stem cells into adipocytes in vitro. *Journal of Cell Science,* **110**, 1279–1285.

Dean, W., Bowden, L., Aitchison, A., Klose, J., Moore, T., Meneses, J.J., Reik, W., and Feil, R. (1998) Altered imprinted gene methylation and expression in completely ES cell-derived mouse fetuses: association with aberrant phenotypes. *Development*, **125**, 2273–2282.

Doetschman, T.C., Eistetter, H., Katz, M., Schmidt, W., and Kemler, R. (1985) The in vitro development of blastocyst-derived embryonic stem cell lines: formation of visceral yolk sac, blood islands and myocardium. *Journal of Embryology & Experimental Morphology*, **87**, 27–45.

Evans, M.J. and Kaufman, M.H. (1981) Establishment in culture of pluripotential cells from mouse embryos. *Nature*, **292**, 154–156.

Faust, N., Bonifer, C., Wiles, M.V., and Sippel, A.E. (1994) An in vitro differentiation system for the examination of transgene activation in mouse macrophages. *DNA & Cell Biology*, **13**, 901–907.

Folger, K.R., Wong, E.A., Wahl, G., and Capecchi, M.R. (1982) Patterns of integration of DNA microinjected into cultured mammalian cells: evidence for homologous recombination between injected plasmid DNA molecules. *Molecular & Cellular Biology*, **2**, 1372–1387.

Forlani, S. and Nicolas, J. (1997) Gene activity in the preimplantation mouse embryo. In *Transgenic animals, generation and use*, L.M. Houdebine (ed), Harwood Academic Press, Amsterdam, pp. 345–359.

Forrester, L.M., Nagy, A., Sam, M., Watt, A., Stevenson, L., Bernstein, A., Joyner, A.L., and Wurst, W. (1996) An induction gene trap screen in embryonic stem cells: Identification of genes that respond to retinoic acid in vitro. *Proceedings of the National Academy of Sciences of the United States of America*, **93**, 1677–1682.

Fulka, J., Jr., First, N.L., and Moor, R.M. (1996) Nuclear transplantation in mammals: remodelling of transplanted nuclei under the influence of maturation promoting factor. *Bioessays*, **18**, 835–840.

Gossler, A., Doetschman, T., Korn, R., Serfling, E., and Kemler, R. (1986) Transgenesis by means of blastocyst-derived embryonic stem cell lines. *Proceedings of the National Academy of Sciences of the United States of America*, **83**, 9065–9069.

Gridley, T., Soriano, P., and Jaenisch, R. (1987) Insertional mutagenesis in mice. *Trends in Genetics*, **3**, 162–166.

Hogan, B., Beddington, R., and Costantini, F. (1994) *Manipulating the mouse embryo: a laboratory manual*, Cold Spring Harbor, NY.

Hole, N., Graham, G.J., Menzel, U., and Ansell, J.D. (1996) A limited temporal window for the derivation of multilineage repopulating hematopoietic progenitors during embryonal stem cell differentiation in vitro. *Blood*, **88**, 1266–1276.

Jaenisch, R., Jahner, D., Nobis, P., Simon, I., Lohler, J., Harbers, K., and Grotkopp, D. (1981) Chromosomal position and activation of retroviral genomes inserted into the germ line of mice. *Cell*, **24**, 519–529.

Jahner, D., Stuhlmann, H., Stewart, C.L., Harbers, K., Lohler, J., Simon, I., and Jaenisch, R. (1982) De novo methylation and expression of retroviral genomes during mouse embryogenesis. *Nature*, **298**, 623–628.

Kono, T., Kwon, O.Y., Watanabe, T., and Nakahara, T. (1992) Development of mouse enucleated oocytes receiving a nucleus from different stages of the second cell cycle. *Journal of Reproduction & Fertility*, **94**, 481–487.

Kuehn, M.R., Bradley, A., Robertson, E.J., and Evans, M.J. (1987) A potential animal model for Lesch-Nyhan syndrome through introduction of HPRT mutations into mice. *Nature*, **326**, 295–298.

Laker, C., Meyer, J., Schopen, A., Friel, J., Heberlein, C., Ostertag, W., and Stocking, C. (1998) Host cis-mediated extinction of a retrovirus permissive for expression in embryonal stem cells during differentiation. *Journal of Virology*, **72**, 339–348.

Lindenbaum, M.H. and Grosveld, F. (1990) An in vitro globin gene switching model based on differentiated embryonic stem cells. *Genes & Development*, **4**, 2075–2085.

Martin, G.R. (1981) Isolation of a pluripotent cell line from early mouse embryos cultured in medium conditioned by teratocarcinoma stem cells. *Proceedings of the National Academy of Sciences of the United States of America*, **78**, 7634–7638.

Martin, G.R. and Evans, M.J. (1975) Differentiation of clonal lines of teratocarcinoma cells: formation of embryoid bodies in vitro. *Proceedings of the National Academy of Sciences of the United States of America*, **72**, 1441–1445.

McGrath, J. and Solter, D. (1983) Nuclear transplantation in the mouse embryo by microsurgery and cell fusion. *Science*, **220**, 1300–1302.

McWhir, J., Schnieke, A.E., Ansell, R., Wallace, H., Colman, A., Scott, A.R., and Kind, A.J. (1996) Selective ablation of differentiated cells permits isolation of embryonic stem cell lines from murine embryos with a non-permissive genetic background. *Nature Genetics*, **14**, 223–226.

Nagy, A., Gocza, E., Diaz, E.M., Prideaux, V.R., Ivanyi, E., Markkula, M., and Rossant, J. (1990) Embryonic stem cells alone are able to support fetal development in the mouse. *Development*, **110**, 815–821.

Nagy, A., Rossant, J., Nagy, R., Abramow-Newerly, W., and Roder, J.C. (1993) Derivation of completely cell culture-derived mice from early-passage embryonic stem cells. *Proceedings of the National Academy of Sciences of the United States of America*, **90**, 8424–8428.

Page, R.L., Canseco, R.S., Russell, C.G., Johnson, J.L., Velander, W.H., and Gwazdauskas, F.C. (1995) Transgene detection during early murine embryonic development after pronuclear microinjection. *Transgenic Research,* **4**, 12–17.

Potocnik, A.J., Nielsen, P.J., and Eichmann, K. (1994) In vitro generation of lymphoid precursors from embryonic stem cells. *EMBO Journal,* **13**, 5274–5283.

Rao, P.N., Wilson, B., and Puck, T.T. (1977) Premature chromosome condensation and cell cycle analysis. *Journal of Cellular Physiology,* **91**, 131–141.

Robertson, E.J. (1987) Embryo-derived stem cell lines. In *Teratocarcinomas and embryonic stem cells, a practical approach,* E.J. Robertson (ed), IRL Press, Oxford, pp. 71–112.

Sandmeyer, S.B., Hansen, L.J., and Chalker, D.L. (1990) Integration specificity of retrotransposons and retroviruses. *Annual Review of Genetics,* **24**, 491–518.

Schnieke, A.E., Kind, A.J., Ritchie, W.A., Mycock, K., Scott, A.R., Ritchie, M., Wilmut, I., Colman, A., and Campbell, K.H. (1997) Human factor IX transgenic sheep produced by transfer of nuclei from transfected fetal fibroblasts. *Science,* **278**, 2130–2133.

Schoonjans, L., Albright, G.M., Li, J.L., Collen, D., and Moreadith, R.W. (1996) Pluripotential rabbit embryonic stem (ES) cells are capable of forming overt coat color chimeras following injection into blastocysts. *Molecular Reproduction & Development,* **45**, 439–443.

Schwartzberg, P.L., Goff, S.P., and Robertson, E.J. (1989) Germ-line transmission of a c-abl mutation produced by targeted gene disruption in ES cells. *Science,* **246**, 799–803.

Taketo, M., Schroeder, A.C., Mobraaten, L.E., Gunning, K.B., Hanten, G., Fox, R.R., Roderick, T.H., Stewart, C.L., Lilly, F., Hansen, C.T., *et al.* (1991) FVB/N: an inbred mouse strain preferable for transgenic analyses. *Proceedings of the National Academy of Sciences of the United States of America,* **88**, 2065–2069.

Thomson, J.A., Kalishman, J., Golos, T.G., Durning, M., Harris, C.P., Becker, R.A., and Hearn, J.P. (1995) Isolation of a primate embryonic stem cell line. *Proceedings of the National Academy of Sciences of the United States of America,* **92**, 7844–7848.

Thomson, J.A., Kalishman, J., Golos, T.G., Durning, M., Harris, C.P., and Hearn, J.P. (1996) Pluripotent cell lines derived from common marmoset (*Callithrix jacchus*) blastocysts. *Biology of Reproduction,* **55**, 254–259.

Townes, T.M., Lingrel, J.B., Chen, H.Y., Brinster, R.L., and Palmiter, R.D. (1985) Erythroid-specific expression of human beta-globin genes in transgenic mice. *EMBO Journal,* **4**, 1715–1723.

Ueda, O., Jishage, K., Kamada, N., Uchida, S., and Suzuki, H. (1995) Production of mice entirely derived from embryonic stem (ES) cell with many passages by coculture of ES cells with cytochalasin B induced tetraploid embryos. *Experimental Animals,* **44**, 205–210.

Various Authors (1997) Part II. Section A. Gene transfer into mammal embryos. In *Transgenic animals, generation and use,* L.M. Houdebine (ed), Harwood Academic Press, Amsterdam, pp. 7–44.

Vittet, D., Prandini, M.H., Berthier, R., Schweitzer, A., Martin-Sisteron, H., Uzan, G., and Dejana, E. (1996) Embryonic stem cells differentiate in vitro to endothelial cells through successive maturation steps. *Blood,* **88**, 3424–3431.

Wakayama, T., Perry, A.C., Zuccotti, M., Johnson, K.R., and Yanagimachi, R. (1998) Full-term development of mice from enucleated oocytes injected with cumulus cell nuclei. *Nature,* **394**, 369–374.

Wang, Z.Q., Kiefer, F., Urbanek, P., and Wagner, E.F. (1997) Generation of completely embryonic stem cell-derived mutant mice using tetraploid blastocyst injection. *Mechanisms of Development,* **62**, 137–145.

Warren, A.J., Colledge, W.H., Carlton, M.B., Evans, M.J., Smith, A.J., and Rabbitts, T.H. (1994) The oncogenic cysteine-rich LIM domain protein rbtn2 is essential for erythroid development. *Cell,* **78**, 45–57.

Willadsen, S.M. (1983) Nuclear transplantation in sheep embryos. *Nature,* **320**, 63–65.

Willecke, K. and Ruddle, F.H. (1975) Transfer of the human gene for hypoxanthine-guanine phosphoribosyltransferase via isolated human metaphase chromosomes into mouse L-cells. *Proceedings of the National Academy of Sciences of the United States of America,* **72**, 1792–1796.

Wilmut, I., Schnieke, A.E., McWhir, J., Kind, A.J., and Campbell, K.H. (1997) Viable offspring derived from fetal and adult mammalian cells. *Nature,* **385**, 810–813.

Withers-Ward, E.S., Kitamura, Y., Barnes, J.P., and Coffin, J.M. (1994) Distribution of targets for avian retrovirus DNA integration in vivo. *Genes & Development,* **8**, 1473–1487.

Further reading

General

Houdebine, L.M. (ed) (1997) *Transgenic animals, generation and use,* Harwood Academic Press, Amsterdam.

Nuclear transfer (background)

Fulka, J., Jr., First, N.L., and Moor, R.M. (1996) Nuclear transplantation in mammals: remodelling of transplanted nuclei under the influence of maturation promoting factor. *Bioessays,* **18**, 835–840.

Grana, X. and Reddy, E.P. (1995) Cell cycle control in mammalian cells: role of cyclins, cyclin dependent kinases (CDKs), growth suppressor genes and cyclin-dependent kinase inhibitors (CKIs). *Oncogene,* **11**, 211–219.

YACs

Lamb, B.T. and Gearhart, J.D. (1995) YAC transgenics and the study of genetics and human disease. *Current Opinion in Genetics & Development,* **5**, 342–348.

BACs

Kim, U.J., Shizuya, H., de Jong, P.J., Birren, B., and Simon, M.I. (1992) Stable propagation of cosmid sized human DNA inserts in an F factor based vector. *Nucleic Acids Research,* **20**, 1083–1085.

Cosmids

Hohn, B., Koukolikova-Nicola, Z., Lindenmaier, W., and Collins, J. (1988) Cosmids. *Bio/Technology,* **10**, 113–127.

PACs

Sternberg, N.L. (1992) Cloning high molecular weight DNA fragments by the bacteriophage P1 system. *Trends in Genetics,* **8**, 11–16.

CHAPTER EIGHT

Random DNA integration

Techniques for generating transgenic multicellular organisms had to await developments in molecular biology. The requirements are simple but stringent. The first is for a fragment of DNA, produced in some quantity and small in size relative to the genome of the organism, in the case of mammals typically a few millionths of the genome. Such fragments became available only when cloning technology had reached a fairly advanced stage with the development of a range of DNA restriction and modification enzymes, sophisticated bacterial vectors and screening techniques. The second is for some understanding, the more the better, of the structure of genes and its relationship to function, which depends to a large extent upon nucleotide sequencing methods and other methods, such as Northern blotting, for assessing RNA levels in cells. Once these tools were available a meaningful DNA fragment could be chosen for introduction into the genome, and methods could be devised to monitor its presence and assess its activity.

Success in introducing genes into the mouse genome was first reported in 1980 (Gordon *et al.* 1980) and transmission of transgenes through the genome in 1981 (Gordon and Ruddle 1981; Wagner *et al.* 1981). Within a few years success had been reported with rabbits, sheep and pigs (Hammer *et al.* 1985) and by 1990 with rats (Mukha *et al.* 1990; Hammer *et al.* 1990; Mullins and Ganten 1990). All of the earliest transgenic mammals were generated by directly injecting DNA into one of the two haploid pronuclei present in newly fertilised oocytes (1-cell embryos). When DNA is introduced in this way, it integrates at random chromosomal sites, irrespective of whether or not it possesses regions of identity with one of the chromosomes. This is to say that we have no way of directing the insertion to a chosen location and simply have to accept what we get.

Somewhat later, mouse ES cells became available (see Chapter 3). These are an invaluable tool for generating transgenic mice by inserting DNA at a chosen chromosomal location by homologous recombination, often with the objective of inactivating a particular gene (see Chapter 9). However, when foreign DNA is introduced into ES cells, or indeed many other cell types, it integrates into random chromosomal sites in a minority of cells. Even when it possesses regions of identity with one of the chromosomes, it more frequently integrates randomly than by homologous recombination.

As we shall see, methods that lead to random DNA integration are basically flawed. New methods are currently being developed to achieve similar experimental objectives more efficiently and with a much greater degree of precision. It is likely that within a few years the older methods will no longer be employed.

Design of DNA constructs

Promoters and reporters

A basic construct consists of a promoter sequence and polyadenylation signals and a reporter sequence (Figure 8.1) . The promoter, at the 5'-end of the construct, contains sequences that bind protein factors which initiate RNA transcription. Some promoters direct transcription generally, though not uniformly, throughout the organism while others show tissue specificity to varying degrees. The polyadenylation signals, once transcribed, initiate

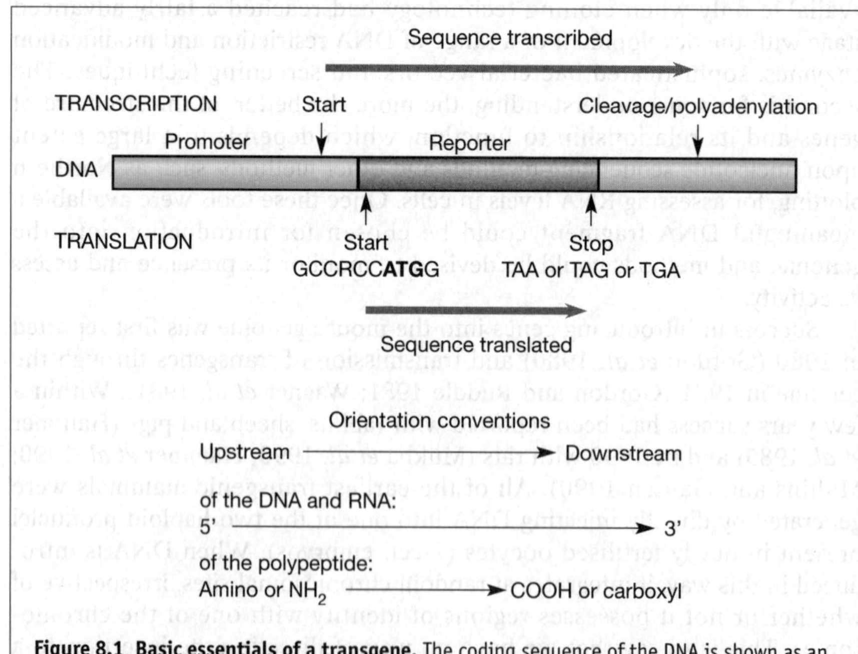

Figure 8.1 Basic essentials of a transgene. The coding sequence of the DNA is shown as an open box, flanking sequences as shaded boxes. The consensus translational start sequence and the stop codon (see text) are shown as DNA sequences (R = A or G). Additional elements, not shown here, which are commonly added to transgenes are discussed in Chapter 10.

the cleavage of the RNA transcript at a nearby site that terminates it at its 3' end and attracts polyadenylation. Polyadenyl*yl*ation, as it is correctly but seldom termed, is the post-transcriptional addition of a string of 200 to 300 adenylic acid residues. The reporter sequence usually directs the synthesis of a protein. This may variously be one that contributes to the biochemistry of the cell or the physiology of the organism, or a selective marker protein, or one that allows visualisation of cells in which it is expressed. In most cases the reporter requires a normal open reading frame, beginning with a start (ATG) codon and ending with one or more stop (TAA, TAG, TGA) codons.

The most frequently used selective reporter gene is the bacterial *neo* gene. The marker most commonly used to visualise cells, whether within cultured colonies, embryos or tissues of animals, has been the bacterial *lacZ* gene that codes for the enzyme β-galactosidase. This is usually visualised through its conversion of a colourless soluble substrate such as X-gal (5-bromo-4-chloro-3-indolyl β-D-galactoside) to an insoluble blue product. The *β-geo* gene was constructed from parts of the *neo* and *lacZ* genes and codes for a hybrid protein that performs both functions (Friedrich and Soriano 1991). A recently developed reporter is the green fluorescent protein (GFP) gene from the jellyfish *Aequoria victoria*. Unlike other fluorogenic proteins, GFP requires no substrate or cofactor (Ikawa *et al.* 1995; Kain *et al.* 1995). Commonly used reporter genes are listed in Table 8.1.

Table 8.1 Commonly used reporter transgenes.

Gene	*Origin and function*	*Reporter properties*
h*gh*	Human growth hormone	Secreted into blood plasma, immunoassay
neo	Bacterial neomycin resistance gene	Positive selection
hph	Bacterial hygromycin resistance gene	Positive selection
pur	*Streptomyces alboniger* puromycin resistance gene	Positive selection
tk	Herpesvirus thymidine kinase gene	Positive (in tk^- cells) and negative selection
HyTK	Hybrid of *hph* and *tk* genes	Positive and negative selection
m*hprt*	Mouse hypoxanthine-guanine phosphoribosyl transferase gene	Positive and negative selection in $hprt^-$ cells
lacZ	Bacterial β-galactosidase gene	Visual reporter
GFP	Jellyfish green fluorescent protein gene	Visual reporter
β-geo	Hybrid of *lacZ* and *neo* genes	Positive selection and visual reporter

A construct will often carry additional elements, such as introns, enhancers, locus control regions, matrix attachment regions and internal ribosome entry sites, intended to affect the level, tissue specificity and timing of the expression of the reporter sequence. These are considered later in the text, mainly in Chapter 10.

Directed ectopic gene expression

While ectopic gene expression due to chromosomal position effect can be very troublesome when it is unwanted (Chapter 10), in some cases ectopic expression can be the desired objective. This can often be arranged by coupling a promoter that is expressed predominantly or exclusively in the target tissue with a cDNA or other construct that codes for the protein that we wish to be expressed there. A well known example is the synthesis of human therapeutic proteins in the milk of sheep or cows. Here a mammary gland-specific promoter sequence (e.g. the promoter of the β-lactoglobulin gene or a casein gene) is coupled to a DNA sequence coding for the human therapeutic protein (e.g. α1-antitrypsin for the treatment of cystic fibrosis and emphysema or Factor IX for the treatment of haemophilia C).

In some cases this has worked out well by the simplest route of coupling a few kilobases of DNA containing the promoter to the cDNA sequence. In other cases it has proved more difficult, due to additional requirements for high-level activity of the promoter. These can include a requirement for enhancer sequences that may be located at some distance from the gene or for elements within introns, removed and replaced by the human cDNA in the above example. Thus to be sure of achieving high-level ectopic expression requires an understanding of the specific properties of the promoter in question (Chapter 10).

Integration of foreign DNA

Configuration

In most instances multiple copies of the transgene are integrated into a chromosomal site as a transgene **array**. The array is almost certainly generated extrachromosomally, prior to integration. The evidence for this comes from the study, not of embryos, but of cultured cells. In early studies of the chromosomal integration of cloned DNA in cultured cells the DNA was usually introduced together with an excess of a DNA 'carrier', such as commercially available salmon sperm DNA. This is not equivalent to the introduction of cloned DNA into embryos by microinjection or into ES cells by electroporation, where carrier DNA is not employed, and data obtained with this procedure can largely be ignored here. Cloned DNA without carrier also became integrated into the chromosomes of cultured cells, and the general configurational properties of the integrated transgenes were essentially the same as in injected embryos (Huttner *et al.* 1981;

Folger *et al.* 1982), whether the DNA was introduced by transfection or microinjection. The predominant configuration in both cases is the same in the following respects:

- the DNA is invariably found to be integrated at only one or at most a very few chromosomal sites in a given embryo or clone of cells
- usually multiple copies of the foreign DNA are integrated together (in tandem) at each site
- most of the multiple tandem copies at a given site are arranged in direct (i.e. the same) orientation (see Figure 8.2)
- most importantly, these direct tandem copies are perfect or near-perfect copies of the input DNA, usually with, at the most, minor imperfections at the junctions between neighbouring copies
- more rarely, two copies are found in inverted orientation, and these copies sometimes have a terminal deletion

The similarity in the configuration of foreign DNA when integrated into the chromosomes of cultured cells embryos argues that the arrays are formed by very similar mechanisms.

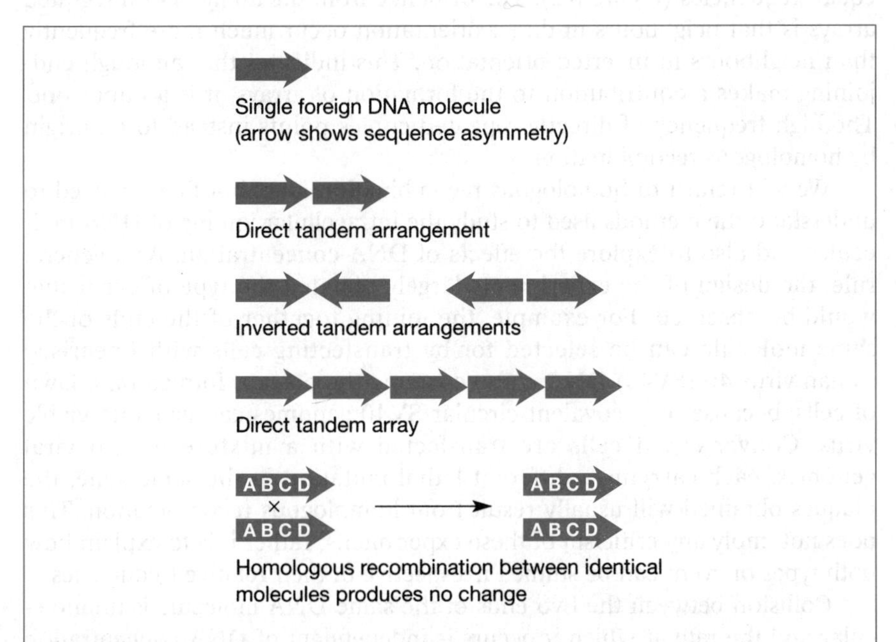

Figure 8.2 Direct and inverted gene arrangements. The blue arrow signifies a DNA molecule, e.g. a transgene, and its 5′→3′ orientation. Bottom: homologous recombination between identical linear molecules does not alter them.

Formation of arrays

It was argued previously that foreign DNA arrays are generated extrachromosomally, mainly by a process of recombination prior to integration into the chromosomes (Bishop and Smith 1989; Bishop 1996). The argument is based on two considerations:

- Productive collisions between two extrachromosomal DNA molecules are likely to be more frequent than those between an extrachromosomal molecule and one that is already integrated.
- Recombinatorial extrachromosomal interactions have been studied extensively in transfected cells (Lin *et al.* 1984; Wake *et al.* 1985; Folger *et al.* 1985). The results show that two different mechanisms are deployed, namely end-joining and homologous recombination.

End-joining means the joining of DNA duplexes end-to-end, and can occur by blunt-end ligation, or with pairing between complementary single-strand sequences exposed by a restriction enzyme, or by illegitimate recombination (i.e. recombination between imperfectly matched or even unmatched DNA duplexes). When the ends of the input DNA molecules are blunt, or have identical single-strand extensions, we would expect neighbouring copies to be joined in the same (direct) and opposite (inverted) orientation with equal frequencies (Figure 8.2). The evidence from the analysis of integrated arrays is that neighbours in direct orientation occur much more frequently than neighbours in inverted orientation. This indicates that although end-joining makes a contribution to the formation of arrays, it is a minor one. The high frequency of directly repeated copies points instead to an origin by homologous recombination.

We will return to homologous recombination later, but first we need to understand the methods used to study the intracellular joining of DNA molecules and also to explore the effects of DNA concentration. As a general rule, the design of the experiments largely dictated the type of event that would be observed. For example, the joining together of the ends of the same molecule can be selected for by transfecting cells with linearised simian virus 40 (SV40) DNA and recovering the plaques formed on a lawn of cells, because only covalent circular SV40 genomes can generate viable virus. Conversely, if cells are transfected with a mixture of two viral genomes, each carrying a different lethal mutation in the same gene, the plaques obtained will usually result from homologous recombination. This does not imply any criticism of these experiments; rather it is to explain how both types of event can be studied irrespective of their relative frequencies.

Collision between the two ends of the same DNA molecule is unimolecular and the rate at which it occurs is independent of DNA concentration (although it is of course influenced by salt concentration, protein binding, viscosity of the medium and so on). In contrast, the collision of ends of two DNA molecules is bimolecular and the rate is concentration-dependent.

Given a sufficiently high DNA concentration, the rate of the bimolecular reaction can exceed that of the unimolecular reaction. In the transfection procedure in which cloned and carrier DNA are introduced into cells together, copies of the cloned DNA become integrated in a complex together with fragments of carrier DNA. These complexes are almost certainly formed by end-joining. Very great amounts of carrier DNA are normally employed in this procedure, and an excess of carrier over cloned DNA of the order of 1000 or more (Wigler *et al.* 1977; Pellicer *et al.* 1978; Wigler *et al.* 1979). Under these circumstances it is not surprising that the cloned DNA becomes end-joined to the excess of genomic DNA fragments. Very much smaller amounts of DNA are normally introduced into ES cells and it is usual to introduce only from 50 to 1000 DNA molecules into the nucleus of a 1-cell embryo. It is therefore not surprising if the unimolecular reaction predominates. A discussion of DNA concentration in relation to intra- and inter-molecular ligation of DNA molecules *in vitro* can be found in Sambrook *et al.* 1989.

Now consider a population of identical linear DNA molecules, which is what is usually introduced into an embryo or ES cell. We have seen that bimolecular end-joining very probably makes only a minor contribution to the arrays of integrated DNA. The alternative pathway to arrays is homologous recombination. However, homologous recombination between such identical molecules would simply generate different copies of the same molecule by reciprocal exchange (Figure 8.2). How then are we to explain the extrachromosomal generation of direct tandem arrays?

When, say, 100 molecules are introduced into a cell they can be expected to participate in end-joining. End-joining of input DNA molecules has been observed within minutes of microinjecting mouse embryo nuclei (Burdon and Wall 1992). The model proposes that some of the linear molecules rapidly become circularised in transfected cells by unimolecular end-joining. Another reaction that occurs in transfected cells is that circular DNA molecules are randomly cleaved by cellular nucleases, generating a population of molecules best described as 'circularly permuted linear' molecules (Figure 8.3) (Bishop and Smith 1989). If two circularly permuted molecules undergo homologous recombination, or if one of them recombines with one of the original DNA molecules, a linear molecule that can be up to nearly twice as long will be produced. Importantly for the argument, the longer molecule will usually incorporate at least one of the end-joins that were formed unimolecularly. Because of this, the model accommodates damage at the junctions between the tandem DNA repeats (Hamada *et al.* 1993; Chen *et al.* 1995; Kopchick and Stacey 1984); much less internal damage should occur, and any that does occur will imply multiple rounds of circularisation and linearisation by cleavage. Perhaps the most telling feature of the model is that it allows for the assembly of perfect tandem arrays (apart from the junction points) even from fragments of the circularised molecules that are shorter than unit length.

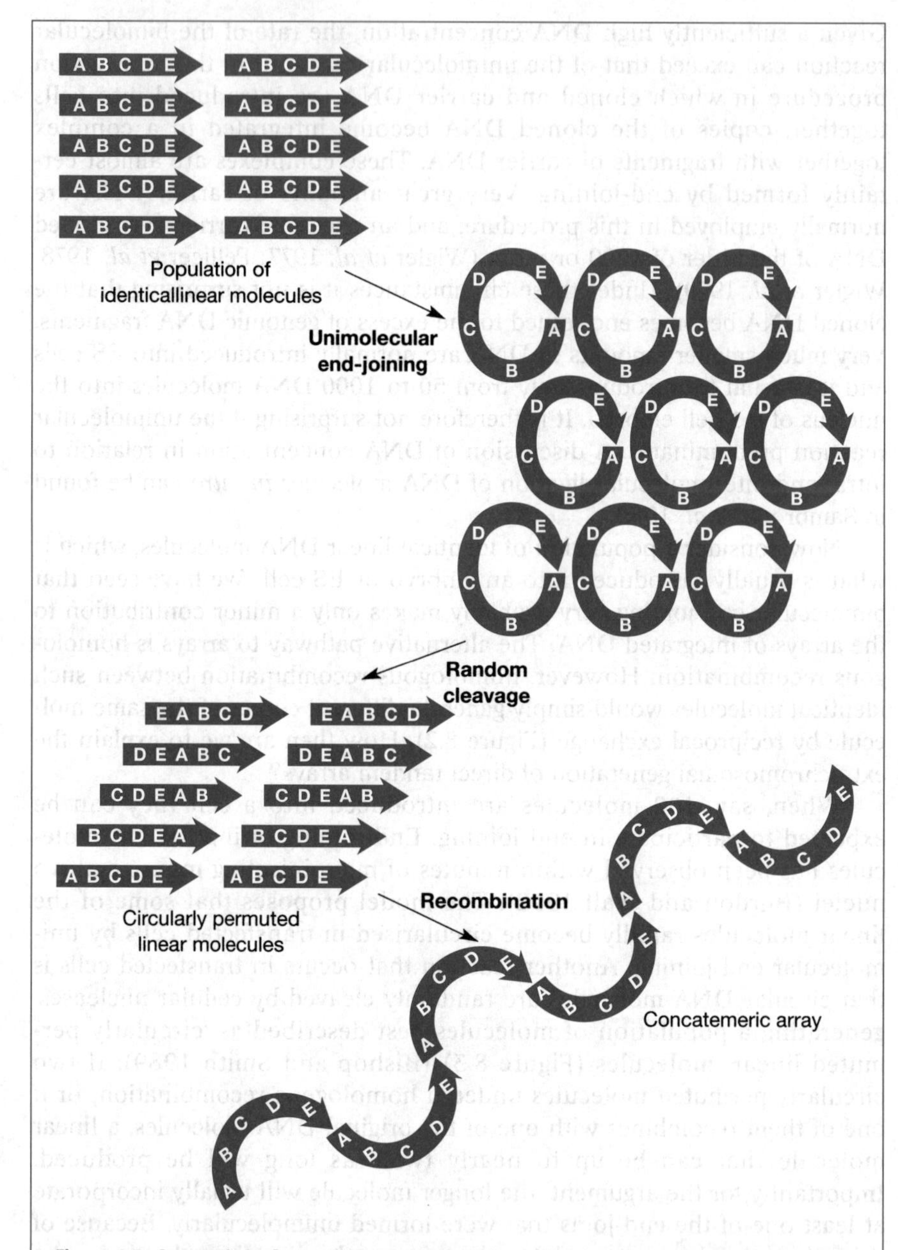

Figure 8.3 Generation of a tandem repeat array from identical linear molecules. The blue arrows indicate copies of a DNA molecule and its 5′→3′ orientation. Random cleavage of circularised molecules generates circularly permuted linear molecules. Recombination between these produces concatemeric direct tandem arrays.

Irrespective of whether the input DNA molecules are linear or circular, they should generate concatemers of a size that will in general be proportional to the square of the DNA input. An important implication of the model is that if overlapping fragments of a larger DNA molecule are introduced into the same cell, the larger DNA molecule should be reconstituted intracellularly by recombination. Such reconstituted genes have been reported in transgenic animals generated by pronuclear microinjection of overlapping fragments (Shimoda *et al.* 1991; Pieper *et al.* 1992) and in one of the studies a fragmented gene was reconstituted so faithfully that it was correctly transcribed and spliced and the mRNA was correctly translated (Pieper *et al.* 1992).

Which homologous recombinatorial mechanism is in effect is inconsequential to the model. However a substantial body of work shows that, in transfected cultured cells, the mechanism of DNA recombination is *nonconservative* (Lin *et al.* 1984; Wake *et al.* 1985; Folger *et al.* 1985). According to this mechanism (Lin *et al.* 1984), single-stranded ends are exposed in linear DNA molecules by exonuclease action. Collisions between exposed complementary regions lead to base-pairing, very likely followed by branch-migration. The strands in the junction region are covalently closed by well characterised repair reactions involving further nuclease action, ligation and probably DNA synthesis (Figure 8.4). In L cells 3'-exonuclease activity seems to dominate the recombination reaction (Lin *et al.* 1987) while there may be more 5'-exonuclease activity in ES cells (Henderson and Simons 1997).

Integration into the chromosome

DNA molecules, generally catenated, are inserted into the DNA duplex of one or, more rarely, more than one chromosome. Some understanding of the processes involved may be gleaned by examining changes in the structure of the chromosome around the site of integration. The structure of the foreign DNA is usually known and a restriction enzyme map of the chromosome around the integration site can be generated by Southern blotting, using the foreign DNA as a probe, but determining the structure of the chromosome in this region prior to integration requires the isolation of a new chromosomal probe. Consequently, the number of studies that provide all of this information is small. An overall picture of the results of DNA integration does emerge from the data, although it is not very illuminating. There is considerable diversity among the few well studied integrants, suggesting that chance plays a large part in the outcome. Most commonly the chromosome is altered by the deletion of a short or long segment at the integration site (Covarrubias *et al.* 1987; Naora *et al.* 1994; Xiang *et al.* 1990; Brown *et al.* 1994; Chen *et al.* 1995). Other effects observed are duplication and inversion (Wilkie and Palmiter 1987), complex disturbances of the original chromosome at some distance from the point at which the foreign DNA is inserted (Covarrubias *et al.* 1986) and multiple insertions in close proximity to each other (Covarrubias *et al.* 1986; Michalova *et al.* 1988).

We can hope to learn more by comparing the nucleotide sequences of the junctions between the foreign DNA and the chromosome with those of the input foreign DNA and the chromosomal site as it was prior to integra-

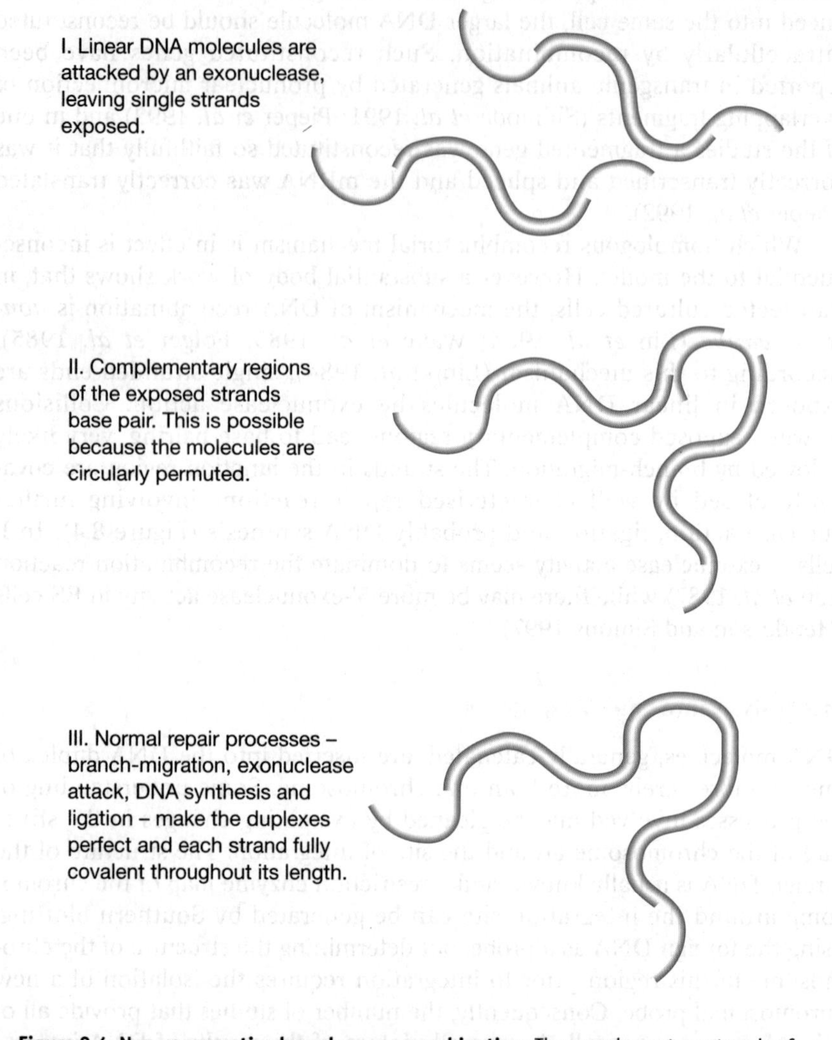

Figure 8.4 Non-conservative homologous recombination. The complementary strands of two circularly permuted DNA molecules are shown. Exonuclease attack exposes single strands at the ends of the molecules. These find a complement in another molecule because the molecules are circularly permuted. Repair generates a continuous duplex.

tion. The sequence of the foreign DNA is usually known and junctions can usually be isolated with ease, but as a general rule determining the chromosomal nucleotide sequence prior to integration is more laborious. However, a more coherent picture emerges.

- Several instances of low-level homology between the foreign DNA and the chromosomal integration site (Rohan *et al.* 1990; Allen *et al.* 1994; Hamada *et al.* 1993), and the frequent integration of a minisatellite transgene into a resident satellite sequence (Allen *et al.* 1994), both indicate that integration often occurs through illegitimate recombination between poorly matched sequences. It is likely that the preferential integration of some other transgenes into regions of repeated DNA is due to low-level homology between part of the transgene and one or other of the commoner genomic repeat sequences, rather than a sequence-independent preference for sites in repeat DNA.

- The most remarkable feature of the junctions between foreign and chromosomal DNA is the inclusion of short (so-called 'filler') sequences that do not derive in an obvious way from either the chromosome or the foreign DNA. The shorter the sequence the more difficult it is to identify its origin, since very short sequences occur frequently in DNA. Mammalian DNA is so complex that any 16-mer can be expected to occur in the genome by chance and the chance probability of any 18-mer occurring is about 0.05. Longer sequences that derive from elsewhere in the genome (Wilkie and Palmiter 1987) or have no known homology with either the foreign DNA or the resident chromosome complement (Wilkie and Palmiter 1987; Chen *et al.* 1995) are also found at some junctions.

Quite recently a number of devices have been introduced to make systematic studies of integration less laborious. For instance, if the foreign DNA contains part of a selectable gene and the cells the remainder, insertion events that bring the two parts together in a productive configuration can be selected for. If the break between the two parts lies within an intron, and provided they do not overlap, productive configurations arise by random integration into the chromosomal part of the intron. At least one junction will be easily amplified by PCR. Such approaches allow the nucleotide sequences of junctions to be determined rather easily, although at the risk that the nature of the junctions may be biased by the requirement for a productive configuration. Figure 8.5 shows a summary of data relating to sequence identity at more than 100 junctions. These show that sequence homology may have played no part in the formation of about half of the junctions, while it does seem to have been involved in the remainder.

As mentioned above, junctions between foreign and resident DNA often incorporate filler DNA sequences. A brief summary of some of the relevant data (Table 8.2) shows that about a third of the fillers consist of a single nucleotide, about a third are between 2 and 15 nucleotides long (and therefore their origin is non-attributable), and about a third contain up to several hundred nucleotides. The distribution of insert sizes is as smooth as could be expected with such a range of lengths and relatively small number of incidences, with a strong bias towards shorter lengths; thus the data give no indication that more than one mechanism might be involved. The same phenomenon is found at the junctions between translocated chromosome regions (Roth *et al.* 1989) and at junctions formed by extrachromosomal end-joining (Wake *et al.* 1984; Roth *et al.* 1985).

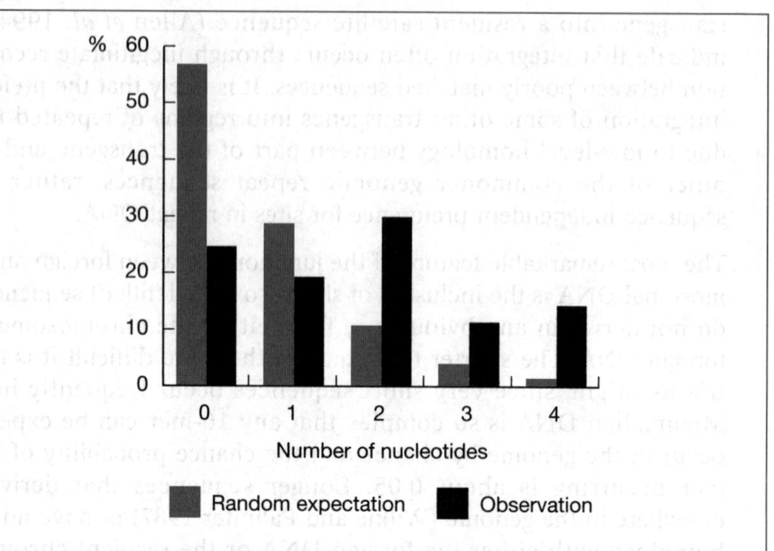

Figure 8.5 Short regions of identity at junctions between foreign and chromosomal DNA. Of 129 junctions, 127 were between regions with no identity or with from one to four sequentially identical terminal nucleotides. The percentages of junctions in these five groups are shown (black) and compared with the frequencies expected if sequence identity had no effect on the joining reaction (blue). This was calculated from $P(x) = (1+x)\cdot(0.25)^x\cdot(0.75)^2$, where x is the number of sequentially identical nucleotides and $P(x)$ is the probability that these will be present at the junction by chance (Roth *et al.* 1985). Data from Stary and Sarasin 1992; Phillips and Morgan 1994; Lehman *et al.* 1994; Merrihew *et al.* 1996.

It is tempting to think that single-stranded ends of extrachromosomal DNA molecules, exposed by exonuclease action, invade DNA duplexes to initiate the process of integration. The invasion of a DNA duplex by two exposed ends of an extrachromosomal molecule at two different points, followed by replacement of the intervening chromosomal DNA by the foreign DNA, could explain deletions of resident DNA. However this explanation would be insufficient to explain the filler DNA or junctions that occur in the absence of a region of identity between the participating sequences. Consequently additional or alternative mechanisms, such as blunt-end ligation, need to be considered. It seems that blunt-end ligation occurs both chromosomally and extrachromosomally (Wake *et al.* 1984; Roth *et al.* 1985; Rouet *et al.* 1994a). However, while selective chromosome breaks increase the frequency of homologous recombination (Smih *et al.*, 1995; Rouet *et al.*, 1994b, and see Chapter 9) they do not increase the frequency of random integration in *Dictyostelium* (Kuspa and Loomis 1992) and there is as yet no evidence that they do so in mice. Filler DNA sequences can be explained in a number of ways. Short filler sequences could be produced by the random addition of nucleotides. Longer sequences could be introduced if a recombinogenic DNA end can make an abortive invasion of the chromosome during which it picks up a new terminus, and can subsequently integrate elsewhere in the genome (Merrihew *et al.* 1996).

Table 8.2 Lengths of filler DNA sequences at junctions between foreign and chromsomal DNA. (From Phillips and Morgan 1994[a], Merrihew *et al.* 1996[b] and Roth *et al.* 1989[c].)

	Length of filler sequence		
Experimental procedure	1	2–15	>15
Inactivation of the *aprt* gene after *in vivo* treatment with a blunt-end restriction enzyme[a]	16	7	6
Reconstitution of an active *aprt* gene from non-overlapping segments in the chromosome and foreign DNA[b]	2	4	6
Compilation[c]	7	16	8

Transgenic mosaicism and the time of DNA integration

An animal is mosaic when it is made up of at least two cell types that differ in their genomes and that derive from a single zygote. Transgenic mosaics consist of a mixture of transgenic and non-transgenic cells (or, less frequently, two or more transgenic cell types). The concept is relevant to transgenic animals produced by embryo microinjection, but not to those produced by the ES cell route. In principle, transgenic mosaics can arise either by insertion of the transgene into a region of the chromosome that has replicated at least once, or by the loss of the transgene from one of the daughter (or grand-daughter, etc.) chromosomes following insertion. Because integration of the foreign DNA is likely to occur before compaction and because as a general rule only three morula cells are selected to form the precursor of the embryo, a mosaic will most commonly be one-third or two-thirds transgenic. Other things being equal, a one-third transgenic animal will transmit the transgene to only one-sixth of its progeny. This means that to have a 95% probability of transmission to at least one offspring requires 17 to be produced, which means taking two to three litters from a founder female mouse.

Early estimates of the frequency of mosaicism among transgenic mice produced by embryo microinjection suggested a figure of 15–30% (Brinster *et al.* 1985; Wilkie *et al.* 1986). More recent estimates have been much higher, around 65% (Whitelaw *et al.* 1993; Ellison *et al.* 1995; Bishop 1997), probably due to the use of more sensitive methods rather than differences in the frequency of mosaicism between laboratories.

Such a high level of mosaicism is consistent with the interpretation that DNA integration frequently occurs within regions of the chromosome that have already divided, or are in the process of division. In mice the foreign DNA is introduced around the time that S-phase begins, so that both duplicated and duplicating chromosomal DNA is immediately available. PCR methods (Burdon and Wall 1992; Cousens *et al.* 1994) and FISH (Echelard 1997) have been used to detect foreign DNA in blastomeres following the early divisions of injected 1-cell embryos. Foreign DNA was found in every blastomere in only 6% (PCR) and 5% (FISH) of 8-cell embryos. The interpretation of the experiment is somewhat clouded by the possible persistence of extrachromosomal DNA to the 8-cell stage. There is a further complication in that only 15% of injected embryos survive to birth, and we do not know what bias there might be towards survival among embryos in which the number of blastomeres carrying DNA differs. However, an examination of these data is interesting. In a model mouse blastocyst three of the eight blastomeres will form the inner cell mass at the time of compaction (Chapter 3). Assuming that the process is random, we can define the probabilities, for each of the nine blastomere classes (Figure 8.6), that the embryo proper will be transgenic and that it will be mosaic. Frequency predictions are made by combining the probabilities with the numbers of blastocysts in each class. These calculations indicate that 70% (PCR) and 46% (FISH) of the embryos would be transgenic, whereas in practice about 20% of pups born are transgenic. The difference is quite consistent with the lethal effect of DNA injection (62%) (Canseco *et al.* 1994; Page *et al.* 1995). Of the transgenic pups 70% (PCR) and 75% (FISH) would be mosaic, in good agreement with the frequency observed, which suggests that mortality is much the same in uniformly transgenic and mosaic embryos.

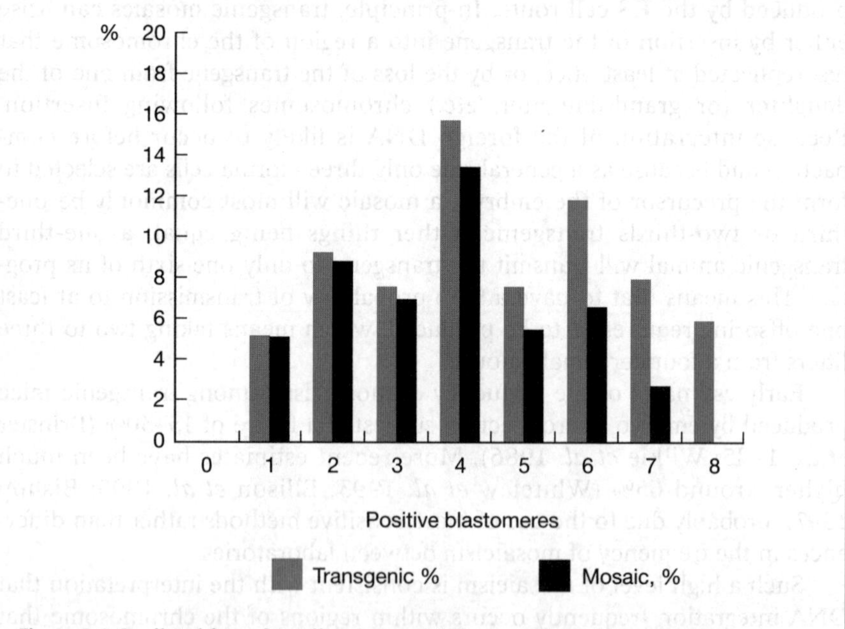

Figure 8.6 Predicted fate of 8-cell embryos with from one to eight transgenic blastomeres. Numbers of blastomeres in each class are pooled data of Burdon & Wall (1992) and Cousens *et al.* (1994). The figure shows predicted frequencies of transgenic (blue) and mosaic (black) embryos as a percentage of all embryos. The calculations assume that a blastomere that is positive at the 8-cell stage has integrated the foreign DNA and that three blastomeres out of eight form the embryo proper at compaction.

To provide an explanation for one particular well studied integration event, Wilkie and Palmiter proposed that a replication 'eye' (formed by the replication forks that diverge from a chromosomal origin of replication) had been invaded by the extrachromosomal DNA (Wilkie and Palmiter 1987). This provides an attractive *general* model for DNA insertion for several reasons. For one thing, we can imagine that the replication eye will be more accessible to invasion than a non-replicating DNA duplex. In addition, regions where replication is under way are expected to be populated by enzymes that carry out DNA synthesis and repair. Replication eyes also offer an explanation of how invading DNA might initiate duplications of resident DNA (Wilkie and Palmiter 1987; Hamada 1986). The high level of mosaicism observed does not prove that integration occurs during DNA replication, but it is certainly consistent with the idea.

Transgenic founders and transgenic lines

Each transgenic pup produced by embryo microinjection is called a transgenic founder (or G_0 mouse). The same transgene will be configured uniquely in each founder and will also be integrated at a unique chromosomal site. In general transgenes are faithfully transmitted through the germ line and consequently each founder, provided it is fertile, is capable of giving rise to a unique transgenic line. Thus a transgenic line may be defined as a pedigree, all descended from the same transgenic founder, and all carrying the same transgene in the same unique configuration at the same unique chromosomal site.

When transgenic mice are produced by the ES cell route, each unique transgenic ES cell clone is the equivalent of the transgenic founder. A distinct advantage of this route is that each clone, once expanded, can be aliquoted and cryopreserved. The transgenic cells of the chimeras carry the same transgene as the clone, and of course so do the transgenic offspring of the chimera. Barring later loss of transgenes, all transgenic mice derived from a given clone constitute a unique transgenic line.

Insertional mutation and gene trapping

The integration of foreign DNA into any gene is liable to disrupt its function, and random insertion of DNA into the mouse genome follows the general rule. Gene disruption is usually recessive, detected when homozygous transgenic animals fail to appear among the offspring of mated heterozygotes or die soon after birth. At one time it was anticipated that an approach based on random DNA integration would allow the identification of numbers of genes important in development; the foreign DNA would provide a target sequence that would allow the inactivated gene to be readily recovered from a DNA library. In the event the expectation was not realised and only a few genes have been isolated in this way (Rijkers *et al.*

1994). One obstacle is that widespread disruption of the chromosome tends to accompany gene inactivation. The tidy integration of foreign DNA into much of the mammalian genome (including intergenic regions, repetitive DNA and introns) will generally not inactivate a gene. Consequently integration events that cause chromosomal rearrangements are almost certainly over-represented among those that cause gene inactivation.

In contrast, retroviral proviruses invariably integrate neatly into the genome and have provided a more economical source of insertionally inactivated genes (Soriano *et al.* 1987; Stacey *et al.* 1987). However this approach has not been widely used, possibly because most insertional inactivation has been a by-product of the introduction of transgenes into the genome for other reasons; some of these lines would attract attention when homozygous transgenic animals turned out to have a lethal or otherwise unexpected phenotype.

Gene trap methods were first applied to cultured mammalian cells. The principle is to introduce a reporter gene, usually the bacterial *neo* gene, that will be expressed only when integrated in a particular juxtaposition with one of a chosen class of chromosomal targets. When selection is applied colonies develop mainly from cells in which this has occurred. The reporter also provides a tag which allows at least part of each trapped gene to be isolated from a genomic DNA library prepared from the affected cells. The gene sequence recovered in this way acts as a second tag which allows the native gene to be isolated from a library of normal genomic DNA, if so desired.

The design of the transgene is most important, since it determines the class of genomic element that will be trapped. Examples of three types of gene trap are illustrated in Figure 8.7:

- a promoterless reporter gene, i.e. a coding region, acts as a promoter trap (Pulm and Knippers 1985; Hiller *et al.* 1988)
- a reporter gene with a minimal promoter (Chapter 10) acts as an enhancer trap (Hamada 1986; Bhat *et al.* 1988)
- an intron trap has neither promoter nor enhancer, and usually carries at its 5' end the 3' part of an intron and a splice acceptor sequence so that the expression of the reporter gene depends on integration into an intron in the correct orientation (Gossler *et al.* 1989)

Reference has been made to the isolation of trapped and insertionally inactivated genes by probing libraries to identify vector sequences. Gene traps also provide alternative, more direct and less laborious approaches to the identification and isolation of the normal allele of a trapped gene. The procedure, 5'-RACE (rapid amplification of cDNA ends), is made possible by the fact that the 5' part of the mRNA transcript of the trapped gene is from the normal gene and its 3' part from the gene trap construct. Since the sequence of the latter is known, an oligonucleotide primer can be designed that allows a DNA molecule containing the sequence of the hybrid mRNA to be selectively synthesised by reverse transcriptase and DNA polymerase and amplified by the polymerase chain reaction (see Box 8.1). When the nucleotide sequence of the PCR product is determined, the part of it that derives from the gene trap vector serves to identify it positively. The gene and EST databases can be searched for sequences identical or similar to the fragment of mRNA sequence, often allowing the trapped gene to be identified immediately. The PCR product also serves as a probe that can be used for various purposes: to identify mRNA transcripts on Northern blots and to screen cDNA and genomic libraries.

Gene trapping with ES cells

ES cells have allowed gene trapping to be developed into a more powerful technique. The favourable properties of ES cells are as follows:

- Under appropriate culture conditions they will differentiate *in vitro* and can be encouraged to follow particular differentiation pathways by manipulating the growth medium.
- Their capacity to contribute to embryos allows mice carrying an ES cell genome to be produced.
- The gene inactivated by the trapping event can be made homozygous by breeding or by mitotic recombination (Mortensen *et al.* 1992), or the second allele of the gene can be inactivated by homologous recombination once the gene has been isolated. This allows genes essential for normal development to be identified, and the phenotype of homozygous null animals to be studied.
- The effects of the homozygous null recessive lethal mutations can be studied in chimeric embryos and, if the reporter is chromogenic, the expression pattern of the trapped gene can be ascertained.

Constraints

The brief treatment of gene expression at the beginning of Chapter 10 may help some readers with this section.

Different levels of constraint can be placed on a gene trap fragment. If the reporter sequence has no effective translational start codon at its 5' end, it will be translated only when integrated into an intron that lies downstream of the start codon of the gene (Figure 8.7-3). Furthermore, the correct reporter polypeptide sequence will be translated only when RNA splicing between the splice donor of the gene and the splice acceptor of the trap places the reporter sequence in the same reading frame as the upstream coding sequence of the gene. Under these circumstances a fusion protein will be translated which will usually exhibit at least some reporter enzyme activity. Otherwise a nonsense polypeptide will be produced. Other

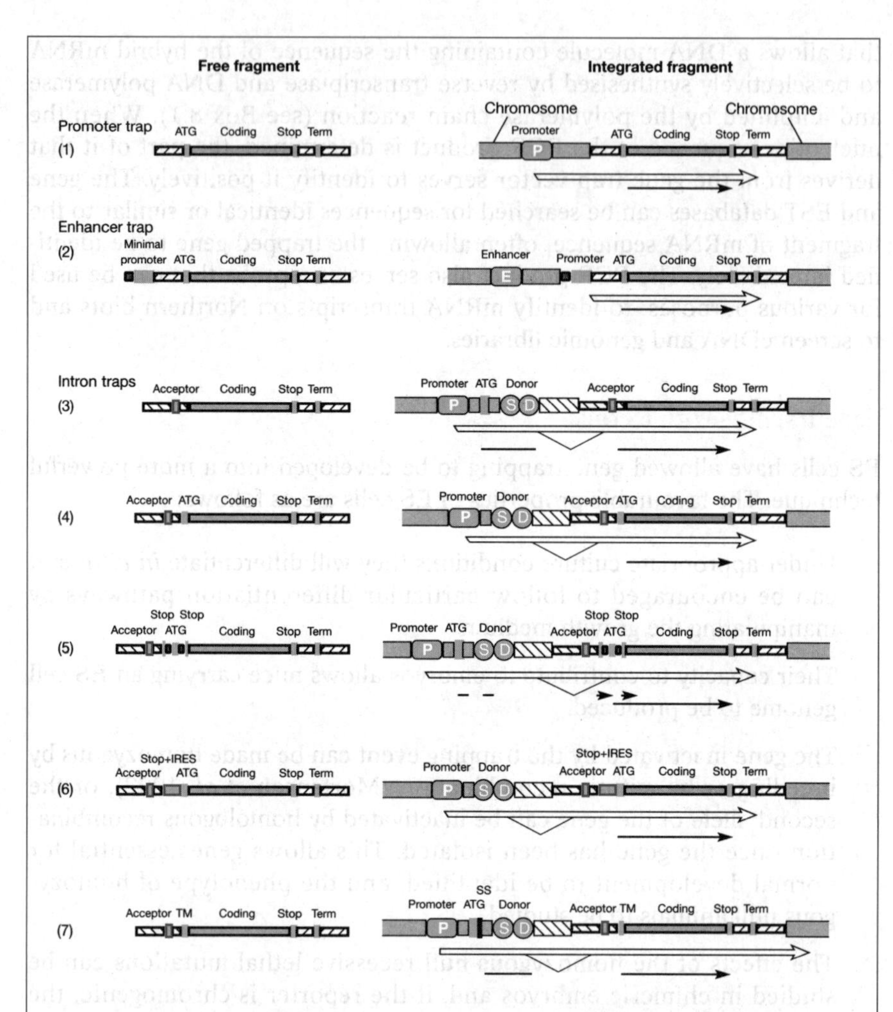

Figure 8.7 Diagrammatic representation of gene trap vector DNA fragments both free and when productively integrated into the genome. Coding regions in the vectors are shown as thin open boxes, flanking regions as thin hatched boxes, notable sequence elements in blue. Term, transcriptional termination (cleavage/polyadenylation) site; ATG, translational start codon; acceptor, splice acceptor site; IRES, internal ribosome entry site; TM, region coding for transmembrane protein domain. On the right chromosomal elements are shown as fatter boxes, the flanking chromosome as open boxes, introns hatched. Donor (S-D), splice donor site; SS, signal sequence. Open arrows show the regions transcribed, and the attached triangles signify RNA splicing. Continuous and broken line arrows show respectively the regions of the vectors and the chromosomes that are translated, interrupted across regions that are spliced out.

things being equal, only one-sixth of the integrations of such a vector into an intron will be productive (one-third when the trap is in the same 5′ –3′ orientation as the gene and none when it is not).

If the reporter sequence is given an effective ATG start codon of its own (Figure 8.7–4), some translation may be initiated from it, but this will depend on the 'strength' of the upstream start codon(s) and the length of mRNA intervening (Kozak 1986; Kozak 1987b; Kozak 1987a). However, the polypeptide translated in this way is not encumbered by a novel amino-terminal protein sequence, and should possess full enzyme activity. More efficient translational initiation can be obtained at an internal start codon by placing translation stop codons in its vicinity (Figure 8.7–5). Some of the ribosomes that terminate translation at these codons are able to reinitiate at the nearby start codon. Placing a stop codon in both of the reading frames not used by the reporter ensures the termination of all out-of-frame translation initiated upstream.

A more potent device is to place an internal ribosome entry site (IRES) upstream of the reporter (Figure 8.7–6). The IRES is a sequence isolated from a picornavirus such as encephalomyocarditis virus or poliovirus that permits efficient translational initiation at an internal start codon in an mRNA molecule (see Chapter 10). An IRES also contains upstream stop codons in all three reading frames. Consequently there is high-level expression of the reporter protein and it is not fused to an upstream polypeptide.

One reporter or two?

Some gene trap fragments have two reporter genes, others one. When there are two reporters, one of them, such as *neo*, is coupled to a dedicated mammalian promoter (such as the β-actin or phosphoglycerate kinase (*pgk*) promoter). This allows the selection of colonies in which the fragment is integrated at any chromosomal position that does not positively inhibit its expression. The other reporter (such as *lacZ*) will have one of the configurations discussed in the previous section and serves to identify the subset of the G418-resistant colonies that carry the fragment integrated productively into an active transcription unit.

The most commonly used single reporter is the hybrid *β-geo* gene which serves both for selection and as a visual indicator. The two configurations seem to give much the same overall result; in one case (Forrester *et al.* 1996) the two-reporter configuration gave 50 neo^R colonies per 10^6 electroporated cells, of which 1/18 expressed β-galactosidase, i.e. three expressing cells per 10^6 overall. In another study (Skarnes *et al.* 1995) a comparable single reporter construct gave 1.6 neo^R cells per 10^6 cells, 95% of which expressed β-galactosidase. These frequencies, which appear to be very reproducible (Friedrich and Soriano 1991), indicate that 5% of integrations are productive, equivalent to 10% of integrations in one orientation.

The gene trapping strategies described so far identify only genes that are expressed in the cell population. This is potentially a serious limitation, for example where the intention is to trap and identify genes involved in development. There are indications that transcription in ES cells may be especially promiscuous or leaky, and if so the problem will be less serious. Even so, it is unlikely to be negligible. However there are ways in which it may be addressed. One way to broaden the spectrum of

BOX 8.1: MAKING DOUBLE-STRANDED DNA FROM CELLULAR RNA

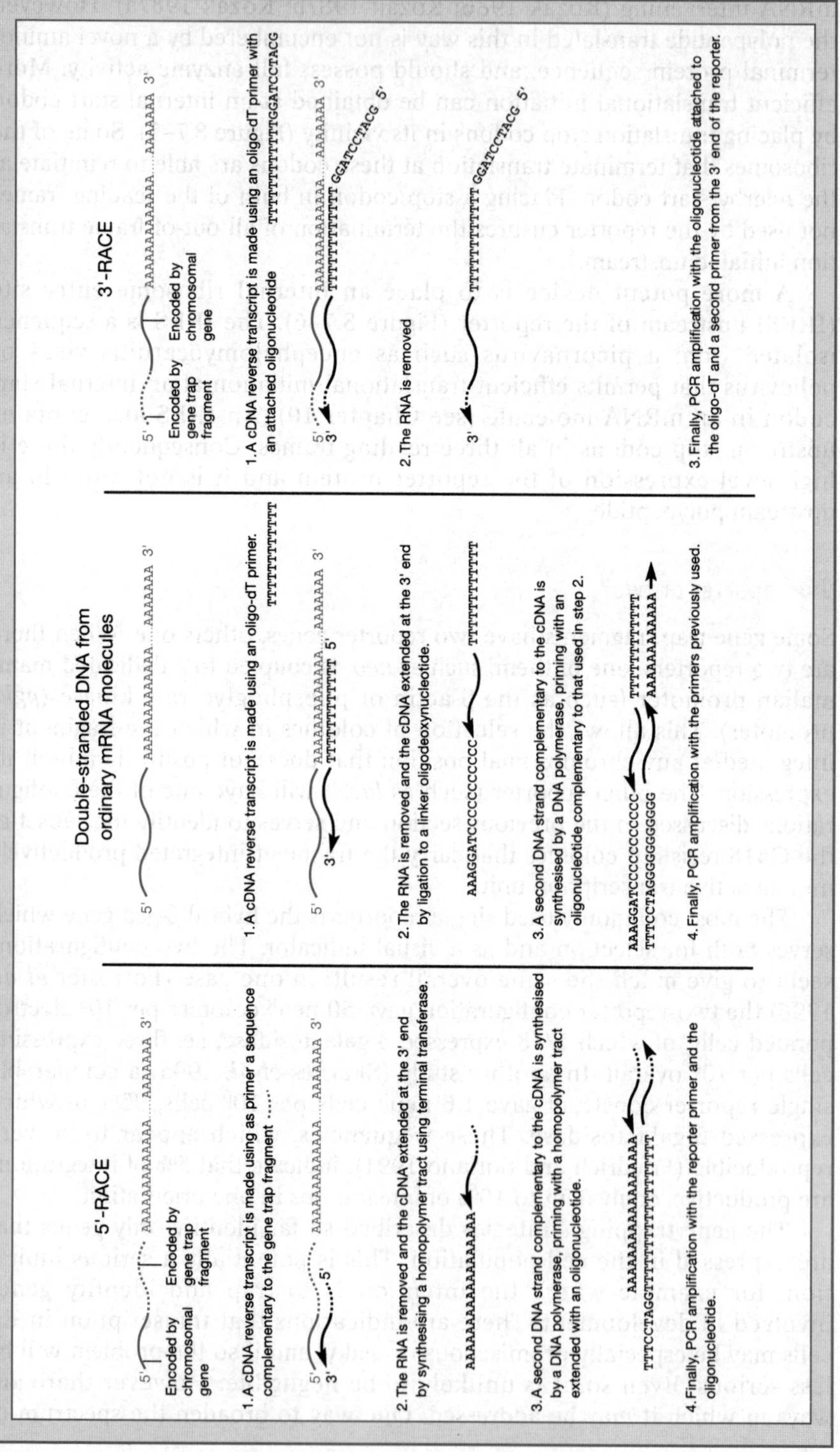

5'-RACE

5' AAAAAAAAAAAAA.....AAAAAAA 3'

Encoded by chromosomal gene — Encoded by gene trap fragment

1. A cDNA reverse transcript is made using as primer a sequence complementary to the gene trap fragment

5' AAAAAAAAAAAAA.....AAAAAAA 3'
3' 5'

2. The RNA is removed and the cDNA extended at the 3' end by synthesising a homopolymer tract using terminal transferase.

AAAAAAAAAAAAAAAAAA

3. A second DNA strand complementary to the cDNA is synthesised by a DNA polymerase, priming with a homopolymer tract extended with an oligonucleotide.

AAAAAAAAAAAAAAAAAA
TTTCCTAGGTTTTTTTTTTTTTTTTTT

4. Finally, PCR amplification with the reporter primer and the oligonucleotide.

Double-stranded DNA from ordinary mRNA molecules

5' AAAAAAAAAAAAA.....AAAAAAA 3'

1. A cDNA reverse transcript is made using an oligo-dT primer.

TTTTTTTTTTTT

5' AAAAAAAAAAAAA.....AAAAAAA 3'
3' TTTTTTTTTTTT 5'

2. The RNA is removed and the cDNA extended at the 3' end by ligation to a linker oligodeoxynucleotide.

AAAGGATCCCCCCCCCCC TTTTTTTTTTTT

3. A second DNA strand complementary to the cDNA is synthesised by a DNA polymerase, priming with an oligonucleotide complementary to that used in step 2.

AAAGGATCCCCCCCCCCC TTTTTTTTTTTT
TTTCCTAGGGGGGGGGGG AAAAAAAAAAAA

4. Finally, PCR amplification with the primers previously used.

3'-RACE

5' AAAAAAAAAAAAA.....AAAAAAA 3'

Encoded by gene trap fragment — Encoded by chromosomal gene

1. A cDNA reverse transcript is made using an oligo-dT primer with an attached oligonucleotide TTTTTTTTTTTTGGATCCTACG

5' AAAAAAAAAAAAA.....AAAAAAA 3'
3' TTTTTTTTTTTT GGATCCTACG 5'

2. The RNA is removed

3' TTTTTTTTTTTT GGATCCTACG 5'

3. Finally, PCR amplification with the oligonucleotide attached to the oligo-dT and a second primer from the 3' end of the reporter.

The Figure shows an example of a route to making double-stranded copy DNA from a cellular mRNA population and examples of 5'-RACE and 3'-RACE modifications.

In each case copy DNA is first synthesised using reverse transcriptase, but the primers differ. When the oligo-dT primer is used, whether or not with an attached oligonucleotide, most of the different mRNA molecules in the population will be copied. In contrast, the primer that is complementary to the gene trap fragment will lead to preferential copying of the trapped transcript.

Alternative ways of introducing a chosen restriction enzyme site at the 5' end of the copy DNA are shown. One way (shown under 5′-RACE) is to extend the 3′ end of the copy DNA with an oligonucleotide homopolymer by synthesis with terminal transferase, and to introduce the new site attached to a complementary homopolymer primer at the next step. The other way (shown under ordinary RNA molecules) is to ligate an oligonucleotide containing the site to the 3′ end of the copy DNA and then to prime synthesis of the reverse strand with a complementary oligonucleotide. Either method can of course be used either to indiscriminately amplify an entire mRNA population or in the 5′-RACE procedure.

When copies of the entire mRNA population are wanted, amplification of the double-stranded copy DNA is not necessary unless the supply of RNA is severely limiting. In the RACE examples the PCR step is the only (3'-RACE) or most powerful (5'-RACE) discriminatory step that selectively amplifies the DNA copy of the trapped mRNA. It is sometimes found necessary to follow this step with a second amplification step, seeded with part of the product of the first and primed by a second pair of primers ‘nested’ between the first pair.

trapped genes is to employ a variety of differentiated cell types. Another, which can be applied to ES cells, is to eliminate cells that express the reporter while still undifferentiated, by selection or fluorescence-activated cell sorting (Gogos *et al.* 1997), and identify among the remainder those that express the reporter when they differentiate. Another promising approach is to provide the reporter gene with its own promoter but to deprive it of a transcription terminator; this should enrich for integration into transcriptionally inactive genes provided they supply a termination signal (Niwa *et al.* 1993; Yoshida *et al.* 1995). Partial nucleotide sequences of genes trapped in this way can be determined by a variation of the RACE procedure called 3'-RACE (Box 8.1).

Retroviral gene trap vectors

Retroviral vectors offer some distinct advantages for gene trapping purposes:

- the provirus is integrated neatly into the chromosome without causing rearrangements
- a single copy of the provirus is inserted, never a tandem array
- multiple integrations can readily be minimised by regulating the virion/cell ratio at the time of infection

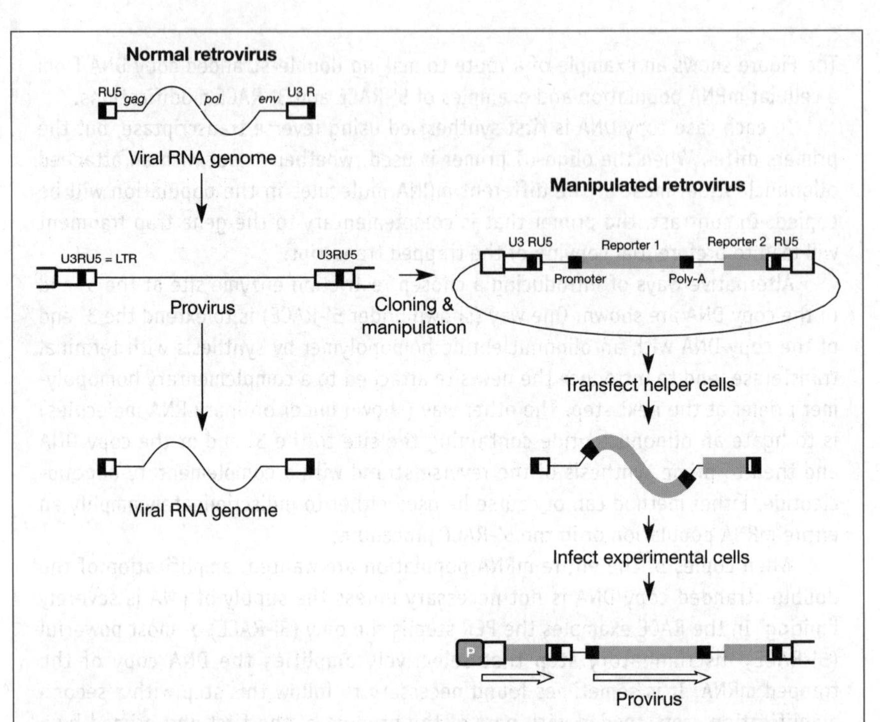

Figure 8.8 Manipulation of the genome of a retrovirus to make a U3-type vector. The *gag*, *pol* and *env* genes of the cloned provirus are replaced by the Reporter 1 gene and most of the 3′ U3 region is replaced by the Reporter 2 gene, promoterless in this case. Virus is recovered from helper cells transfected with the manipulated DNA. When the new provirus integrates into the chromosome the manipulated U3 region is present in both LTRs. The 5′ copy of the Reporter 2 gene is transcribed only when the provirus integrates into a chromosomal transcription unit. For more details see the text and Chapter 7.

More conventional retroviral vectors contain the same range of reporter elements employed in gene trap fragments and produce much the same results (Friedrich and Soriano 1991). The development of U3 vectors (von Melchner and Ruley 1989) marks a new approach that takes full advantage of the retrovirus life cycle. In these vectors a reporter gene is inserted into the U3 region of the 3' LTR of the provirus leaving in place the short inverted repeat that is necessary for integration (Reporter 2 in Figure 8.8). The manipulated retrovirus is introduced into helper cells in the usual way (Chapter 7) producing virus particles carrying the RNA genome illustrated in the figure. When this integrates into the experimental cells one copy of the Reporter 2 gene is in close proximity to the 5' junction with the chromosome, in a position to be transcribed if insertion is within a transcription unit. Reporter 1 can be a mammalian cellular reporter with its own promoter and a poly-A site, when the virion will act in much the same way as a two-reporter gene trap fragment. Alternatively, Reporter 1 can be a bacterial drug resistance gene, with a bacterial promoter, coupled to a plasmid replication origin. In this case plasmid rescue can be used to clone the 5' copy of Reporter 2 together with part of the transcription unit into which the provirus has integrated (Hicks *et al.* 1997). This is a viable alternative to 5'-RACE as a way of obtaining a sequence tag with which to search databases and probe libraries and blots.

Special screens

Methods can be devised to select particular classes of trapped genes. Some of these are listed in Table 8.3. Those that have not yet been used in conjunction with ES cells could probably be adapted to that purpose. Only one is described here, as an example.

Table 8.3 Properties shared by groups of genes that have been selected for by special gene trap protocols.

Property	*Method*	*Cells*	*Reference*
Response to retinoic acid	Replica plating, RA administration	ES	Forrester *et al.* 1996
Response to oestradiol	Negative then positive selection	3T3	Gogos *et al.* 1997
Radiation inducible	FACS	CHO	Menichini *et al.* 1997
Involved in apoptosis	Induction of CRE protein	FLOXIL3	Russ *et al.* 1996
Secretory and membrane	Special vector	ES	Skarnes *et al.* 1995
Lineage-specific expression	Replica plating and immunofluorescence with lineage-specific antibodies	ES	Baker *et al.* 1997

Cells can be selected in which the trapped genes specify secreted and membrane-bound proteins. Such proteins carry signals in their amino acid sequences that lead them into the appropriate processing pathway, among which are the signal sequence (SS) at the amino-terminal end and an internal transmembrane region (TM). In addition to the intron and splice acceptor and the *β-geo* gene, the trap fragment codes for a TM (Figure 8.7–7). When the gene trap integrates into an intron just 3' of a sequence that codes for an SS, the β-geo hybrid protein locates to the cytoplasmic side of the endoplasmic reticulum and is enzymatically active. When the gene trap integrates productively into one of the majority of genes that code for cytoplasmic proteins without an SS, the protein locates to the lumen side of the reticulum and this inactivates its β-galactosidase activity.

Consequently, clones of the desired class are those that exhibit β-galactosidase activity. The neomycin phosphotransferase moiety of the hybrid protein seems to be active in both configurations, or at least more generally active, because only 20% of G418-resistant clones exhibited β-galactosidase activity (Skarnes *et al.* 1995).

The sledgehammer approach

Modern developmental biology stems to a large extent from exercises in saturation mutagenesis in the fruit fly and the nematode. Driven largely by biomedicine a vast effort is under way to determine the nucleotide sequences of the human and mouse genomes, among others. At the same time, libraries of expressed sequence tags are being accumulated. These are of immense value to researchers, although to what extent their expense is justified is difficult to assess. Gene trapping in ES cells is generating a similar database, with the unique property that the individual DNA sequences are linked to a co-ordinated set of tagged ES cell clones. Many of these are already carrying a null mutation in the tagged gene and many, although unfortunately not all, retain the capacity to colonise the germ line.

One moderately large gene trap exercise employed a gene trap fragment (Chowdhury *et al.* 1997), another a retrovirus trap (Hicks *et al.* 1997). Both are characterised by making their first point of reference a database search with a short nucleotide sequence from the trapped gene, determined in one case by 5'-RACE and in the other by plasmid rescue. As shown in Table 8.4, the results of the screens were quite different; in one case 83% of the trapped sequences were novel, in the other 51%. This may be due in part to the considerable differences in the trapping methods employed and may also owe something to the adoption of different database search strategies. In both cases a wide spectrum of genes was trapped, but the expectation is that only genes expressed in ES cells will be trappable by these approaches. If as suggested ES cells express 10 000–20 000 different genes, perhaps 80% of genes will be inaccessible to these approaches, and some of these will certainly be important in development. Another problem is likely to be the long-term viability of ES cells in storage. Again, the task of determining the

Table 8.4 Results of two large general gene trap exercises directed first at gene identification through nucleotide sequence determination and database searches. ESTs – expressed sequence tags.

		Identification			
Trap	*Recovery*	*Known genes*	*ESTs*	*Neither*	*Reference*
Provirus	Plasmid rescue	42	21	337 (83%)	Hicks *et al.* 1997
Fragment	5'-RACE	16	11	28 (51%)	Chowdhury *et al.* 1997

functional significance of even a single important trapped gene is fairly daunting. Thus, while this approach will undoubtedly generate useful information, the saturation of the mouse developmental gene map is likely to involve a very long-term programme.

References

Allen, M.J., Jeffreys, A.J., Surani, M.A., Barton, S., Norris, M.L., and Collick, A. (1994) Tandemly repeated transgenes of the human minisatellite MS32 (D1S8), with novel mouse gamma satellite integration. *Nucleic Acids Research*, **22**, 2976–2981.

Baker, R.K., Haendel, M.A., Swanson, B.J., Shambaugh, J.C., Micales, B.K., and Lyons, G.E. (1997) In vitro preselection of gene-trapped embryonic stem cell clones for characterizing novel developmentally regulated genes in the mouse. *Developmental Biology*, **185**, 201–214.

Bhat, K., McBurney, M.W., and Hamada, H. (1988) Functional cloning of mouse chromosomal loci specifically active in embryonal carcinoma cells. *Molecular & Cellular Biology*, **8**, 3251–3259.

Bishop, J.O. (1996) Chromosomal insertion of foreign DNA. *Reproduction, Nutrition, Development*, **36**, 607–618.

Bishop, J.O. (1997) Chromosomal insertion of foreign DNA. In *Transgenic animals, generation and use*, L.M. Houdebine (ed), Harwood Academic Press, Amsterdam, pp. 219–223.

Bishop, J.O. and Smith, P. (1989) Mechanism of chromosome integration of microinjected DNA. *Molecular Biology and Medicine*, **6**, 283–298.

Brinster, R.L., Chen, H.Y., Trumbauer, M.E., Yagle, M.K., and Palmiter, R.D. (1985) Factors affecting the efficiency of introducing foreign DNA into mice by microinjecting eggs. *Proceedings of the National Academy of Sciences of the United States of America*, **82**, 4438–4442.

Brown, A., Copeland, N.G., Gilbert, D.J., Jenkins, N.A., Rossant, J., and Kothary, R. (1994) The genomic structure of an insertional mutation in the dystonia musculorum locus. *Genomics*, **20**, 371–376.

Burdon, T.G. and Wall, R.J. (1992) Fate of microinjected genes in preimplantation mouse embryos. *Molecular Reproduction and Development*, **33**, 436–442.

Canseco, R.S., Sparks, A.E., Page, R.L., Russell, C.G., Johnson, J.L., Velander, W.H., Pearson, R.E., Drohan, W.N., and Gwazdauskas, F.C. (1994) Gene transfer efficiency during gestation and the influence of co-transfer of non-manipulated embryos on production of transgenic mice. *Transgenic Research*, **3**, 20–25.

Chen, C.M., Choo, K.B., and Cheng, W.T. (1995) Frequent deletions and sequence aberrations at the transgene junctions of transgenic mice carrying the papillomavirus regulatory and the SV40 TAg gene sequences. *Transgenic Research*, **4**, 52–59.

Chowdhury, K., Bonaldo, P., Torres, M., Stoykova, A., and Gruss, P. (1997) Evidence for the stochastic integration of gene trap vectors into the mouse germline. *Nucleic Acids Research*, **25**, 1531–1536.

Cousens, C., Carver, A.S., Wilmut, I., Colman, A., Garner, I., and O'Neill, G.T. (1994) Use of PCR-based methods for selection of integrated transgenes in preimplantation embryos. *Molecular Reproduction & Development*, **39**, 384–391.

Covarrubias, L., Nishida, Y., and Mintz, B. (1986) Early postimplantation embryo lethality due to DNA rearrangements in a transgenic mouse strain. *Proceedings of the National Academy of Sciences of the United States of America*, **83**, 6020–6024.

Covarrubias, L., Nishida, Y., Terao, M., D'Eustachio, P., and Mintz, B. (1987) Cellular DNA rearrangements and early developmental arrest caused by DNA insertion in transgenic mouse embryos. *Molecular & Cellular Biology*, **7**, 2243–2247.

Echelard, Y. (1997) Genetic mosaicism in the generation of transgenic mice. In *Transgenic animals, generation and use*, L.M. Houdebine (ed), Harwood Academic Press, Amsterdam, pp. 233–236.

Ellison, A.R., Wallace, H., Al-Shawi, R., and Bishop, J.O. (1995) Different transmission rates of herpesvirus thymidine kinase reporter transgenes from founder male parents and male parents of subsequent generations. *Molecular Reproduction and Development*, **41**, 425–434.

Folger, K.R., Wong, E.A., Wahl, G., and Capecchi, M.R. (1982) Patterns of integration of DNA microinjected into cultured mammalian cells: evidence for homologous recombination between injected plasmid DNA molecules. *Molecular & Cellular Biology*, **2**, 1372–1387.

Folger, K.R., Thomas, K., and Capecchi, M.R. (1985) Nonreciprocal exchanges of information between DNA duplexes coinjected into mammalian cell nuclei. *Molecular & Cellular Biology*, **5**, 59–69.

Forrester, L.M., Nagy, A., Sam, M., Watt, A., Stevenson, L., Bernstein, A., Joyner, A.L., and Wurst, W. (1996) An induction gene trap screen in embryonic stem cells: Identification of genes that respond to retinoic acid in vitro. *Proceedings of the National Academy of Sciences of the United States of America*, **93**, 1677–1682.

Friedrich, G. and Soriano, P. (1991) Promoter traps in embryonic stem cells: a genetic screen to identify and mutate developmental genes in mice. *Genes & Development*, **5**, 1513–1523.

Gogos, J.A., Lowry, W., and Karayiorgou, M. (1997) Selection for retroviral insertions into regulated genes. *Journal of Virology*, **71**, 1644–1650.

Gordon, J.W., Scangos, G.A., Plotkin, D.J., Barbosa, J.A., and Ruddle, F.H. (1980) Genetic transformation of mouse embryos by microinjection of purified DNA. *Proceedings of the National Academy of Sciences of the United States of America*, **77**, 7380–7384.

Gordon, J.W. and Ruddle, F.H. (1981) Integration and stable germ line transmission of genes injected into mouse pronuclei. *Science*, **214**, 1244–1246.

Gossler, A., Joyner, A.L., Rossant, J., and Skarnes, W.C. (1989) Mouse embryonic stem cells and reporter constructs to detect developmentally regulated genes. *Science*, **244**, 463–465.

Hamada, H. (1986) Random isolation of gene activator elements from the human genome. *Molecular & Cellular Biology*, **6**, 4185–4194.

Hamada, T., Sasaki, H., Seki, R., and Sakaki, Y. (1993) Mechanism of chromosomal integration of transgenes in microinjected mouse eggs: sequence analysis of genome-transgene and transgene-transgene junctions at two loci. *Gene*, **128**, 197–202.

Hammer, R.E., Pursel, V.G., Rexroad, C.E., Jr., Wall, R.J., Bolt, D.J., Ebert, K.M., Palmiter, R.D., and Brinster, R.L. (1985) Production of transgenic rabbits, sheep and pigs by microinjection. *Nature*, **315**, 680–683.

Hammer, R.E., Maika, S.D., Richardson, J.A., Tang, J.P., and Taurog, J.D. (1990) Spontaneous inflammatory disease in transgenic rats expressing HLA-B27 and human beta 2m: an animal model of HLA-B27-associated human disorders. *Cell*, **63**, 1099–1112.

Henderson, G. and Simons, J.P. (1997) Processing of DNA prior to illegitimate recombination in mouse cells. *Molecular & Cellular Biology*, **17**, 3779-3785.

Hicks, G.G., Shi, E.G., Li, X.M., Li, C.H., Pawlak, M., and Ruley, H.E. (1997) Functional genomics in mice by tagged sequence mutagenesis. *Nature Genetics*, **16**, 338–344.

Hiller, S., Hengstler, M., Kunze, M., and Knippers, R. (1988) Insertional activation of a promoterless thymidine kinase gene. *Molecular & Cellular Biology*, **8**, 3298–3302.

Huttner, K.M., Barbosa, J.A., Scangos, G.A., Pratcheva, D.D., and Ruddle, F.H. (1981) DNA-mediated gene transfer without carrier DNA. *Journal of Cell Biology*, **91**, 153–156.

Ikawa, M., Kominami, K., Yoshimura, Y., Tanaka, K., Nishimune, Y., and Okabe, M. (1995) A rapid and non-invasive selection of transgenic embryos before implantation using green fluorescent protein (GFP). *FEBS Letters*, **375**, 125–128.

Kain, S.R., Adams, M., Kondepudi, A., Yang, T.T., Ward, W.W., and Kitts, P. (1995) Green fluorescent protein as a reporter of gene expression and protein localization. *BioTechniques*, **19**, 650–655.

Kopchick, J.J. and Stacey, D.W. (1984) Differences in intracellular DNA ligation after microinjection and transfection. *Molecular & Cellular Biology*, **4**, 240–246.

Kozak, M. (1986) Point mutations define a sequence flanking the AUG initiator codon that modulates translation by eukaryotic ribosomes. *Cell*, **44**, 283–292.

Kozak, M. (1987a) Effects of intercistronic length on the efficiency of reinitiation by eucaryotic ribosomes. *Molecular & Cellular Biology*, **7**, 3438–3445.

Kozak, M. (1987b) At least six nucleotides preceding the AUG initiator codon enhance translation in mammalian cells. *Journal of Molecular Biology*, **196**, 947–950.

Kuspa, A. and Loomis, W.F. (1992) Tagging developmental genes in Dictyostelium by restriction enzyme-mediated integration of plasmid DNA. *Proceedings of the National Academy of Sciences of the United States of America*, **89**, 8803–8807.

Lehman, C.W., Trautman, J.K., and Carroll, D. (1994) Illegitimate recombination in Xenopus: characterization of end-joined junctions. *Nucleic Acids Research*, **22**, 434–442.

Lin, F.L., Sperle, K., and Sternberg, N. (1984) Model for homologous recombination during transfer of DNA into mouse L cells: role for DNA ends in the recombination process. *Molecular & Cellular Biology*, **4**, 1020–1034.

Lin, F.L., Sperle, K.M., and Sternberg, N.L. (1987) Extrachromosomal recombination in mammalian cells as studied with single- and double-stranded DNA substrates. *Molecular & Cellular Biology*, **7**, 129–140.

Menichini, P., Viaggi, S., Gallerani, E., Fronza, G., Ottaggio, L., Comes, A., Ellwart, J.W., and Abbondandolo, A. (1997) A gene trap approach to isolate mammalian genes involved in the cellular response to genotoxic stress. *Nucleic Acids Research*, **25**, 4803–4807.

Merrihew, R.V., Marburger, K., Pennington, S.L., Roth, D.B., and Wilson, J.H. (1996) High-frequency illegitimate integration of transfected DNA at preintegrated target sites in a mammalian genome. *Molecular & Cellular Biology*, **16**, 10–18.

Michalova, K., Bucchini, D., Ripoche, M.A., Pictet, R., and Jami, J. (1988) Chromosome localization of the human insulin gene in transgenic mouse lines. *Human Genetics*, **80**, 247–252.

Mortensen, R.M., Conner, D.A., Chao, S., Geisterfer-Lowrance, A.A., and Seidman, J.G. (1992) Production of homozygous mutant ES cells with a single targeting construct. *Molecular & Cellular Biology*, **12**, 2391–2395.

Mukha, D.V., Gorodetskii, Ignat'eva, T.V., and Dyban, A.P. (1990) The integration and expression of the beta-casein gene of *Bos taurus* in transgenic rats and mice. *Ontogenez*, **21**, 215–218. [Russian]

Mullins, J.J. and Ganten, D. (1990) Transgenic animals: new approaches to hypertension research. *Journal of Hypertension – Supplement*, **8**, S35–7.

Naora, H., Kimura, M., Otani, H., Yokoyama, M., Koizumi, T., Katsuki, M., and Tanaka, O. (1994) Transgenic mouse model of hemifacial microsomia: cloning and characterization of insertional mutation region on chromosome 10. *Genomics*, **23**, 515–519.

Niwa, H., Araki, K., Kimura, S., Taniguchi, S., Wakasugi, S., and Yamamura, K. (1993) An efficient gene-trap method using poly A trap vectors and characterization of gene-trap events. *Journal of Biochemistry*, **113**, 343–349.

Page, R.L., Canseco, R.S., Russell, C.G., Johnson, J.L., Velander, W.H., and Gwazdauskas, F.C. (1995) Transgene detection during early murine embryonic development after pronuclear microinjection. *Transgenic Research*, **4**, 12–17.

Pellicer, A., Wigler, M., Axel, R., and Silverstein, S. (1978) The transfer and stable integration of the HSV thymidine kinase gene into mouse cells. *Cell*, **14**, 133–141.

Phillips, J.W. and Morgan, W.F. (1994) Illegitimate recombination induced by DNA double-strand breaks in a mammalian chromosome. *Molecular & Cellular Biology*, **14**, 5794–5803.

Pieper, F.R., de Wit, I.C., Pronk, A.C., Kooiman, P.M., Strijker, R., Krimpenfort, P.J., Nuyens, J.H., and de Boer, H.A. (1992) Efficient generation of functional transgenes by homologous recombination in murine zygotes. *Nucleic Acids Research*, **20**, 1259–1264.

Pulm, W. and Knippers, R. (1985) Transfection of mouse fibroblast cells with a promoterless herpes simplex virus thymidine kinase gene: number of integrated gene copies and structure of single and amplified gene sequences. *Molecular & Cellular Biology*, **5**, 295–304.

Rijkers, T., Peetz, A., and Ruther, U. (1994) Insertional mutagenesis in transgenic mice. *Transgenic Research*, **3**, 203–215.

Rohan, R.M., King, D., and Frels, W.I. (1990) Direct sequencing of PCR-amplified junction fragments from tandemly repeated transgenes. *Nucleic Acids Research*, **18**, 6089–6095.

Roth, D.B., Porter, T.N., and Wilson, J.H. (1985) Mechanisms of nonhomologous recombination in mammalian cells. *Molecular & Cellular Biology*, **5**, 2599–2607.

Roth, D.B., Chang, X.B., and Wilson, J.H. (1989) Comparison of filler DNA at immune, nonimmune, and oncogenic rearrangements suggests multiple mechanisms of formation. *Molecular & Cellular Biology*, **9**, 3049–3057.

Rouet, P., Smih, F., and Jasin, M. (1994a) Introduction of double-strand breaks into the genome of mouse cells by expression of a rare-cutting endonuclease. *Molecular & Cellular Biology*, **14**, 8096–8106.

Rouet, P., Smih, F., and Jasin, M. (1994b) Expression of a site-specific endonuclease stimulates homologous recombination in mammalian cells. *Proceedings of the National Academy of Sciences of the United States of America*, **91**, 6064–6068.

Russ, A.P., Friedel, C., Ballas, K., Kalina, U., Zahn, D., Strebhardt, K., and von Melchner, H. (1996) Identification of genes induced by factor deprivation in hematopoietic cells undergoing apoptosis using gene-trap mutagenesis and site-specific recombination. *Proceedings of the National Academy of Sciences of the United States of America*, **93**, 15279–15284.

Sambrook, J., Fritsch, E.F., and Maniatis, T. (1989) *Molecular cloning: a laboratory manual*, C. Nolan (ed), Cold Spring Harbor Laboratory Press, Cold Spring Harbor, NY.

Shimoda, K., Cai, X., Kuhara, T., and Maejima, K. (1991) Reconstruction of a large DNA fragment from coinjected small fragments by homologous recombination in fertilized mouse eggs. *Nucleic Acids Research*, **19**, 6654.

Skarnes, W.C., Moss, J.E., Hurtley, S.M., and Beddington, R.S. (1995) Capturing genes encoding membrane and secreted proteins important for mouse development. *Proceedings of the National Academy of Sciences of the United States of America*, **92**, 6592–6596.

Smih, F., Rouet, P., Romanienko, P.J., and Jasin, M. (1995) Double-strand breaks at the target locus stimulate gene targeting in embryonic stem cells. *Nucleic Acids Research*, **23**, 5012–5019.

Soriano, P., Gridley, T., and Jaenisch, R. (1987) Retroviruses and insertional mutagenesis in mice: proviral integration at the Mov 34 locus leads to early embryonic death. *Genes & Development*, **1**, 366–375.

Stacey, A., Mulligan, R., and Jaenisch, R. (1987) Rescue of type I collagen-deficient phenotype by retroviral-vector-mediated transfer of human pro alpha 1(I) collagen gene into Mov-13 cells. *Journal of Virology*, **61**, 2549–2554.

Stary, A. and Sarasin, A. (1992) Molecular analysis of DNA junctions produced by illegitimate recombination in human cells. *Nucleic Acids Research*, **20**, 4269–4274.

von Melchner, H. and Ruley, H.E. (1989) Identification of cellular promoters by using a retrovirus promoter trap. *Journal of Virology*, **63**, 3227–3233.

Wagner, T.E., Hoppe, P.C., Jollick, J.D., Scholl, D.R., Hodinka, R.L., and Gault, J.B. (1981) Microinjection of a rabbit beta-globin gene into zygotes and its subsequent expression in adult mice and their offspring. *Proceedings of the National Academy of Sciences of the United States of America*, **78**, 6376–6380.

Wake, C.T., Gudewicz, T., Porter, T., White, A., and Wilson, J.H. (1984) How damaged is the biologically active subpopulation of transfected DNA? *Molecular & Cellular Biology*, **4**, 387–398.

Wake, C.T., Vernaleone, F., and Wilson, J.H. (1985) Topological requirements for homologous recombination among DNA molecules transfected into mammalian cells. *Molecular & Cellular Biology*, **5**, 2080–2089.

Whitelaw, C.B., Springbett, A.J., Webster, J., and Clark, J. (1993) The majority of G0 transgenic mice are derived from mosaic embryos. *Transgenic Research*, **2**, 29–32.

Wigler, M., Silverstein, S., Lee, L.S., Pellicer, A., Cheng Yc., and Axel, R. (1977) Transfer of purified herpes virus thymidine kinase gene to cultured mouse cells. *Cell*, **11**, 223–232.

Wigler, M., Sweet, R., Sim, G.K., Wold, B., Pellicer, A., Lacy, E., Maniatis, T., Silverstein, S., and Axel, R. (1979) Transformation of mammalian cells with genes from procaryotes and eucaryotes. *Cell*, **16**, 777–785.

Wilkie, T.M., Brinster, R.L., and Palmiter, R.D. (1986) Germline and somatic mosaicism in transgenic mice. *Developmental Biology*, **118**, 9–18.

Wilkie, T.M. and Palmiter, R.D. (1987) Analysis of the integrant in MyK-103 transgenic mice in which males fail to transmit the integrant. *Molecular & Cellular Biology*, **7**, 1646–1655.

Xiang, X., Benson, K.F., and Chada, K. (1990) Mini-mouse: disruption of the pygmy locus in a transgenic insertional mutant. *Science*, **247**, 967-969.

Yoshida, M., Yagi, T., Furuta, Y., Takayanagi, K., Kominami, R., Takeda, N., Tokunaga, T., Chiba, J., Ikawa, Y., and Aizawa, S. (1995) A new strategy of gene trapping in ES cells using 3'RACE. *Transgenic Research*, **4**, 277–287.

Further reading

Insertional mutagenesis

Gridley, T., Soriano, P., and Jaenisch, R. (1987) Insertional mutagenesis in mice. *Trends in Genetics*, **3**, 162–166.

Rijkers, T., Peetz, A., and Ruther, U. (1994) Insertional mutagenesis in transgenic mice. *Transgenic Research*, **3**, 203–215.

Gene trapping

Joyner, A.L. (1991) Gene targeting and gene trap screens using embryonic stem cells: new approaches to mammalian development. *Bioessays*, **13**, 649–656.

Evans, M.J., Carlton, M.B., and Russ, A.P. (1997) Gene trapping and functional genomics. *Trends in Genetics*, **13**, 370–374.

CHAPTER NINE

DNA integration by homologous recombination

Two regions of double-stranded DNA that are essentially identical, but not necessarily completely so, are said to be homologous. Recombination in bacteria, and both meiotic and mitotic recombination in eukaryotes, take place by reciprocal exchange between homologous duplexes. Provided that it contains appropriate regions of homology, foreign DNA can become integrated into eukaryotic chromosomes by a similar process of homologous recombination. DNA integration by homologous recombination provides a way of introducing a predetermined mutation into a chosen gene. This is very often referred to as gene targeting. In prokaryotes and robust unicellular eukaryotes like yeast, sophisticated procedures allow the selection of subgroups of mutations that affect chosen cellular functions, and these have been very successfully applied following random mutagenesis. The same sort of approach has had some success with mammalian cells in culture, despite their relative complexity and fragility and the fact that they usually contain at least two copies of most genes. However, the genes that relate to interactions between cells and tissues are virtually all inaccessible to this approach. Mutations induced in mice by radiation or chemical mutagens have been useful, but the expense and labour involved in identifying and characterising them is enormous. Gene targeting in ES cells permits mice carrying new mutations to be produced quite efficiently. The mutated gene can be inactive (a null allele) or a novel allele encoding a protein that may differ from the wild-type protein by as little as a single amino acid.

Homologous recombination can be conservative or non-conservative. During the process of non-conservative recombination some part of one or both duplexes is destroyed. On the other hand during the process of conservative recombination the lengths of both participating duplexes are maintained or restored, although in some circumstances one of them may subsequently be destroyed or lost (see below). We saw in the previous chapter that extrachromosomal DNA concatemers are formed mainly by a process of homologous recombination that is non-conservative. In contrast, the process of homologous recombination by which DNA is integrated into the chromosome is regarded as conservative. Because of the DNA loss that occurs, the distinction verges on the semantic. It is more meaningful however if we keep in mind the related process of reciprocal exchange between chromosomes.

Transgene integration

The need for selection

We saw in Chapter 8 that foreign DNA becomes integrated into the chromosomes at random sites by illegitimate recombination. This happens whether or not the foreign DNA shares regions of identity with the chromosome. The relative frequencies of integration by illegitimate and homologous recombination vary greatly between cell types and also depend upon properties of the DNA (discussed below). However, random DNA integration is invariably much more frequent than homologous recombination and consequently it is necessary to distinguish the two types of event.

The absolute frequencies of integration by both recombination routes also vary between modes of delivery. Direct microinjection of foreign DNA gives integration frequencies perhaps three orders of magnitude higher than transfection by calcium phosphate coprecipitation or electroporation (Capecchi 1980). However, even microinjection generates only about ten random integrations and one homologous integration per 1000 injected cells. Thus selective techniques are desirable if the very laborious microinjection delivery route is employed and absolutely necessary when delivery is by electroporation, as is usually the case.

Selection has two possible alternative aims. One is to select the minority of cells that have integrated the foreign DNA either randomly or by homologous recombination, killing off the majority that have failed to integrate it. The other is to select only the cells that have integrated the foreign DNA by homologous recombination. These aims can be met in different ways, depending on the objective of the experiment, the structure of the foreign DNA and the properties of the target gene. Methods used to select targeted genes are described in Box 9.3.

Modes of transgene integration

Two modes of transgene integration by homologous recombination have been recognised for some years (Thomas and Capecchi 1987). In one mode the transgene is added to the chromosome without any loss of the pre-existing chromosomal DNA; in effect the chromosome becomes a little longer. This is the insertion mode (Figure 9.1a). In the other mode, called replacement (Figure 9.1b), foreign DNA replaces part of the chromosomal DNA;

the chromosome may become longer or shorter depending on the lengths of the two stretches of DNA. Which form of integration occurs is determined to a very large extent by the structure of the foreign DNA, and is consequently under experimental control. Box 9.1 shows how the same DNA fragment may be manipulated with a view to insertion or replacement. Similar targeting frequencies are reported for the two types of vector when they contain the same regions of homology with the chromosome (Deng and Capecchi 1992).

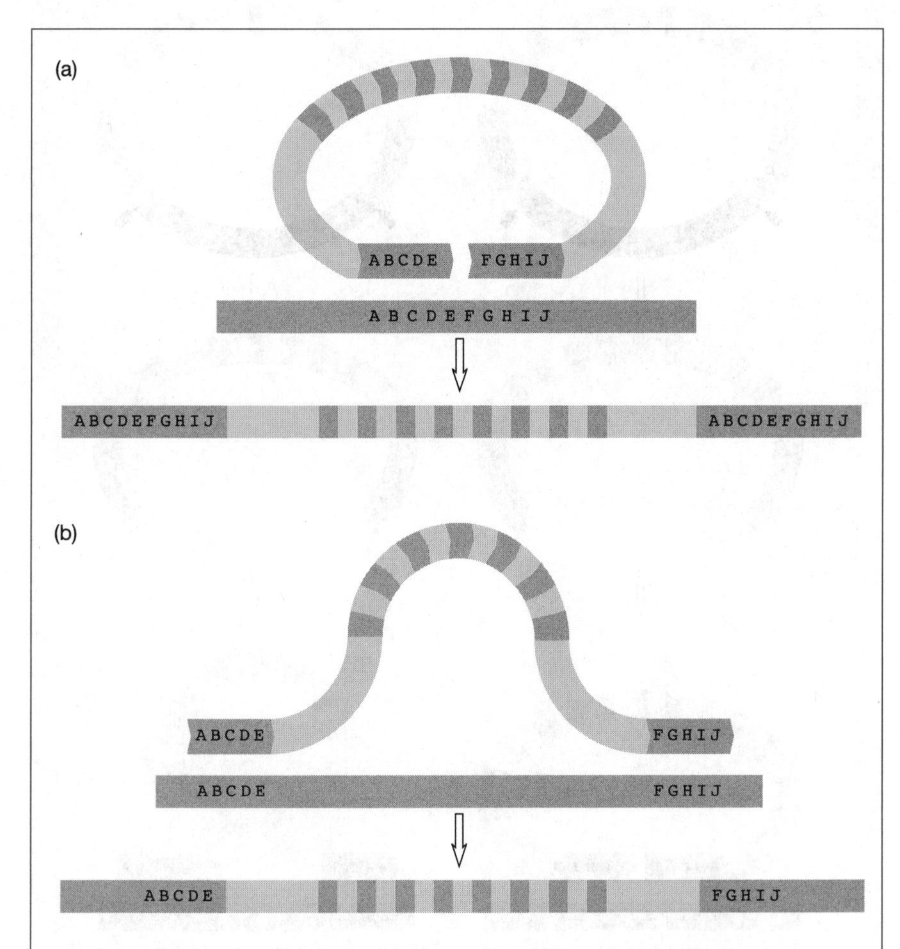

Figure 9.1 Integration of foreign DNA into chromosomal DNA by homologous recombination. The chromosomal DNA and homologous sequences present in the foreign DNA are shown as large grey boxes, anonymous foreign DNA in blue, and a notional selective reporter gene in blue and grey. (a) *Insertion* into a chromosomal sequence. The region of homology (A–J) is repaired and duplicated, and the duplicated regions lie on either side of the reporter. (b) *Replacement* of a chromosomal sequence. The chromosomal DNA between the regions of homology (A–E and F–J) is replaced by the reporter.

BOX 9.1 MANIPULATING A CLONED GENE TO MAKE VECTORS FOR INSERTION AND REPLACEMENT BY HOMOLOGOUS RECOMBINATION

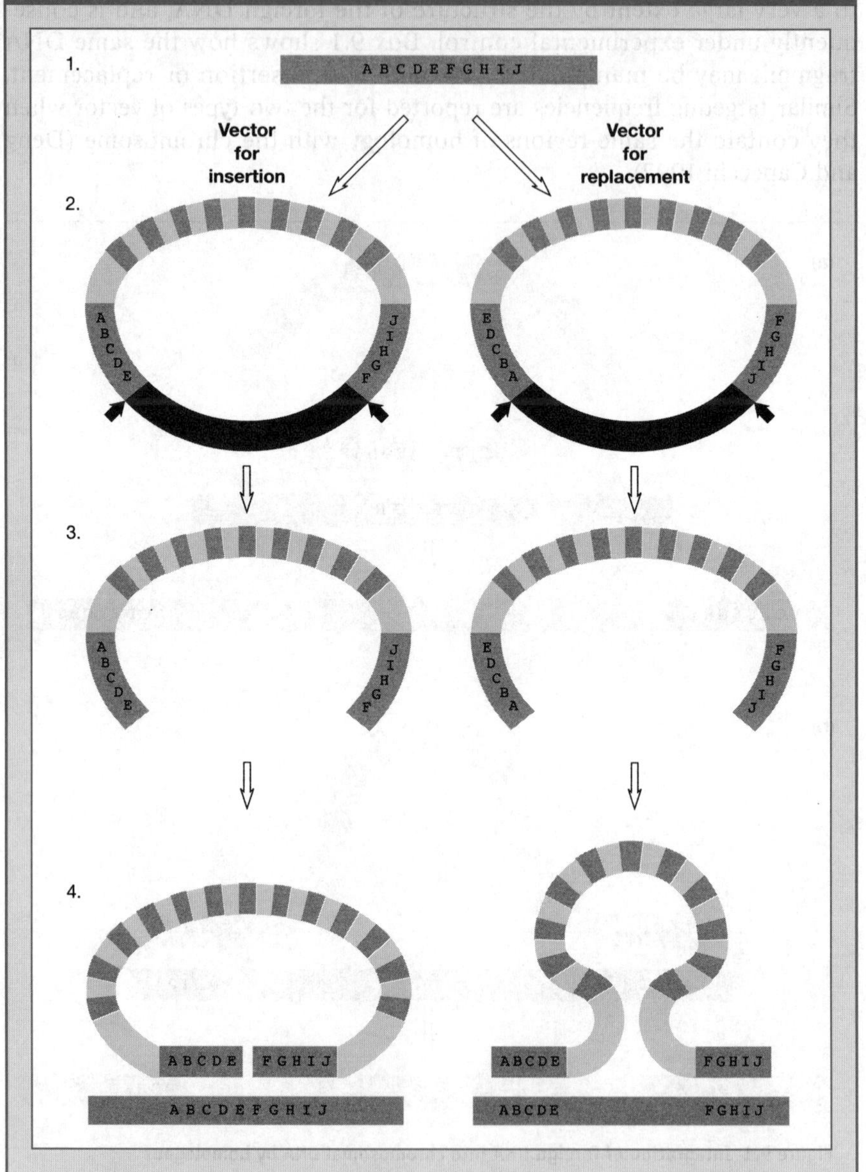

Cloned parts of the chromosome are represented by large grey boxes and a gene by the letters A–E and F–J within them. Plasmid DNA is in black, and a selective reporter gene in blue and grey with anonymous flanking regions in blue. The second row shows how the two types of vector can be constructed from the same gene fragments; the difference is in the orientation of these with respect to the plasmid DNA and the reporter sequence. Row 3 shows the result of cleaving the two vectors with restriction enzymes at unique sites that require to be chosen in advance, or newly introduced if appropriate sites are not already present at these locations (filled arrows in row 2). The bottom row shows how the vector fragments align with the homologous chromosome region.

Some years ago it was observed that the frequency of recombination between tandem genes on the same chromosome required around 200 base pairs of uninterrupted perfect homology (Waldman and Liskay 1988); reducing the uninterrupted length from 232 to 134 base pairs while keeping the overall length the same reduced the recombination frequency 20-fold. The frequency of homologous recombination between a transgene and the chromosome is similarly affected by mismatched base pairs. Several studies have employed pairs of targeting vectors that contain the same regions of DNA from two different inbred strains of mice, one being the strain from which the cells to be targeted were isolated. In each case the vectors that were prepared from the same strain (called isogenic vectors) gave significantly higher targeting frequencies (te Riele *et al.* 1992; van Deursen and Wieringa 1992; Deng and Capecchi 1992). The observed differences between different pairs of vectors were not the same, but this is not surprising since genetic differences between mouse strains vary and are not distributed uniformly across the genome. The importance of this observation is that the use of isogenic DNA increases the frequency of homologous recombination relative to random integration (Figure 9.2).

The *E. coli* MutS protein identifies mismatched base pairs in DNA and forms part of the multiprotein 'repairosome' complex which corrects the mismatch (Modrich 1991; Modrich 1994). ES cells with two inactivated

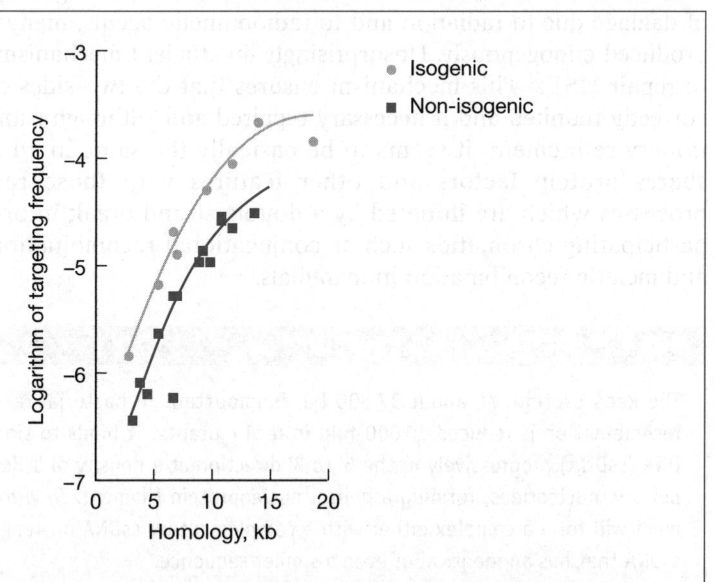

Figure 9.2 Dependency of frequency of targeting on the length of homology between the foreign and chromosomal DNA. The highest frequencies shown are in the region of 2 per 10 000 cells. The frequency is about five times greater when isogenic DNA (blue circles, see text) is employed. Each curve incorporates data obtained with both insertion and replacement vectors. Data are from Deng and Capecchi (1992).

copies of the *msh2* gene, a mouse homologue of *MutS*, eliminated the targeting deficit of non-isogenic DNA (de Wind *et al.* 1995). This indicates a parallel in the mouse to the involvement of the *E. coli* repairosome in preventing recombination between evolutionarily diverged sequences (Rayssiguier *et al.* 1989). Assuming that the process of mismatch repair is much the same in the mouse and *E. coli,* the extensive strand degradation that accompanies mismatch repair can be expected to abort the integration of foreign DNA.

The lengths of the regions of homology with the target sequence have a very strong influence on the frequency of homologous recombination (Figure 9.2). Increasing the homologous length from 2 to 14 kb gives an increase of more than 100-fold in targeting frequency (Deng and Capecchi 1992). Perhaps surprisingly, insertion and replacement vectors with the same homologous length gave essentially the same targeting frequencies.

Whether introduced by replacement or insertion the newly integrated DNA contains very few alterations, other than those introduced by gene conversion or by the deletion of terminal inhomologies (see below). An examination of 80 kb of DNA introduced into 44 clones in human and mouse cells by a sensitive chemical method revealed only two base-substitutions (Zheng *et al.* 1991).

Repair of double-strand breaks

Double-strand breaks (DSBs) occur in the DNA of all organisms as a result of damage due to radiation and to radiomimetic agents, many of which are produced endogenously. Unsurprisingly an efficient mechanism has evolved to repair DSBs. This mechanism ensures that the two sides of a DSB are correctly reunited and if necessary repaired and, although subject to evolutionary refinement, it seems to be basically the same in all life forms. It shares protein factors and other features with those recombination processes which are initiated by a double-strand break in one of the two participating chromatids such as conjugational recombination in bacteria and meiotic recombination in mammals.

BOX 9.2 BACTERIAL RecA PROTEIN

The RecA protein, M_r about 37 800 Da, is important in bacterial recombination; recombination is reduced 10 000-fold in *recA* mutants. It binds to single-stranded DNA (ssDNA) progressively in the 5' to 3' direction at a density of 1 RecA monomer per 3–4 nucleotides, forming a helical nucleoprotein filament. *In vitro* such a filament will form a complex either with a complementary ssDNA or, remarkably, with ssDNA that has an identical or even a similar sequence.

Accordingly, the filament is thought to form a three-stranded structure with a normal DNA duplex *in vivo* provided that the sequence of one of the strands of the duplex is similar or identical to that of the filamentous DNA (the other being complementary to it). The triplex structure is transient and quickly resolves to give, on

▶

the one hand, a new heteroduplex made up of the filament strand still bound up with recA protein and the complementary strand of the original duplex, and on the other the like strand of the original duplex, now a single strand. The net effect is the displacement of one strand of the duplex by the single strand, often called **strand invasion.**

In recombination *in vivo* single strands on either side of a double strand break in duplex DNA are exposed by exonucleolytic attack from recA filaments and invade the homologous regions of another DNA duplex. As indicated above, RecA-mediated strand invasion seems not to be highly discriminating and recombination between mismatched regions is detected and aborted at later stages. This inaccuracy may be a price that has to be paid in order to achieve the tricky task of bringing together like chromosome regions.

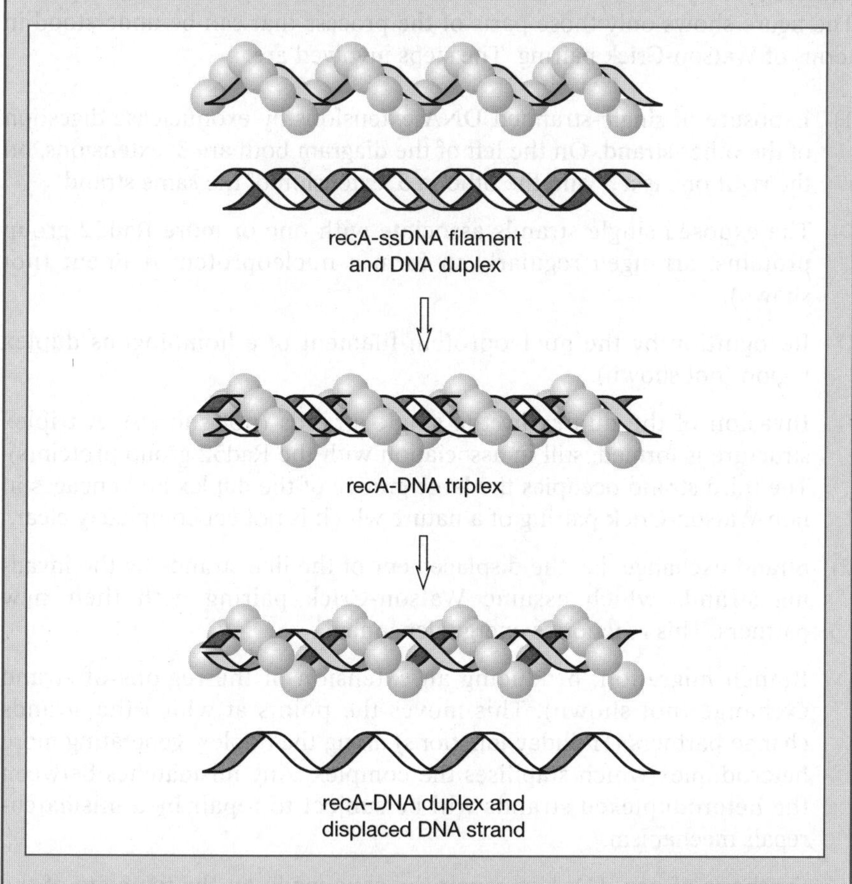

DNA strands are represented by helical tapes and RecA protein by blue spheres.The top diagram shows the helical RecA-ssDNA filament and a DNA duplex, one strand of which is similar in sequence to the DNA in the filament. The filament owes its configuration to the properties of RecA protein. The diagram does not show that the filament DNA is actually more extended than duplex DNA. The centre diagram shows a hypothetical triple-stranded filament. The bottom diagram illustrates the situation following displacement of the like strand of the duplex by the filament.

Although many other proteins are also involved in the repair of DSBs in *E. coli,* the RecA protein lies at the heart of the process. This remarkable protein facilitates recognition between a single-stranded DNA (ssDNA) and the region of a DNA duplex that contains the same sequence (and its complement). *In vivo,* ssDNA is exposed on both sides of a DSB by exonuclease action. RecA protein binds to the exposed single strands and this allows them to locate to the homologous region of an intact duplex, displace the like strands from the duplex and base-pair with the complementary sequences (Box 9.2). Eukaryote genomes each specify a number of proteins related to recA, collectively called the Rad52 epistasis group of proteins, which perform the same function.

The diagram in Figure 9.3 shows how a homologous duplex is used to provide a template for the repair of a DSB or to bring about crossing-over. The figure shows only those parts of the process that can be understood in terms of Watson-Crick pairing. The steps involved are:

1) Exposure of single-stranded DNA extensions by exonuclease digestion of the other strand. On the left of the diagram both are 3' extensions, on the right one is a 3' and the other a 5' extension of the same strand.

○ The exposed single strands associate with one or more Rad52 group proteins, arranged regularly to form a nucleoprotein filament (not shown).

○ Recognition by the nucleoprotein filament of a homologous duplex region (not shown).

○ Invasion of the duplex by the single strands (not shown). A triplex structure is formed, still in association with the Rad52 group protein(s). The third strand occupies the large groove of the duplex and engages in non-Watson-Crick pairing of a nature which is not yet completely clear.

2) Strand exchange, i.e. the displacement of the like strands by the invading strands which assume Watson-Crick pairing with their new partners. This reaction is energy-dependent.

○ Branch migration, producing an extension of the regions of strand exchange (not shown). This moves the points at which the strands change partners (Holliday junctions) along the duplex, generating more heteroduplex which stabilises the complex. Any mismatches between the heteroduplexed strands will be subject to repair by a mismatch-repair mechanism.

3) Synthesis of new DNA to repair the gaps made on the template of the new partner strand and ligation to restore covalent continuity.

4) Resolution. Cleavage and religation to separate out two molecules from the complex. This can result either in the ends of each initial duplex remaining as part of the same molecule (non-crossover outcome) or in an exchange of partners between the ends (crossover outcome).

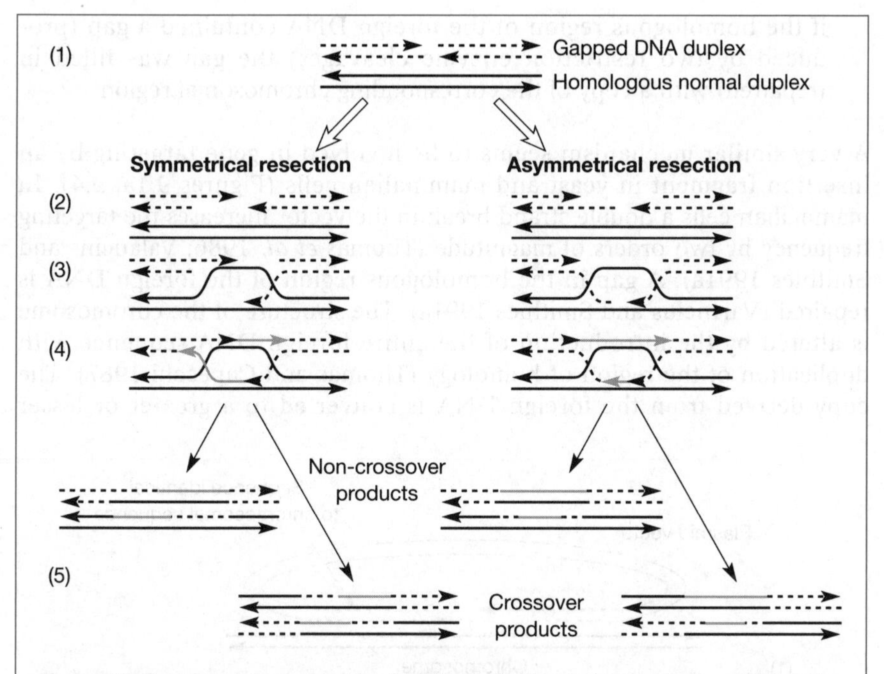

Figure 9.3 Homologous recombination by DSB repair. The single strands of one DNA duplex are shown as continuous lines and of the other by broken lines, with arrowheads at their notional 3' ends. Newly synthesised DNA is in blue. Lines that are not horizontal are out of scale with the rest of the diagram. (1) The upper duplex is represented as having the DSB. (2) Resection of the open ends of the duplex. The difference between the left and right sides of the diagram is determined at this point. On the left both ends are resected by a 5' exonuclease; on the right one end is resected by a 3' exonuclease. (3) Invasion of the intact duplex by the exposed single strands. (4) Repair synthesis and ligation to form a double Holliday junction. The Holliday junctions branch-migrate (not shown) and isomerise, presumably assisted by protein effectors. (5) In each case two modes of resolution are possible, one leading only to gene conversion, the other to conversion and crossing-over. In places where heteroduplex is present at this point (where a broken line and a continuous line lie together) repair of any inhomologies might already have occurred, or may occur later.

Gene targeting by insertion

The importance of DSB repair in transgene insertion first came to light in studies in which budding yeast were transformed with circular duplex DNA containing a region of a yeast chromosome (Orr-Weaver *et al.* 1981; Szostak *et al.* 1983). The main findings were as follows:

- when the foreign DNA became integrated into the chromosome the region of homology was duplicated
- an alternative outcome was gene conversion, usually of the plasmid sequence by the chromosomal sequence
- the frequency of transformation was increased when a DSB was introduced into the foreign DNA within the homologous region

- if the homologous region of the foreign DNA contained a gap (produced by two restriction enzyme cleavages) the gap was filled in (repaired) with a copy of the corresponding chromosomal region

A very similar mechanism seems to be involved in gene targeting by an insertion fragment in yeast and mammalian cells (Figures 9.1a, 9.4). In mammalian cells a double strand break in the vector increases the targeting frequency by two orders of magnitude (Thomas *et al.* 1986; Valancius and Smithies 1991a). A gap in the homologous region of the foreign DNA is repaired (Valancius and Smithies 1991a). The structure of the chromosome is altered by the introduction of the entire foreign DNA sequence with duplication of the region of homology (Thomas and Capecchi 1987). The copy derived from the foreign DNA is converted to a greater or lesser

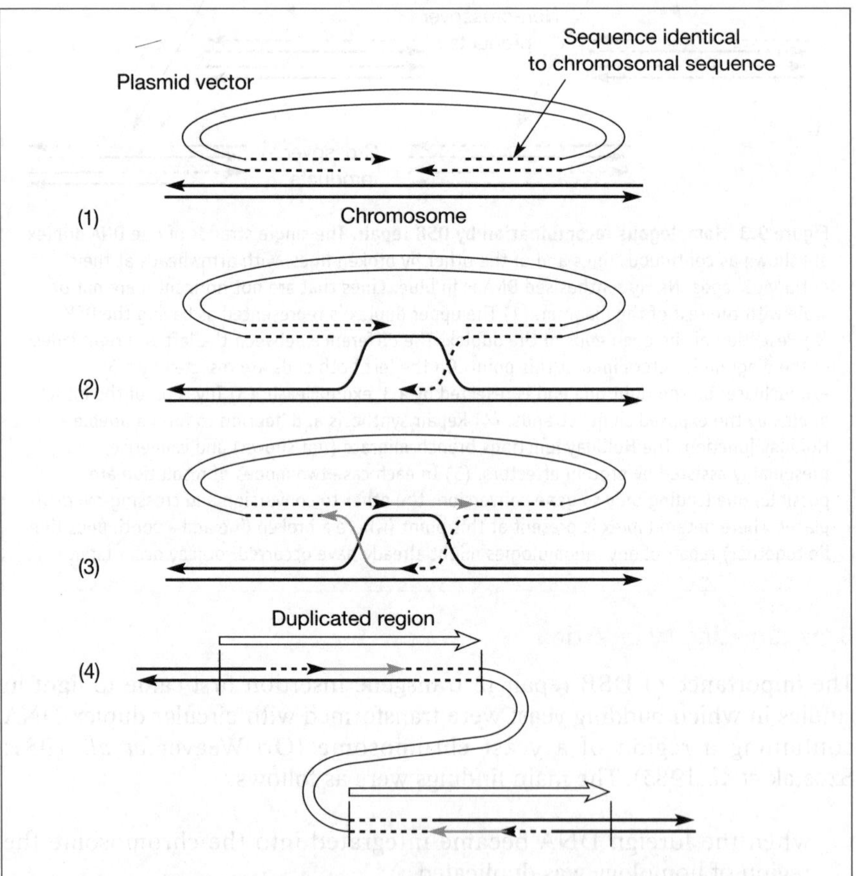

Figure 9.4 Insertion of foreign DNA *via* a DSB. Foreign and vector DNA is shown by thin lines, chromosomal DNA by thick lines. Chromosomal sequences attached to the vector are shown as broken lines. Newly synthesised DNA is in blue. (1) Resection of the open ends by a 5' exonuclease; (2) strand invasion; (3) synthesis and ligation; (4) resolution product with duplication of a region of the chromosome. This diagram corresponds to the left part of Figure 9.3. A path corresponding to the right part is equally plausible.

extent by the chromosomal sequence (Doetschman *et al.* 1987; Hasty *et al.* 1995; Hasty *et al.* 1991b), both by extension of the invading strand and by mismatch repair of heteroduplexed regions. Conversion of the chromosomal sequence by the vector sequence occurs much less frequently (Hasty *et al.* 1995). The extent and direction of conversion are both variable because they depend on exonuclease degradation of the non-invading strands and on branch migration, both of which are variable in extent.

When the objective is to inactivate the chromosomal gene these observations constrain the targeting strategy to some extent. The region of homology should lie within the coding region of the target gene and the selective marker should interrupt an exon. Following insertion the 5' copy of the gene is truncated at its 3' end at the point to which the 3' targeting fragment extends. Adjacent to this is the selective reporter, and adjacent to that the second copy of the gene, truncated at its 5' end at the point to which the 3' targeting fragment extends (Figures 9.1a, 9.4).

Two potential problems arise. One is that if a transcript arising from the promoter of the targeted gene traverses both gene copies, RNA splicing could generate an unaltered mRNA from it. This will be avoided if the selective reporter has an effective transcriptional termination/poly-A addition site, preventing transcription of the second gene copy. The second problem stems from the fact that the original gene can be restored by intrachromosomal recombination between the tandem copies (Figure 9.5). Any recombinants that arise in cell cultures will be eliminated by maintaining selection for the reporter, and the frequency of intrachromosomal recombination is low in any event (Hasty *et al.* 1991a; Zheng *et al.* 1991). However, it is not clear what effects it might have *in vivo* on animals carrying an ES cell genome, where selection is not maintained.

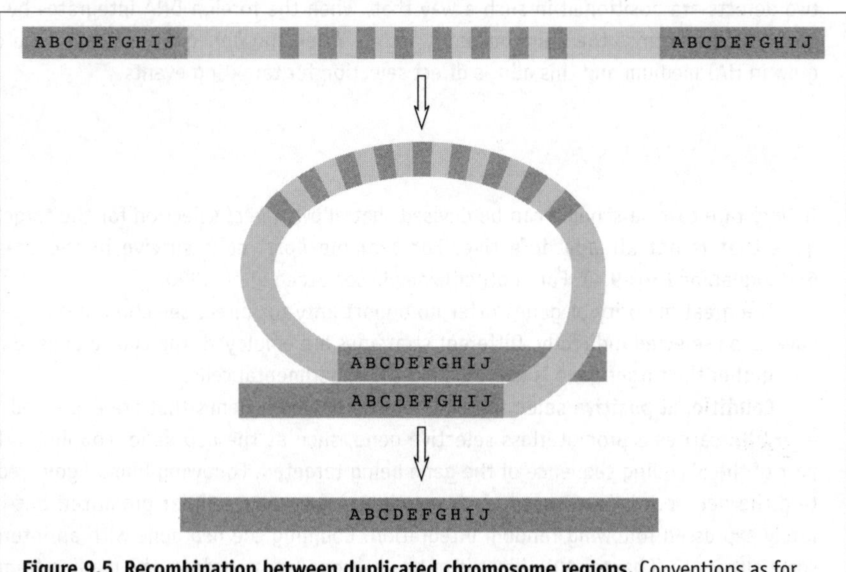

Figure 9.5 Recombination between duplicated chromosome regions. Conventions as for Figure 9.1.

BOX 9.3 SELECTION FOR HOMOLOGOUS RECOMBINATION AND GENE TARGETING

Selectable reporter genes commonly used in gene targeting studies. Sources: b, bacterium; s, streptomycete; v, human Herpes simplex virus; m, mammals.

Gene	*Source*	*Selection*	*Agent*
neo	b	Positive	G418
hph	b	Positive	Hygromycin
pur	s	Positive	Puromycin
aprt	m	Positive	HAT medium
hprt	m	Positive	HAT medium
		Negative	6-thioguanine
HSVtk	v	Negative	Ganciclovir, FIAU[1]
Dt	b	Negative	None (toxic)

[1]Antiherpetic agents kill cells expressing the herpesvirus thymidine kinase but not those expressing the cellular enzyme; consequently negative selection against a *tk* transgene can be employed in cells with active cellular thymidine kinase genes.

Studies of targeting frequency and the conditions that affect it

The foreign DNA usually carries two reporters. One of these, usually the bacterial *neo* gene under the control of a mammalian promoter, confers resistance to G418 irrespective of the mode of integration. Both random integration and homologous recombination produce G418 resistant colonies. The other reporter is a defective gene, frequently the X-linked *hprt* gene, a copy of which already present in the recipient cells and inactivated by a different defect. The two defects are positioned in such a way that, when the foreign DNA integrates by homologous recombination, the gene becomes active. When the *hprt* gene is active, the cells can grow in HAT medium and this allows direct selection for targeting events.

Targeting new genes

In very rare cases a scheme can be devised that allows *direct* selection for the targeting of a gene that is not already defective. For example *hprt*⁻ cells survive in the presence of 6–thioguanine (Box 9.4). For another example see Steeg *et al.* 1990.

The great majority of genes offer no opportunity for direct selection and targeted cells have to be selected indirectly. Different strategies are employed, the choice depending partly on whether the target gene is expressed in the experimental cells.

Conditional positive selection can be used to target genes that are expressed. The foreign DNA carries a promoterless selective gene, such as the *neo* gene, coupled in frame to part of the 5' coding sequence of the gene being targeted. Following homologous recombination the *neo* gene is expressed *via* transcription from the cellular promoter, but it is only rarely expressed following random integration. Coupling the *neo* gene with an internal ribosome entry site (Chapter 10) allows it to be positioned further down the reading frame.

A device called **positive–negative selection** (Mansour *et al.* 1988) is frequently used to enrich for targeted cells relative to cells that have integrated the foreign DNA randomly, and is especially useful when the target gene is not expressed in the recipient cells. A gene that allows selection against cells that express it (usually the *HSVtk* gene) is added at one or both ends of the foreign DNA. When this integrates by homologous recombination the *HSVtk* gene or genes, which lie outside the regions of homology, are lost (see Figure). When it integrates randomly, any terminal *HSVtk* gene is likely to be incorporated into the chromosome, all the more so because random integrants so frequently take the form of tandem arrays (Chapter 8). Cells carrying random integrants can consequently be selected against.

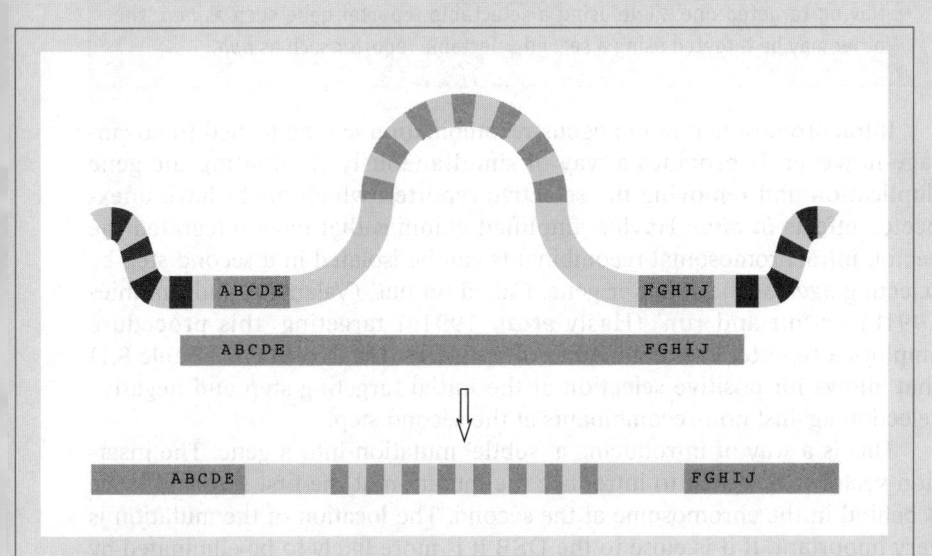

Arrangement of DNA sequences in a targeting fragment for DNA replacement by homologous recombination using positive–negative selection. The chromosome and homologous regions in the targeting fragment are shown by grey boxes and the letters A–E and F–J. Other targeting fragment sequences contain an element of blue: the positive selection reporter gene is shown in blue and grey and two copies of a negative selection reporter are shown in blue and black; solid blue represents anonymous flanking sequences. Upon homologous recombination with the target sequence the positive selection reporter is retained and the negative reporters are lost. When integration of either end occurs by illegitimate recombination (not shown) the corresponding negative selection reporter will usually be retained.

Targeting both alleles

Most genes in mammalian cells are present in two copies, if the cells have a normal diploid chromosome complement. Targeting one of the two copies of a gene produces a heterozygous cell, a heterozygous null in the case of an integration that inactivates the gene. Several devices have been employed to isolate homozygous nulls:

- In the case of ES cells the most obvious is to generate mice with the ES cell genome and produce homozygotes by mating. This approach is obviously unavailable when the gene is essential for the survival of embryos (i.e. when the mutation is an embryonic recessive lethal).
- During cell proliferation parts of the genome become homozygous spontaneously at a very low frequency due to mitotic recombination. Cells homozygous for gene targeting events that confer resistance to a drug have been selected by increasing the drug concentration by about an order of magnitude (Mortensen *et al.* 1992; de Wind *et al.* 1995).
- Having targeted one allele using a selectable reporter gene such as *neo*, the other may be targeted using a second selectable reporter such as *hph*.

Intrachromosomal homologous recombination can be turned to advantage however. It provides a way of simultaneously eliminating the gene duplication and removing the selective reporter, which might have unexpected effects *in vivo*. Having amplified colonies that have integrated the vector, intrachromosomal recombinants can be isolated in a second step by selecting *against* the reporter gene. Called 'in-out' (Valancius and Smithies 1991b) or 'hit and run' (Hasty *et al.* 1991a) targeting, this procedure employs a reporter or combination of reporters (see Box 9.3 and Table 8.1) that allows for positive selection at the initial targeting step and negative selection against non-recombinants at the second step.

This is a way of introducing a 'subtle' mutation into a gene. The insertion vector is designed to introduce the mutation at the first step and leave it behind in the chromosome at the second. The location of the mutation is very important. If it is close to the DSB it is more likely to be eliminated by conversion at the integration step (Hasty *et al.* 1995) and if it is close to the reporter it is more likely to be eliminated at the second step. After the first step it is necessary to analyse a number of targeted clones in a detailed way so as to identify one that carries the mutation and it is also necessary to analyse a number of clones after the second step to identify one that retains it. As a consequence the method is laborious.

Transgene insertion can be used to duplicate an entire gene, allowing gene dosage effects to be studied. The ends of the vector must span the entire region to be duplicated, and this will include the regulatory region of the gene (promoter and enhancer elements) as well as the coding sequence. Even large gaps are repaired, so that the foreign DNA need consist only of regions flanking the gene and the selective reporter. Smithies and Kim (1994), for example, report the repair of a 16 kb gap and the duplication of 25 kb of DNA.

A further consideration in transgene insertion concerns the fate of non-homologous sequences flanking the DSB or gap. It seems that long non-homologous sequences are eliminated (Figure 9.6b) although they do reduce the gene targeting frequency significantly. Short non-homologous

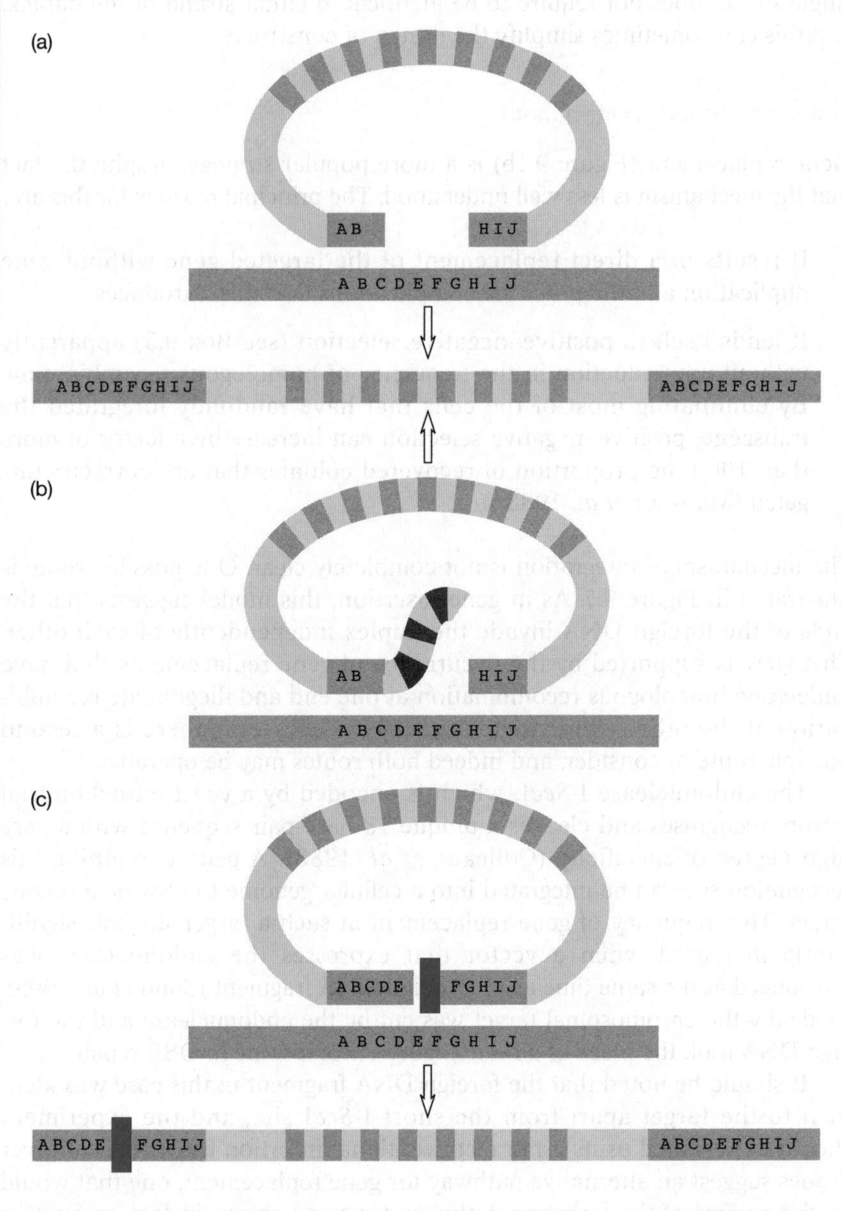

Figure 9.6 Modes of insertion. Conventions as for Figure 9.1. (a) Gap repair. The gap in the foreign DNA fragment is replaced by a copy of the chromosomal sequence; (b) removal of a long non-homologous end, shown in blue and black; (c) incorporation of a short non-homologous end (dark blue).

sequences are sometimes incorporated into the chromosome (Figure 9.6c); in one experiment (Hasty *et al.* 1992) heterologous 13-mers at each end of a DSB did not reduce the targeting frequency significantly, and 25% of the insertions retained at least one of the 13-mers. Thus the end of an invading single strand does not require to be identical to either strand of the duplex, and this can sometimes simplify the design of constructs.

Gene targeting by replacement

Gene replacement (Figure 9.1b) is a more popular strategy, despite the fact that the mechanism is less well understood. The principal reasons for this are:

- It results in a direct replacement of the targeted gene without gene duplication and the potential complications that this introduces.
- It lends itself to positive–negative selection (see Box 9.3) apparently without any reduction in the frequency of homologous recombination. By eliminating most of the cells that have randomly integrated the transgene, positive–negative selection can increase by a factor of more than 1000 the proportion of recovered colonies that are correctly targeted (Mansour *et al.* 1988).

The mechanism of integration is not completely clear. One possible route is illustrated in Figure 9.7. As in gene insertion, this model suggests that the ends of the foreign DNA invade the duplex independently of each other. This view is supported by the occurrence of gene replacements that have undergone homologous recombination at one end and illegitimate recombination at the other (Berinstein *et al.* 1992). However there is a second possible route to consider, and indeed both routes may be operative.

The endonuclease I-*Sce*I, which is encoded by a yeast mitochondrial intron, recognises and cleaves a unique 18 base-pair sequence with a very high degree of specificity (Colleaux *et al.* 1988). A gene containing this recognition site can be integrated into a cellular genome to provide a unique target. The frequency of gene replacement at such a target site was significantly increased when a vector that expresses the endonuclease was introduced at the same time as the foreign DNA fragment (Smih *et al.* 1995). Evidently the chromosomal target was cut by the endonuclease and the foreign DNA took the place of a homologous chromosome in DSB repair.

It should be noted that the foreign DNA fragment in this case was identical to the target apart from the short I-*Sce*I site, and the experiment should be repeated using a more conventional insertion fragment. However it does suggest an alternative pathway for gene replacement, one that would be the reverse of the insertion pathway discussed above, in fact amounting to gap-repair of the chromosome using the foreign DNA as the template rather than another chromosome (Figure 9.8). Resolution of the complex would produce either a converted chromosome or the successful replacement event or a broken chromosome; one fragment of the broken chromosome would have no centromere and the other no telomere.

The plausibility of this model is unclear. The frequency with which DSBs arise in chromosomal DNA is not known with any certainty, and can be expected to differ between cell types, but data of Leonhardt *et al.* (1997)

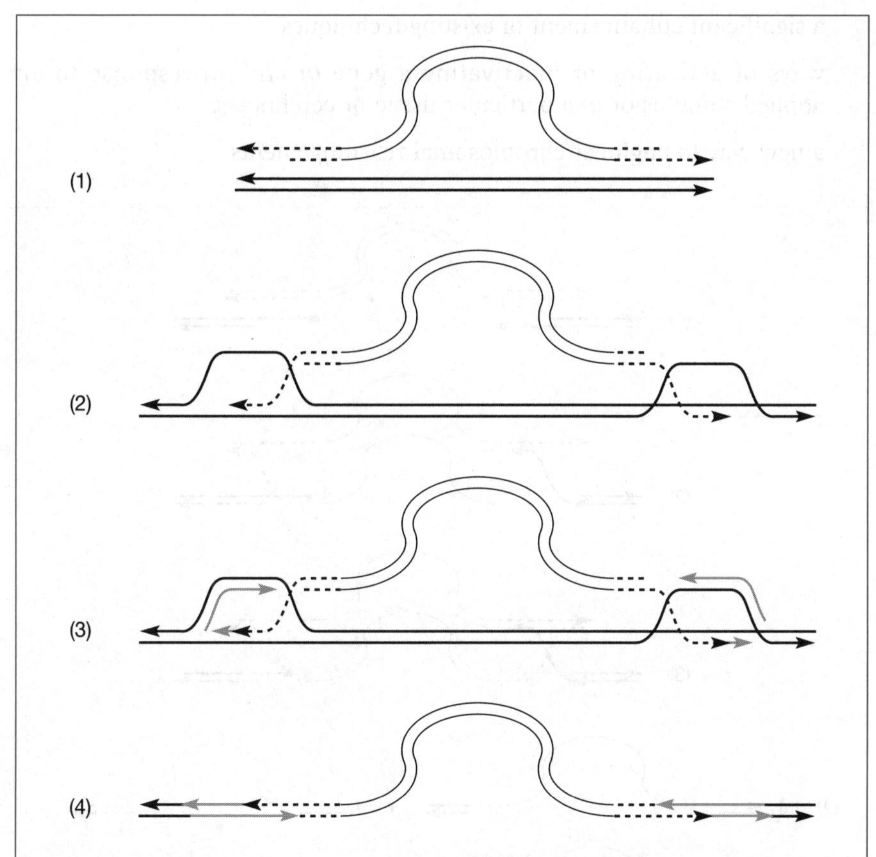

Figure 9.7 Possible mode of integration of a replacement fragment. Conventions as for Figure 9.4. (1) Resection by 5' exonuclease action; (2) invasion of the duplex by the exposed single strands; (3) strand extension by synthesis; (4) product following resolution of the Holliday junctions.

and Schwartz *et al.* (1996) suggest that the spontaneous frequency of occurrence of *misrepaired* DSBs at the *hprt* locus of CHO cells is about 1 in 10^6 cells. If the frequency with which DSBs *originate* is only 100 times higher than this, it comes within the frequency range of the most successful gene replacements. However it is also possible that a mechanism exists to generate a DSB in the chromosome following homologous pairing, induced by proteins involved in the pairing process. If so, the spontaneous frequency of DSBs is, of course, irrelevant.

Site-specific recombination

Yet another new dimension was opened up in transgenic work with the introduction of site-specific recombination (Kilby *et al.* 1993). Used in conjunction with established ways of creating transgenic animals (and indeed plants) site-specific recombination offers:

- a significant enhancement of existing techniques
- ways of activating or inactivating a gene *in vivo* in response to an applied stimulus or in a particular tissue or cell lineage
- a new way to engineer chromosomal rearrangements

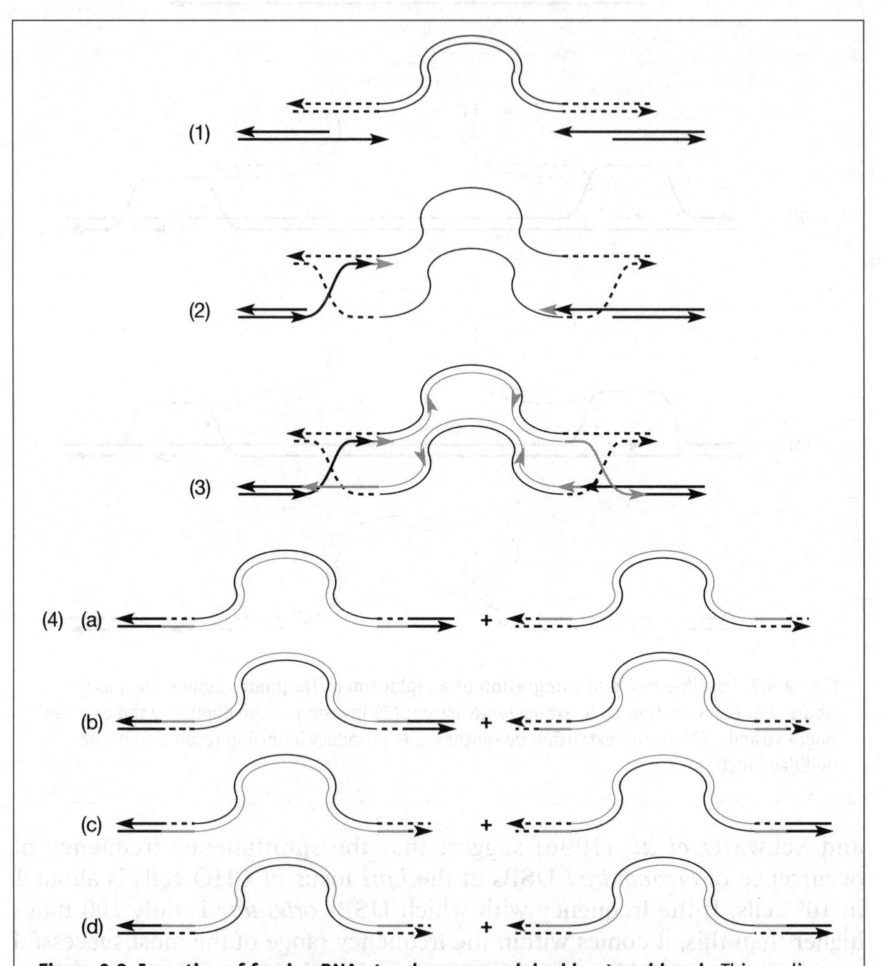

Figure 9.8 Insertion of foreign DNA at a chromosomal double-strand break. Thinner lines show the foreign DNA, thicker lines the chromosomal DNA; foreign DNA homologous to the chromosome is shown by a broken line and a selective reporter gene by continuous lines in an omega shape. Newly synthesised DNA is shown in blue. (1) Resection of the chromosome on either side of the DSB, exposing single strands. (2) Invasion of the foreign DNA by the exposed chromosomal single strands. The complementary strands of the foreign DNA are set apart for the sake of clarity. (3) DNA synthesis proceeding from the 3' ends of the invading strands. (4) The four alternative ways in which the structure may resolve. (a) and (b) show the two ways in which the left and right Holliday junctions resolve to produce an intact chromosome and a free foreign DNA fragment. (c) and (d) show the other two resolution pathways, each of which produces two chromosome fragments each carrying a copy of the foreign DNA; since in each case only one of the fragments will possess a centromere, these configurations are expected to be inviable.

Systems that bring about general recombination between any pair of matching chromosomal sequences involve large numbers of diverse proteins. In contrast, site-specific recombination systems have evolved in quite the opposite direction. These systems promote recombination between members of a very restricted set of similar sequences or, in the most extreme cases, between two identical copies of the same short sequence. This simple objective has produced a simple solution. In these reactions the recombinase binds to eligible DNA sequences and participating sites are brought together by interaction between bound recombinases rather than by the matching of DNA sequences. A small number of proteins participate in some of the reactions; others are effected by the recombinase alone.

Two families of site-specific recombinases are recognised and these employ different molecular mechanisms to cleave, rearrange and join up the strands of the participating DNA duplexes (Stark *et al.* 1992). Two recombinases of the bacteriophage λ-integrase family are now commonly employed in transgenic work: the Cre recombinase of bacteriophage P1 (Austin *et al.* 1981; Hoess and Abremski 1985) and the FLP recombinase of *S. cerevisiae* (Broach *et al.* 1982; Senecoff and Cox 1986). Cre resolves dimers of the plasmid form of the bacteriophage into circular monomers as in Figure 9.9a, and FLP inverts a segment of the yeast 2 micron plasmid as in Figure 9.9c. The R recombinase, which performs the same function as FLP in another yeast, has been employed less frequently.

The mechanism of action of these enzymes is the same. Each recognises a 34 base-pair sequence, called *loxP* in the case of Cre and FRT in the case of FLP (Figure 9.10). Each consists of a core and two flanking sequences. The flanking sequences are identical but inverted and are the enzyme binding sites. The short core sequences are asymmetrical; these confer orientation on the sites, determining which sides of the two participating duplexes are brought together. Two recombinase molecules bind to the recognition sequence in each of two participating DNA duplexes. Strand exchange is by transfer *via* linkage to a tyrosine residue in the recombinase. In the course of the reaction the 3' phosphodiesters at the staggered positions shown in Figure 9.10a are transferred first to the hydroxyl groups of the tyrosine residues and then to the 5' hydroxyl groups of the ribose residues exposed in the other duplex. The recombination process requires four such transfers, catalysed by four recombinase molecules, and no energy input. The rearrangement of the core sequences that this brings about is illustrated in Figure 9.10b. The entire process is clearly illustrated in a review by Stark *et al.* (1992).

Expression of the recombinase

The simplicity of these particular site-specific recombination reactions is illustrated by the fact that the enzymes function well *in vitro* with both linear and circular DNA substrates and in the absence of an energy source

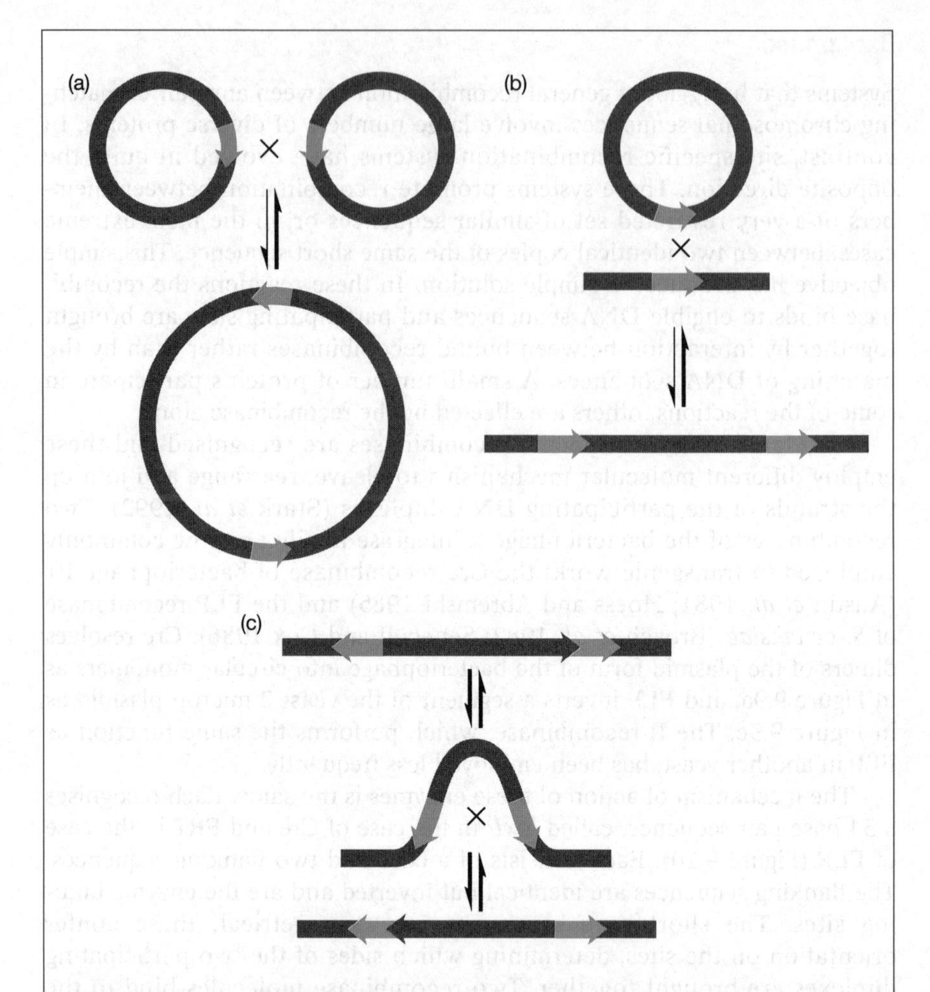

Figure 9.9 Site-specific recombination. The blue arrows represent the recombinase recognition sites (*loxP* or FRT) and their orientation; grey blocks represent other DNA duplex; the half arrows represent the direction of integration and resolution reactions, their width indicating which, if either, is favoured. (a) Cointegration of two closed circular molecules opposed by resolution of the cointegrate; (b) insertion of a closed circular molecule into a linear molecule opposed by its excision; (c) inversion of a sequence located between two recombinase recognition sites in inverted orientation.

(Stark *et al.* 1992). They also function well in all manner of cells, from bacteria through fungi to plant and mammalian cells (Kilby *et al.* 1993). The activity of the isolated recombinases is normally assayed by the breakdown of a cointegrate plasmid, like the large circular molecule illustrated in Figure 9.9a, into the constituent circular molecules represented by the smaller circles. Interlinked circles and complex intermediates are produced in the course of the reaction, so that it is necessary to digest the products with appropriate restriction enzymes in order to obtain quantifiable data (Stark *et al.* 1992). Other methods have been developed to monitor the

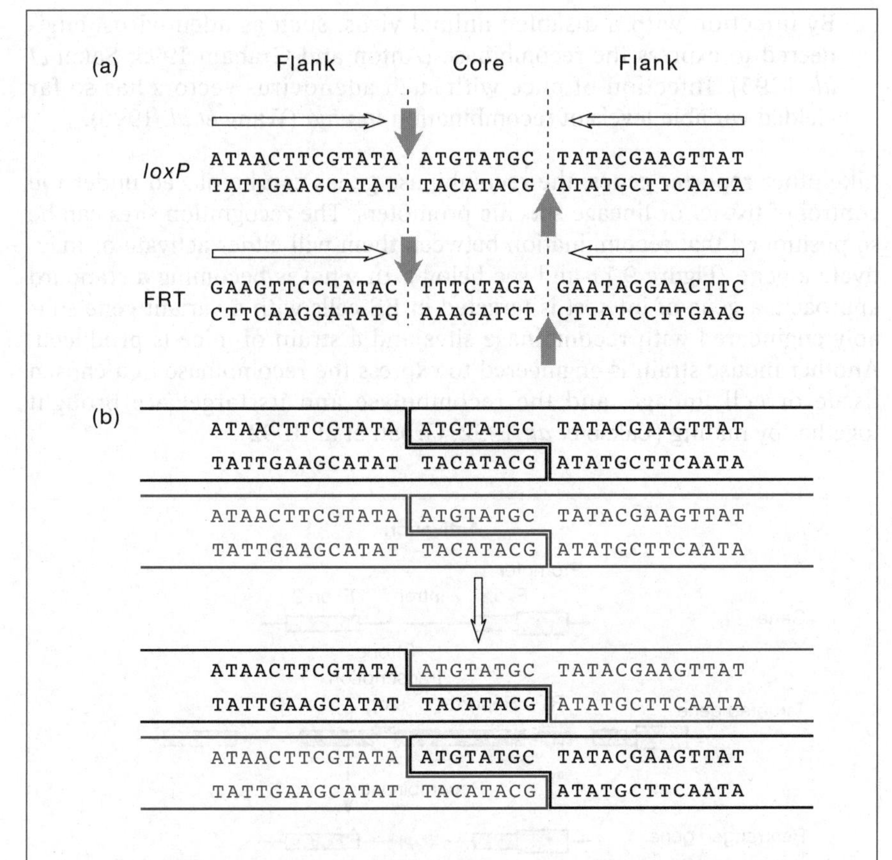

Figure 9.10 (a) Sequences of *loxP* and FRT sites. The open arrows draw attention to the identical inverted flanking regions; the blue arrows show the points at which each of these duplexes exchanges with a second identical duplex; (b) rearrangement of the strands of two participating *loxP* duplexes, one in black and the other in blue, upon recombination between them.

expression of the enzymes *in vivo,* including the use of specific antibodies that can be used in conjunction with protein gel electrophoresis (Schwenk *et al.* 1997), and genes that code for hybrid proteins with both recombinase activity and a visual reporter capability, such as GFP fluorescence (Gagneten *et al.* 1997).

Recombinase can be delivered to cultured cells in a number of ways:

- By lipofection, using the purified enzyme (Baubonis and Sauer 1993).
- By transfecting or electroporating the cells with a circular plasmid that expresses the recombinase under the direction of a cellular promoter (Sauer and Henderson 1990; Gu *et al.* 1993; Fukushige and Sauer 1992). Due to the low frequency of random integration very few of the treated cells incorporate the recombinase-expressing plasmid into their chromosomes.

- By infection with a disabled animal virus, such as adenovirus, engineered to express the recombinase (Anton and Graham 1995; Sakai *et al.* 1995). Infection of mice with such adenovirus vectors has so far yielded variable levels of recombination *in vivo* (Wang *et al.* 1996).

Like other reporter genes, the recombinase genes can be placed under the control of tissue- or lineage-specific promoters. The recognition sites can be so positioned that recombination between them will either activate or inactivate a gene (Figure 9.11 and see below). In what is becoming a standard approach, a gene of interest is targeted in ES cells with a variant gene suitably engineered with recombinase sites and a strain of mice is produced. Another mouse strain is engineered to express the recombinase in a chosen tissue or cell lineage, and the recombinase and its target are brought together by mating (Lakso *et al.* 1992; Orban et al. 1992).

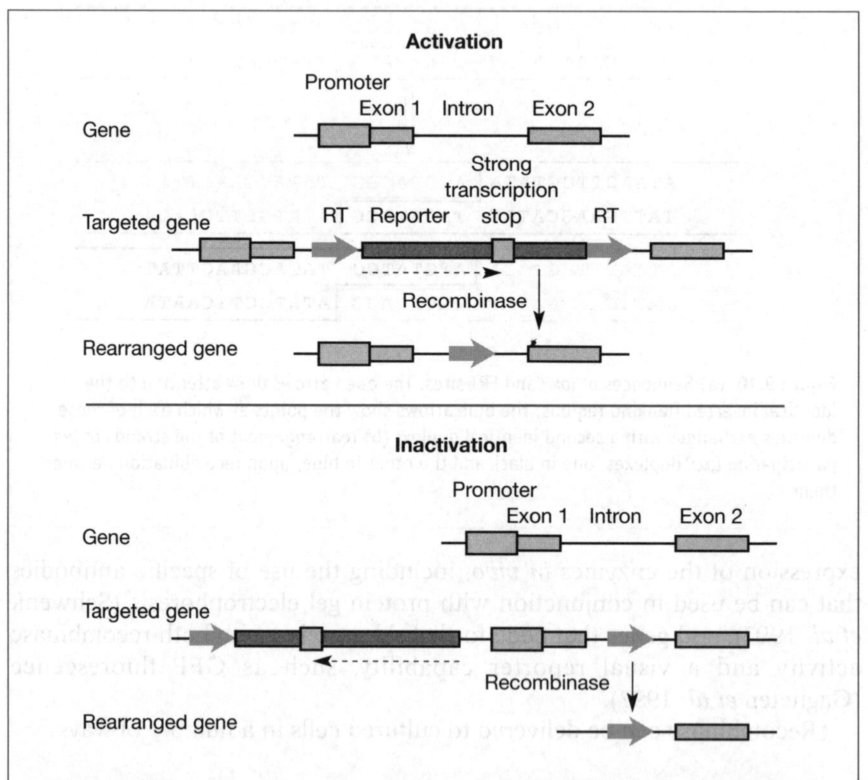

Figure 9.11 Ways in which transgenes may be activated and inactivated by site-specific recombination. The blue arrows represent recombinase recognition sites; the target gene is represented by open boxes and a continuous line, the reporter gene by a shaded box and a strong stop signal by an open box; the broken arrow shows the direction of transcription of the reporter. In each case the top line shows a resident wild-type gene, the middle line the same gene after its alteration by homologous recombination to set up the target for the recombinase, and the bottom line the product of recombination between the recognition sites.

A further variation is to couple the *cre* gene to an inducible promoter, for instance the interferon-inducible promoter of the *Mx1* gene (Kuhn *et al.* 1995); administration of an inducer of interferon such as poly-dI.dC indirectly induces recombination in this system. Yet another variation is to create a gene that codes for a hybrid recombinase which is sequestered in the absence of a specific ligand, such as the Cre-ER^T recombinase, activated *in vivo* by administration of the oestrogen analogue tamoxifen (Feil *et al.* 1996; Brocard *et al.* 1997). So far these approaches have largely failed to produce recombination in every cell of the target tissue (Orban *et al.* 1992; Rickert *et al.* 1997), presumably due to mosaic expression – although when Cre expression was controlled by the interferon-inducible *Mx1* promoter 100% recombination was obtained in the liver (Kuhn *et al.* 1995). Some applications may not require recombination to occur in every cell. Again, some of the published results may appear less promising than they really are since tissues are composed of more than one cell type. Thirdly, as time goes on, no doubt new and improved methods will come forward. The subject of mosaic and variegated expression is taken up in Chapter 10.

Despite the requirement to mate mouse strains together, this is potentially an efficient approach, since the strain carrying the target gene can be mated to strains that express the recombinase in different tissues and likewise a given recombinase-expressing strain can be mated to strains carrying different target genes.

Elimination of reporter genes

One of the commonest uses of site-specific recombination is the removal of a reporter gene from cells once it has served its purpose. This can be achieved simply by bracketing the reporter by recombinase target sites. The recombinase is introduced once suitable clones have been identified. The reaction is the one shown in Figure 9.9b in the upward direction. The excised reporter gene, which does not replicate, will be degraded or simply diluted out during cell proliferation. The procedure is of course facilitated if selection can be applied against the reporter, but the excision reaction can be very efficient, taking place in up to 80% of the treated cells.

A recent enhancement of this method is the development of a strain of mice that express Cre recombinase exclusively during spermatogenesis under the direction of the protamine gene reporter (O'Gorman *et al.* 1997). When ES cells developed from this strain are introduced into a blastocyst, the chromosomes of the ES cell-derived germ cells are exposed to recombinase activity during spermatogenesis. Other strains have been developed which express Cre recombinase in the oocyte, an approach designed to facilitate gene targeting in oocytes (Lewandoski *et al.* 1997).

Both gene activation and inactivation can be brought about by excision of a fragment of DNA bracketed by two recombinase target sites. Examples of the kinds of arrangement that can be used for these purposes are shown in Figure 9.11. In the top diagram a gene has been inactivated by introducing into an intron a reporter that includes a strong transcriptional termination signal. Recombination between the target sites (RT, i.e. *loxP* or FRT) removes the reporter and with it the termination signal, leaving behind in the intron nothing but a single copy of the RT. Provided the RT does not interfere with transcription or RNA splicing, the gene will be activated. The lower part of the figure shows a strategy for gene inactivation. In this case the RTs and the reporter have been positioned so that, with luck, they will not interfere with the transcription of the gene. Subsequently, recombination between the target sites will delete the promoter. The element of luck can be reduced by making sure that the targeted gene is transcriptionally active, for example by transfection of cultured cells.

To generate null cells it will be necessary in most cases to make the recombinase target allele homozygous or else to pair it with a null allele. This will necessarily entail additional matings. In view of the problem of mosaic expression, the better strategy for gene *activation* is probably to make the target allele homozygous, doubling the chance that one or other target gene will be activated in a given cell. Conversely, the better strategy for gene *inactivation* is probably to pair the target allele with a null allele.

Bidirectionality of the recombination reaction

As shown in Figure 9.9 the reactions catalysed by site-specific recombinases can proceed in either direction. The cointegration of two plasmids is opposed by the resolution of the cointegrate and the insertion of a circular molecule into a duplex is opposed by the excision of the circle. The rate at which two molecules become joined depends on their concentrations, while the rate at which a cointegrate resolves into its constituent parts is concentration-independent. Unless the substrate concentration is very high, the rate of resolution will be much higher than the rate of joining. In accordance with this expectation, excision reactions catalysed by recombinase *in vivo* are very efficient, while targeting reactions are by contrast inefficient (Fukushige and Sauer 1992; Rucker and Piedrahita 1997). Furthermore, if the reaction is fully reversible as expected, the equilibrium state will favour the separated DNA molecules at the expense of the cointegrate. Consequently, if more time is provided for a targeting reaction to proceed only a modest increase in frequency can be anticipated.

Two strategies have been proposed to increase the efficiency of insertion by site-specific recombination. The first was originally applied in plant cells (Albert *et al.* 1995) and later in ES cells (Araki *et al.* 1997). Altered *loxP* sites were produced, one (LE) with five base pairs changed at the extreme left of the 34 base-pair sequence (ATAAC → TACCG in the top strand) and the other (RE) with the same alterations at the extreme right (GTTAT → CGGTA, again in the top strand). LE and RE are compatible with each other, consistent with the view that the recombining sequences line up initially with their cores in opposite orientation (Stark *et al.* 1992). The sites produced by recombination between LE and RE are a normal *loxP* site and a site with both alterations (LE+RE), and these are poorly compatible. When a circular foreign DNA carrying RE recombines with a chromosomal LE the foreign DNA is inserted between newly formed *loxP* and LE+RE sites (Figure 9.12). Consequently insertion should be favoured over excision. The efficiency of insertion was in fact about eight times higher than observed in parallel experiments with 'normal' *loxP* targeting, but was nevertheless very low (Table 9.1).

Table 9.1 Efficiencies of insertion of foreign DNA by site-specific recombination with Cre recombinase.

Sites	*Cells*	*Insertions per 10^6 cells*	*Basis for counting total cells*
loxP	CHO	5–45	Colonies formed[a]
loxP	ES	20	Plated cells[b]
LE and RE	ES	Up to 25	Plated cells[c]
loxP and *lox511*	NIH3T3	Up to 8500	Colonies formed[d]
loxP and *lox511*	MEL	Up to 1700	Treated cells[e]

References: [a]Fukushige and Sauer 1992; [b]Jiricny 1994; [c]Araki *et al.* 1997; [d]Bethke and Sauer 1997; [e]Bouhassira *et al.* 1997

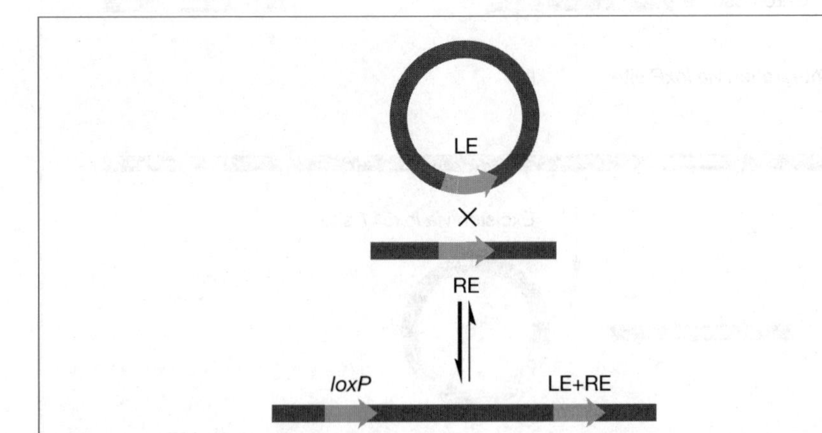

Figure 9.12 The use of altered recombinase recognition sites to favour integration over excision. The altered parts of the sites are shown in grey, unaltered parts in blue. For other details see the text.

The second strategy is based on the observation that a *lox* site called *lox511*, with an alteration in the core sequence (ATGTATGC → ATGTATAC), recombines efficiently with another *lox511* sequence but not with *loxP* (Hoess *et al.* 1986). Both the target site and the foreign DNA are flanked by *loxP* on one side and *lox511* on the other, in each case in the same orientation (Figure 9.13). Insertion may occur in either of two ways:

In a manner analogous to gene replacement by homologous recombination in the sense that both the *loxP* sites and the *lox511* sites may simply recombine, substituting the foreign DNA for the target DNA directly.

If the incoming DNA is circular, recombination between either pair of compatible sites, but not both, is analogous to gene insertion by homologous recombination, and the recombinant chromosome will carry two *loxP* and two *lox511* sites, alternating with each other in the form *loxP-lox511-loxP-lox511*. Recombination between the pair of compatible sites that brought about the insertion will reverse it, but recombination between the other compatible pair will eliminate the target sequence and leave in place the foreign DNA, flanked by incompatible sites and therefore not liable to be excised. In fact this procedure is quite successful, in one case increasing the frequency of insertion events to nearly 1 per 100 surviving cells (Table 9.1).

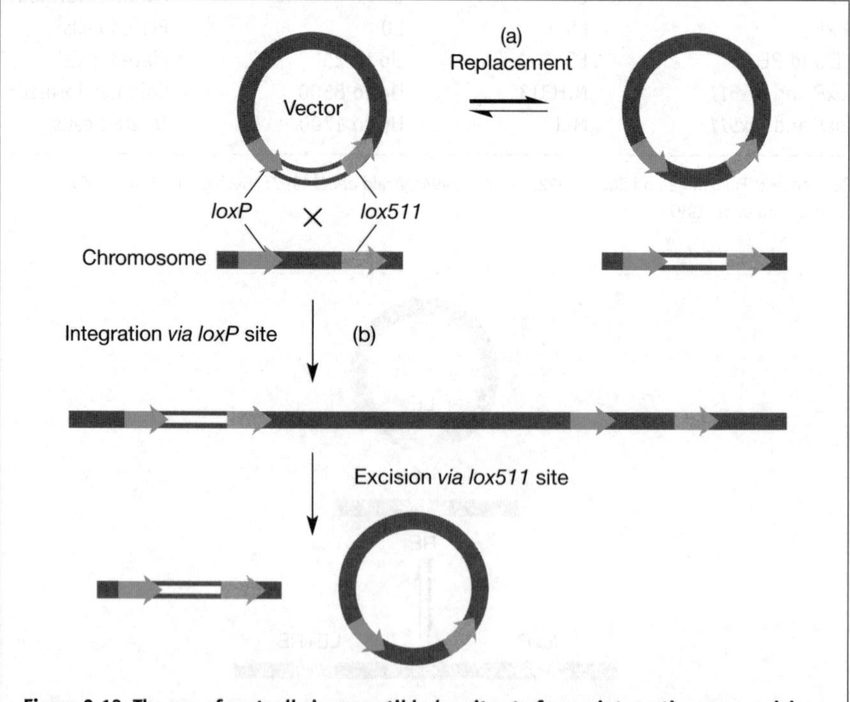

Figure 9.13 The use of mutually incompatible *lox* sites to favour integration over excision. The same outcome is arrived at whether the process is analogous to gene replacement by homologous recombination (a) or, in the first instance, to gene insertion (b). Excision *via* the site not involved in the insertion leads to the integration of the foreign gene (white bar) flanked in the same way by incompatible *lox* sites.

Site-specific recombination offers new opportunities to engineer changes in chromosomes on a grand scale, by producing deletions, inversions and translocations, and even by deleting entire chromosomes. The basic procedure, which has already proved successful in a small number of cases, is first to target two chromosomal sites one after the other by homologous recombination in embryonic stem cells. These sites will be the break-points and can be targeted only if appropriate molecular clones are available. Each site will be altered so as to contain a recombinase recognition sequence in a predetermined orientation and, because of the low frequency with which homologous recombinants are recovered, each will also contain the selectable reporter gene that was used to target the site.

In many cases it will also be necessary to select for the occurrence of the chromosomal rearrangement, for instance if it is brought about by transient expression of the recombinase. If so, a special approach is required. One such approach is illustrated in Figure 9.14 (Smith *et al.* 1995; Ramirez-Solis *et al.* 1995). Here a selectable transgene is split into two parts across an intron. A recombinase target site is added to each part as shown and each is inserted into a different chromosomal location by gene replacement. This can be done of course, only if cloned DNA fragments corresponding to the chosen target sites are available and additional selective reporters need to be included in the constructs. The replacements have to be carried out serially, meaning that after one site has been replaced the cells carrying it become recipients for a second round of replacement at the other site. In the example the *neo* gene and the hygromycin resistance gene (*hph*) are the selective reporters (Figure 9.14b).

The split gene is inactive in this configuration since neither part codes for an active protein. The *hprt* gene, which codes for hypoxanthine phosphoribosyl transferase, has been used for this purpose together with an *hprt*⁻ cell line. *hprt* is a conditionally essential gene, meaning that it is essential only under special conditions of cell culture (see Box 9.4). Cells in which the two parts of the gene have been brought together by site-specific recombination are viable in HAT medium while the non-recombinant cells die. When the replacements are arranged as in Figure 9.14b (i) this selects for a chromosomal translocation as illustrated.

Site-specific recombination can be harnessed to bring about chromosome loss (Figure 9.15). In this example, two recombinase recognition sites were introduced in opposite orientation into the single Y chromosome present in XY ES cells, so that interchromosomal recombination could occur only between replicated daughter chromosomes (i), not between homologues (Lewandoski and Martin 1997). Recombination between daughter copies of the same site exchanges identical downstream chromosome arms, and no alteration results. Recombination between the two sites on one arm leads to an inversion of the intervening stretch of chromosome, probably with no consequences. Recombination between the upstream site on one

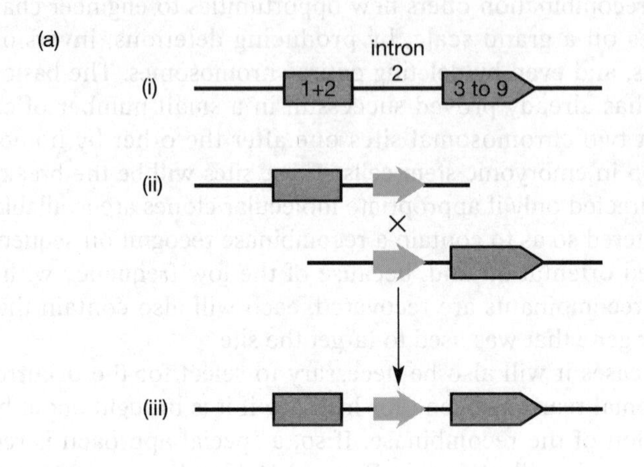

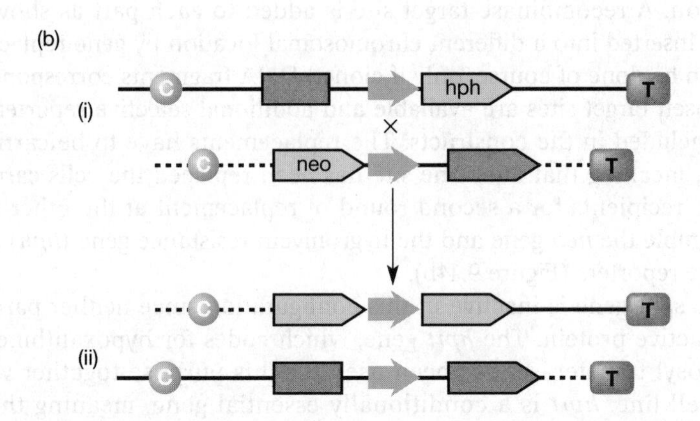

Figure 9.14 Rearrangement of chromosomes by site-specific recombination. (a) Strategy for positive selection of cells carrying the desired rearrangement: (i) a conditionally essential gene (such as the *hprt* gene) containing at least one intron. Two parts of the gene are labelled 1+2 and 3–9, referring to exons, and may or may not have had introns removed; (ii) shows how the two parts of the gene are targeted to different parts of the genome in conjunction with recombinase target sequences (blue arrows); (iii) site-specific recombination reconstitutes the gene with the recognition sequence in the intron so that it is removed from the mRNA when the exons are brought together by RNA splicing. (b) Bringing about a reciprocal translocation by site-specific recombination: (i) two chromosomes are illustrated by continuous and broken lines – C, centromeres; T, telomeres. One has been targeted to contain part of a conditionally essential gene and the hygromycin resistance reporter (*hph*), the other chromosome the second part of the essential gene and the G418 resistance reporter (*neo*); (ii) recombination brings about a reciprocal translocation with the reconstitution of the essential gene.

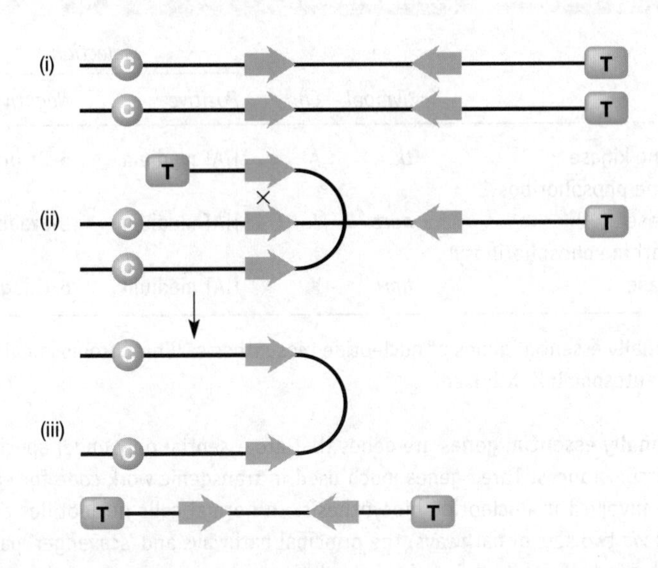

Figure 9.15 Rearrangement of chromosomes by site-specific recombination. Chromosome loss through recombination by site-specific sister-strand exchange; (i) the chromosome is targeted with two recombinase target sites in inverted orientation. Recombination between congruent sites on sister strands brings about no rearrangement; (ii) recombination between site 1 on one sister strand and site 2 on the other produces (iii) acentric and dicentric chromatids, both destined to be lost as proliferation continues.

daughter chromosome and the downstream site on the other produces the result shown in Figure 9.15 (ii and iii), namely dicentric and acentric products, both of which are unstable and are lost during cellular proliferation. In the example, males carrying this arrangement of *loxP* sites on the Y chromosome were mated to females carrying a transgene in which *Cre* is controlled by the β-actin promoter and expressed constitutively and ubiquitously. The majority of male zygotes developed into XO females as a result of loss of the Y chromosome.

A variety of applications are foreseen for this type of methodology. One is to replicate translocations that correlate with physiological or malignant conditions, such as the 12;15 translocation that correlates with murine plasmacytomas (Smith *et al.* 1995), so as to verify the cause and effect relationship between the two, or alternatively to mimic a human condition such as the 6;9 translocation that correlates with acute myeloid leukemia (van Deursen *et al.* 1995) in the expectation of developing a mouse model of the disease. Another is to expand the repertoire of chromosomal deficiencies and deletions that are not lethal when heterozygous. These can be of great assistance in the intensive mapping of chromosomal regions by traditional mutagenesis and in discovering imprinted genes. Yet another is to 'humanise' parts of the genome of another mammal. As pointed out in

BOX 9.4 CONDITIONALLY ESSENTIAL GENES

Name	Symbol	Chr[1]	Selection	
			Positive	*Negative*
Thymidine kinase	*tk*	A	HAT medium	5-Br-uridine
Adenosine phosphoribosyl transferase	*aprt*	A	HAT medium	8-azaadenine
Hypoxanthine phosphoribosyl transferase	*hprt*	X	HAT medium	6-thioguanine

Conditionally essential genes of nucleotide biosynthesis. [1]Chr: chromosomal location: A, autosomal; X, X-linked.

Conditionally essential genes are genes that are essential only under special contrived circumstances. Three genes much used in transgenic work code for enzymes that are involved in nucleotide biosynthesis. In normal cells nucleotides are synthesised *via* two sets of pathways, the principal pathways and 'scavenger' pathways by which products of nucleic acid degradation are recycled. When the principal pathways are inactive for one reason or another the scavenger pathways can supply the cells' needs, provided the medium is supplemented with appropriate biosynthetic intermediates. On the other hand, when the principal pathways are active the scavenger pathways are dispensable and cells that carry an inactivating mutation in one of these pathways are viable under normal conditions. Genes involved in three separate scavenger pathways are listed in the Table.

The enzyme coded for by each of these genes will convert at least one otherwise harmless analogue of its natural substrate to a poisonous derivative by phosphorylating it. This allowed null mutants of the genes to be isolated and also allows negative selection to be applied against each of them. The fact that the *hprt* gene is sex-linked is useful because it means that male-derived cells carry only one copy of the gene.

The principal pathways of AMP, GMP and TMP synthesis are blocked by the folic acid antagonist aminopterin. When aminopterin is present in the medium the scavenger pathways can be supplied by hypoxanthine and thymidine, and under these circumstances the enzymes listed in the Table become essential for survival. HAT medium contains the three compounds. Cells carrying a null mutation in any of the three genes (in both copies if two are present) will die in HAT medium, which can therefore be used to select cells that have integrated a transgene that supplies the missing function (positive selection).

Chapter 5, mammalian genomes contain extensive syntenic regions. It should be possible to flank part or all of a syntenic region by two mutually incompatible recombinase recognition sites in cells of two species, say man and pig. The appropriate human chromosome would be introduced into the pig cells by microcell-mediated transfer and recombination induced by transient expression of recombinase. Pigs might then be generated by nuclear transfer into oocytes (Chapter 7). In such a way the pig major histocompatibility complex, for example, might be humanised at a stroke.

References

Albert, H., Dale, E.C., Lee, E., and Ow, D.W. (1995) Site-specific integration of DNA into wild-type and mutant lox sites placed in the plant genome. *Plant Journal*, **7**, 649–659.

Anton, M. and Graham, F.L. (1995) Site-specific recombination mediated by an adenovirus vector expressing the Cre recombinase protein: a molecular switch for control of gene expression. *Journal of Virology*, **69**, 4600–4606.

Araki, K., Araki, M., and Yamamura, K. (1997) Targeted integration of DNA using mutant *lox* sites in embryonic stem cells. *Nucleic Acids Research*, **25**, 868–872.

Austin, S., Ziese, M., and Sternberg, N. (1981) A novel role for site-specific recombination in maintenance of bacterial replicons. Cell, **25**, 729–736.

Baubonis, W. and Sauer, B. (1993) Genomic targeting with purified Cre recombinase. *Nucleic Acids Research*, **21**, 2025–2029.

Berinstein, N., Pennell, N., Ottaway, C.A., and Shulman, M.J. (1992) Gene replacement with one-sided homologous recombination. *Molecular & Cellular Biology*, **12**, 360–367.

Bethke, B. and Sauer, B. (1997) Segmental genomic replacement by Cre-mediated recombination: genotoxic stress activation of the p53 promoter in single-copy transformants. *Nucleic Acids Research*, **25**, 2828–2834.

Bouhassira, E.E., Westerman, K., and Leboulch, P. (1997) Transcriptional behavior of LCR enhancer elements integrated at the same chromosomal locus by recombinase-mediated cassette exchange. *Blood*, **90**, 3332–3344.

Broach, J.R., Guarascio, V.R., and Jayaram, M. (1982) Recombination within the yeast plasmid 2mu circle is site-specific. *Cell*, **29**, 227–234.

Brocard, J., Warot, X., Wendling, O., Messaddeq, N., Vonesch, J.L., Chambon, P., and Metzger, D. (1997) Spatio-temporally controlled site-specific somatic mutagenesis in the mouse. *Proceedings of the National Academy of Sciences of the United States of America*, **94**, 14559–14563.

Capecchi, M.R. (1980) High efficiency transformation by direct microinjection of DNA into cultured mammalian cells. *Cell*, **22**, 479–488.

Colleaux, L., D'Auriol, L., Galibert, F., and Dujon, B. (1988) Recognition and cleavage site of the intron-encoded omega transposase. *Proceedings of the National Academy of Sciences of the United States of America*, **85**, 6022–6026.

de Wind, N., Dekker, M., Berns, A., Radman, M., and te Riele, H. (1995) Inactivation of the mouse Msh2 gene results in mismatch repair deficiency, methylation tolerance, hyperrecombination, and predisposition to cancer. *Cell*, **82**, 321–330.

Deng, C. and Capecchi, M.R. (1992) Reexamination of gene targeting frequency as a function of the extent of homology between the targeting vector and the target locus. *Molecular & Cellular Biology*, **12**, 3365–3371.

Doetschman, T., Gregg, R.G., Maeda, N., Hooper, M.L., Melton, D.W., Thompson, S., and Smithies, O. (1987) Targetted correction of a mutant HPRT gene in mouse embryonic stem cells. *Nature*, **330**, 576–578.

Feil, R., Brocard, J., Mascrez, B., LeMeur, M., Metzger, D., and Chambon, P. (1996) Ligand-activated site-specific recombination in mice. *Proceedings of the National Academy of Sciences of the United States of America*, **93**, 10887–10890.

Fukushige, S. and Sauer, B. (1992) Genomic targeting with a positive-selection lox integration vector allows highly reproducible gene expression in mammalian cells. *Proceedings of the National Academy of Sciences of the United States of America*, **89**, 7905–7909.

Gagneten, S., Le, Y., Miller, J., and Sauer, B. (1997) Brief expression of a GFP cre fusion gene in embryonic stem cells allows rapid retrieval of site-specific genomic deletions. *Nucleic Acids Research*, **25**, 3326–3331.

Gu, H., Zou, Y.R., and Rajewsky, K. (1993) Independent control of immunoglobulin switch recombination at individual switch regions evidenced through Cre-loxP-mediated gene targeting. *Cell*, **73**, 1155–1164.

Hasty, P., Ramirez Solis, R., Krumlauf, R., and Bradley, A. (1991a) Introduction of a subtle mutation into the Hox-2.6 locus in embryonic stem cells. *Nature*, **350**, 243–246.

Hasty, P., Rivera Perez, J., Chang, C., and Bradley, A. (1991b) Target frequency and integration pattern for insertion and replacement vectors in embryonic stem cells. *Molecular & Cellular Biology*, **11**, 4509–4517.

Hasty, P., Rivera-Perez, J., and Bradley, A. (1992) The role and fate of DNA ends for homologous recombination in embryonic stem cells. *Molecular & Cellular Biology*, **12**, 2464–2474.

Hasty, P., Rivera-Perez, J., and Bradley, A. (1995) Gene conversion during vector insertion in embryonic stem cells. *Nucleic Acids Research*, **23**, 2058–2064.

Hoess, R.H., Wierzbicki, A., and Abremski, K. (1986) The role of the loxP spacer region in P1 site-specific recombination. *Nucleic Acids Research*, **14**, 2287–2300.

Hoess, R.H. and Abremski, K. (1985) Mechanism of strand cleavage and exchange in the Cre-lox site-specific recombination system. *Journal of Molecular Biology,* **181**, 351–362.

Jiricny, J. (1994) Colon cancer and DNA repair: have mismatches met their match? *Trends in Genetics,* **10**, 164–168.

Kilby, N.J., Snaith, M.R., and Murray, J.A. (1993) Site-specific recombinases: tools for genome engineering. *Trends in Genetics,* **9**, 413–421.

Kuhn, R., Schwenk, F., Aguet, M., and Rajewsky, K. (1995) Inducible gene targeting in mice. *Science,* **269**, 1427–1429.

Lakso, M., Sauer, B., Mosinger, B., Lee, E.J., Manning, R.W., Yu, S.H., Mulder, K.L., and Westphal, H. (1992) Targeted oncogene activation by site-specific recombination in transgenic mice. *Proceedings of the National Academy of Sciences of the United States of America,* **89**, 6232–6236.

Leonhardt, E.A., Trinh, M., Forrester, H.B., Johnson, R.T., and Dewey, W.C. (1997) Comparisons of the frequencies and molecular spectra of HPRT mutants when human cancer cells were X-irradiated during G1 or S phase. *Radiation Research,* **148**, 548–560.

Lewandoski, M., Wassarman, K.M., and Martin, G.R. (1997) Zp3-cre, a transgenic mouse line for the activation or inactivation of loxP-flanked target genes specifically in the female germ line. *Current Biology,* **7**, 148–151.

Lewandoski, M. and Martin, G.R. (1997) Cre-mediated chromosome loss in mice. *Nature Genetics,* **17**, 223–225.

Mansour, S.L., Thomas, K.R., and Capecchi, M.R. (1988) Disruption of the proto-oncogene int-2 in mouse embryo-derived stem cells: a general strategy for targeting mutations to non-XX selectable genes. *Nature,* **336**, 348–352.

Modrich, P. (1991) Mechanisms and biological effects of mismatch repair. *Annual Review of Genetics,* **25**, 229–253.

Modrich, P. (1994) Mismatch repair, genetic stability, and cancer. *Science,* **266**, 1959–1960.

Mortensen, R.M., Conner, D.A., Chao, S., Geisterfer-Lowrance, A.A., and Seidman, J.G. (1992) Production of homozygous mutant ES cells with a single targeting construct. *Molecular & Cellular Biology,* **12**, 2391–2395.

O'Gorman, S., Dagenais, N.A., Qian, M., and Marchuk, Y. (1997) Protamine-Cre recombinase transgenes efficiently recombine target sequences in the male germ line of mice, but not in embryonic stem cells. *Proceedings of the National Academy of Sciences of the United States of America,* **94**, 14602–14607.

Orban, P.C., Chui, D., and Marth, J.D. (1992) Tissue- and site-specific DNA recombination in transgenic mice. *Proceedings of the National Academy of Sciences of the United States of America,* **89**, 6861–6865.

Orr-Weaver, T.L., Szostak, J.W., and Rothstein, R.J. (1981) Yeast transformation: a model system for the study of recombination. *Proceedings of the National Academy of Sciences of the United States of America,* **78**, 6354–6358.

Ramirez-Solis, R., Liu, P., and Bradley, A. (1995) Chromosome engineering in mice. *Nature,* **378**, 720–724.

Rayssiguier, C., Thaler, D.S., and Radman, M. (1989) The barrier to recombination between *Escherichia coli* and *Salmonella typhimurium* is disrupted in mismatch-repair mutants. *Nature,* **342**, 396–401.

Rickert, R.C., Roes, J., and Rajewsky, K. (1997) B lymphocyte-specific, Cre-mediated mutagenesis in mice. *Nucleic Acids Research,* **25**, 1317–1318.

Rucker, E.B. and Piedrahita, J.A. (1997) Cre-mediated recombination at the murine whey acidic protein (mWAP) locus. *Molecular Reproduction & Development,* **48**, 324–331.

Sakai, K., Mitani, K., and Miyazaki, J. (1995) Efficient regulation of gene expression by adenovirus vector-mediated delivery of the CRE recombinase. *Biochemical & Biophysical Research Communications,* **217**, 393–401.

Sauer, B. and Henderson, N. (1990) Targeted insertion of exogenous DNA into the eukaryotic genome by the Cre recombinase. *New Biologist,* **2**, 441–449.

Schwartz, J.L., Porter, R.C., and Hsie, A.W. (1996) The molecular nature of spontaneous mutations at the hprt locus in the radiosensitive CHO mutant xrs-5. *Mutation Research,* **351**, 53–60.

Schwenk, F., Sauer, B., Kukoc, N., Hoess, R., Muller, W., Kocks, C., Kuhn, R., and Rajewsky, K. (1997) Generation of Cre recombinase-specific monoclonal antibodies, able to characterize the pattern of Cre expression in cre-transgenic mouse strains. *Journal of Immunological Methods,* **207**, 203–212.

Senecoff, J.F. and Cox, M.M. (1986) Directionality in FLP protein-promoted site-specific recombination is mediated by DNA–DNA pairing. *Journal of Biological Chemistry,* **261**, 7380–7386.

Smih, F., Rouet, P., Romanienko, P.J., and Jasin, M. (1995) Double-strand breaks at the target locus stimulate gene targeting in embryonic stem cells. *Nucleic Acids Research,* **23**, 5012–5019.

Smith, A.J., De Sousa, M.A., Kwabi-Addo, B., Heppell-Parton, A., Impey, H., and Rabbitts, P. (1995) A site-directed chromosomal translocation induced in embryonic stem cells by Cre-loxP recombination [published erratum appears in *Nat. Genet.* 1996 Jan;**12**(1):110]. *Nature Genetics,* **9**, 376–385.

Smithies, O. and Kim, H.S. (1994) Targeted gene duplication and disruption for analyzing quantitative genetic traits in mice. *Proceedings of the National Academy of Sciences of the United States of America,* **91**, 3612–3615.

Stark, W.M., Boocock, M.R., and Sherratt, D.J. (1992) Catalysis by site-specific recombinases. *Trends in Genetics,* **8**, 432–439.

Steeg, C.M., Ellis, J., and Bernstein, A. (1990) Introduction of specific point mutations into RNA polymerase II by gene targeting in mouse embryonic stem cells: evidence for a DNA mismatch repair mechanism. *Proceedings of the National Academy of Sciences of the United States of America,* **87**, 4680–4684.

Szostak, J.W., Orr-Weaver, T.L., Rothstein, R.J., and Stahl, F.W. (1983) The double-strand-break repair model for recombination. *Cell,* **33**, 25–35.

te Riele, H., Maandag, E.R., and Berns, A. (1992) Highly efficient gene targeting in embryonic stem cells through homologous recombination with isogenic DNA constructs. *Proceedings of the National Academy of Sciences of the United States of America,* **89**, 5128–5132.

Thomas, K.R., Folger, K.R., and Capecchi, M.R. (1986) High frequency targeting of genes to specific sites in the mammalian genome. *Cell,* **44**, 419–428.

Thomas, K.R. and Capecchi, M.R. (1987) Site-directed mutagenesis by gene targeting in mouse embryo-derived stem cells. *Cell,* **51**, 503–512.

Valancius, V. and Smithies, O. (1991a) Double-strand gap repair in a mammalian gene targeting reaction. *Molecular & Cellular Biology,* **11**, 4389–4397.

Valancius, V. and Smithies, O. (1991b) Testing an "in-out" targeting procedure for making subtle genomic modifications in mouse embryonic stem cells. *Molecular & Cellular Biology,* **11**, 1402–1408.

van Deursen, J., Fornerod, M., Van Rees, B., and Grosveld, G. (1995) Cre-mediated site-specific translocation between nonhomologous mouse chromosomes. *Proceedings of the National Academy of Sciences of the United States of America,* **92**, 7376–7380.

van Deursen, J. and Wieringa, B. (1992) Targeting of the creatine kinase M gene in embryonic stem cells using isogenic and nonisogenic vectors. *Nucleic Acids Research,* **20**, 3815–3820.

Waldman, A.S. and Liskay, R.M. (1988) Dependence of intrachromosomal recombination in mammalian cells on uninterrupted homology. *Molecular & Cellular Biology*, **8**, 5350–5357.

Wang, Y., Krushel, L.A., and Edelman, G.M. (1996) Targeted DNA recombination in vivo using an adenovirus carrying the cre recombinase gene. *Proceedings of the National Academy of Sciences of the United States of America*, **93**, 3932–3936.

Zheng, H., Hasty, P., Brenneman, M.A., Grompe, M., Gibbs, R.A., Wilson, J.H., and Bradley, A. (1991) Fidelity of targeted recombination in human fibroblasts and murine embryonic stem cells. *Proceedings of the National Academy of Sciences of the United States of America*, **88**, 8067–8071.

CHAPTER TEN

Expression of transgenes

Gene expression is too large a subject to be dealt with in great detail here. Only a bare outline of what seems most important in the context of transgenes is given and the interested reader is referred to the many excellent texts that cover this area and to review articles cited here.

RNA synthesis and processing

The transcription and processing of RNA is conventionally divided into three fairly discrete subtopics, namely (i) synthesis or transcription, (ii) processing and transport and (iii) decay. The amount of each RNA species in a given cell at a given time is determined by the concerted action of these processes. Transcription, processing and transport fix the rate at which new RNA molecules arrive in the cytoplasm, while decay is the rate at which molecules are removed.

Eukaryotes deploy three different systems to transcribe three categories of genes and process the products. Three RNA polymerases sit at the heart of these systems: Pol I transcribes ribosomal RNA genes, Pol III transcribes the genes that specify tRNA and other small stable RNAs, and Pol II transcribes the genes that specify proteins. Only the last of these will be considered here.

The rates of the various processes are individually determined for each gene by the interplay of three sets of influences:

- Signals encoded in DNA within and around the gene and sometimes also at more or less remote chromosomal locations. These interact either as DNA or as RNA sequences with nuclear or cytoplasmic proteins. The signals are features over which we have a large measure of control when constructing transgenes. However, the extent to which we can exploit this opportunity is limited by our incomplete understanding of regulatory processes. Signals that relate to the synthesis and processing of mRNA are shown in Figure 10.1.

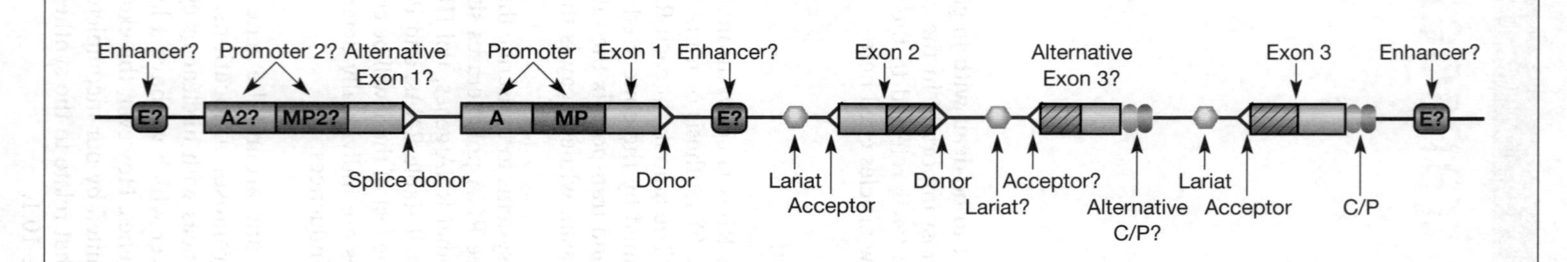

Figure 10.1 Diagram of a gene to illustrate locations of signal elements. For convenience the gene is shown as having only three exons, although most genes have more. The region transcribed to give pre-mRNA is shown in blue, except for regulatory elements and the cleavage and polyadenylation site (C/P) that lie within it; exons are shown as boxes with coding regions hatched and non-coding regions plain blue; introns are shown as lines. Promoters and enhancers are shown as grey shaded boxes. Donor and acceptor splice sites are shown as triangles and the branch (lariat attachment) sites as hexagons. Elements that may or may not be present are shown with question marks. Thus, an enhancer may be beyond the 5' or 3' end of the gene or in an intron, or there may be no known enhancer. Some genes have two promoters, usually utilised in different tissues or at different developmental stages, but most have only one. Many genes have two alternative 3' terminal exons, as shown here, and some have several.

- Epigenetic phenomena modulate the effects of these signals in ways that are inherited by daughter cells. Some but not all instances of DNA methylation, the positioning of nucleosomes and histone acetylation are put in place during the development of the cell lineage, or in the case of imprinting during gametogenesis (Chapter 6). These also depend crucially upon DNA sequence elements in and around the gene in the first instance and, given enough understanding of cellular processes during development, ultimately these should also be controllable.
- Large numbers of regulatory proteins present in the cell interact with DNA sequence signals and signals copied into RNA transcripts. Different cell types deploy different repertoires of active regulatory proteins. The repertoire changes during the diversification and development of cell lineages, and within a given cell it changes during the cell cycle, in response to external stimuli and with time. The same signal can have different effects in conjunction with different regulatory proteins, and in some cells will have none.

Transcription

Although every step in gene expression is subject to regulation it is self-evident that the role of transcription is fundamental. Without transcription there can be no expression, and in the main the subsequent steps in expression, if they act at all, do so to modulate the primary effect of transcriptional control.

Chromatin

It is worth recalling that DNA is intimately associated with the histones, which play a role that is primarily structural. A nucleosome contains two of each of the four core histones, H2A and H2B, which are paired, and H3 and H4, which are also paired. A positively charged N-terminal domain of each histone is exposed. These domains make contact with the DNA, 146 base pairs of DNA being wound around each nucleosome while another 45 base pairs on average extend between successive nucleosomes. Histone H1 links nucleosomes together and participates in chromatin condensation and decondensation.

The condensation of chromatin is thought to be enlisted as an aid to minimising the expression of genes in inappropriate situations (leakiness). If so, it will relate particularly to genes, and there are very many of these, that are required to be active only at a particular time or in a particular cell type.

DNA that is being actively transcribed is more accessible to DNase attack than DNA that is not; this is called DNase sensitivity and it relates to a change in the interaction between DNA and nucleosomes, a decondensation of the chromatin structure. Activated transcriptional regulatory sites, such as enhancers, are frequently even more accessible to DNase, to the extent that cleavages can be produced in fairly localised positions in a preparation of purified nuclei before general DNA degradation proceeds very far. The sites at which this occurs are called DNase hypersensitive sites

(HS; see Figure 10.6). The presence of hypersensitive sites correlates with gene expression, but the correlation is far from perfect. There are constitutive HS, others that arise in a cell lineage in advance of transcription and still others that may be associated with transcriptional repression. However, an HS that arises at about the time of transcriptional activation tends to identify a site of transcriptional regulation.

With the onset of transcription, different changes in chromatin structure occur at the promoter and in the downstream transcribed region. At the promoter there is a net displacement of nucleosomes that comes about through the concerted action of several factors, not all of which necessarily operate at every promoter (Wolffe 1996; Kornberg and Lorch 1995):

- There is an element of sequence preference in nucleosome binding.
- Lysine acetylation and serine phosphorylation reduce the positive charge of the N-terminal histone domains and weaken their interaction with DNA. Acetylation in particular correlates with transcriptional activation.
- *Trans*-acting transcription factors affect histone–DNA interactions; some gain access through the internucleosomal linker regions, others can access nucleosome-bound DNA.
- Specialised multiprotein complexes (remodelling machines), with an affinity for transcription factors and with histone acetyl transferase activity, actively remodel chromatin with the consumption of ATP (Cairns 1998).

The pendulum of opinion has swung about in respect of whether nucleosomes are displaced from the DNA or remain in place as RNA polymerase passes through. Most recently it appears to have settled in mid-position. It seems that the DNA is partially released from the nucleosome as the polymerase passes over it, but resumes contact at the leading end before the trailing end is released so that the relationship between DNA and nucleosome remains essentially the same once the polymerase has gone on downstream.

Minimal promoters

The typical arrangement of signal elements in genes is illustrated in Figure 10.1. A promoter typically has two functionally distinct parts, one sometimes called a minimal promoter (MP) and a second upstream response element or set of response elements (A). The function of the MP is to mark the point in the DNA duplex at which transcription should start so that Pol II can be directed to it. Most vertebrate MP sites contain one of two alternative consensus sequences that perform this function, a few contain both and quite a few contain neither (Novina and Roy 1996). When present, the TATA box (consensus TATA(A/T)A(A/T)) lies about 30 nucleotides upstream of the point at which transcription starts; the Inr sequence (consensus $Y_2AN(T/A)Y_2$), when present, surrounds the site of transcriptional initiation, the embedded adenosine. These short consensus sequences occur much too frequently by chance in vertebrate DNA to mark the initiation sites on their own. This is where the A site comes into the picture, as we shall see shortly. A very large multiprotein complex called the core promoter complex (CPC), of the order of 1 MDa in size, binds to the MP site (Roeder 1996). This consists of a number of components that have been physically purified, carrying acronyms beginning with TFII (transcription factor relating to Pol II), each of which is itself a multiprotein complex. The complex can be assembled on an MP site *in vitro* by sequential addition of purified components (Hoffmann *et al.* 1997). Whether assembly follows this path *in vivo* or whether a preformed core promoter complex associates with the MP site is not yet clear. The linchpin of the CPC is the factor TFIID which contacts the MP site *via* the TATA binding protein (TBP) or an Inr binding protein. This association is modulated in various ways: by the conformation of the MP site, by phosphorylation and dephosphorylation, by the presence or absence of components that repress or activate the complex, and so on. However the most significant modulation is due to transcriptional regulatory proteins that bind to the A site and are considered to be distinct from the CPC.

Activation sites

The A site is often very complex, containing response elements to which a number of different regulatory proteins bind. These act in a concerted way involving protein–DNA and protein–protein interactions to present an acidic activation surface to the CPC, or not as the case may be (Verrijzer and Tjian 1996; Novina and Roy 1996; Hoffmann *et al.* 1997). Interaction between an acidic activation surface and the CPC stabilises the binding of both the CPC and the regulatory proteins to the DNA and activates the CPC transcriptionally. The total number of these regulatory proteins encoded in mammalian DNA is very large. Some are expressed ubiquitously, like the components of the CPC. Others are expressed in a subset of cell types or even, in some cases, in a single cell type. A typical A site is composed of ubiquitous elements or a combination of ubiquitous and tissue-specific elements. Each of the corresponding tissue-specific regulatory proteins may be present in a different subset of cell types. Permutations of response elements at activation sites and of cognate regulatory proteins in different cell types allow for an almost limitless number of different patterns of gene expression.

The regulatory proteins are the principal determinants of tissue-specific gene expression, and also the ultimate agents of stimuli external to the cell that elicit changes in transcription. Many regulatory proteins stimulate transcription. These often exist in two forms, an active form and an inactive form which is activated by chemical modification in response to specific stimuli; others are sequestered in the cytoplasm until released by a specific stimulus; yet others are synthesised *de novo* under the control of other regulatory proteins that respond more directly to a given stimulus.

Proteins that repress transcription also bind to A sites (Cowell 1994). A repressor can simply be deficient in activation and compete with a stimulatory factor for the same binding site, or it can be the inactive form of a stimulatory factor or can actively repress the very low basal level of transcription due to the CPC alone. The active form of many stimulatory factors is a homo- or heterodimer; another class of repressor binds and inactivates such factors in inactive heterodimers. Yet another binds to and blocks the DNA-binding site of the stimulatory factor.

An interesting class of genes, many of which are 'housekeeping genes' responsible for processes carried out by all cells, have neither a TATA box nor an Inf site. Characteristically, transcription of these genes is initiated at a cluster of sites rather than a single site. Most of these genes appear to possess a response element that lies downstream of the initiation site cluster (Ince and Scotto 1995). Details of transcriptional initiation at these promoters are not yet available, though TFIID is known to be involved.

An activated CPC interacts with another gigantic multiprotein complex, the Pol II holoenzyme, which contains factors required for transcriptional initiation, others required for elongation of the transcript and yet others, like Pol II itself, required for both processes. This combined complex can be called the preinitiation complex. Yet another holoenzyme complex is probably involved at this stage in disrupting the nucleosomal organisation of the chromatin in the vicinity of the initiation site (Halle and Meisterernst 1996). Melting of the DNA duplex around the initiation point is effected by the helicase activity of TFIIH, considered to be part of the CPC, and separately involved in the repair of mismatched base pairs in DNA by the excision-repair pathway (Svejstrup *et al.* 1996). TFIIH also has Ser/Thr protein kinase activity which phosphorylates the C-terminal domain of Pol II, dissociating components required for transcriptional initiation from the elongation complex (Koleske and Young 1995; Roeder 1996).

The elongation complex proceeds to transcribe the gene while the initiation-specific factors are available to be recycled. Several additional proteins thought to be involved in elongation assist the polymerase holoenzyme to negotiate obstacles to its progress, such as bound proteins, and extricate it from pauses, some of which may be of its own making; an RNA polymerase stuck in mid-transcription could clearly present the cell with a serious problem.

Enhancers

Enhancers are similar to activator sites in that each contains a cluster of response elements to which a number of ubiquitous and tissue-specific transcriptional regulatory proteins bind. Many of the proteins that bind to enhancers also bind to activators. However, enhancers do differ from activators:

- They do not directly activate CPCs; rather they increase the impact of the activator on its cognate preinitiaton complex.
- While the position of an activator relative to the CPC is relatively inflexible, enhancers are found in different more or less remote locations, upstream, downstream and within introns.
- According to the classical definition, enhancers can be inverted or moved to a different position without abolishing their activity, although it may be modified somewhat. However, there are several examples of sequences within introns which, while otherwise sharing properties with enhancers, are active only in one 5'→3' orientation (see below).

- Enhancers stimulate transcription from promoters in *cis* configuration both during the extrachromosomal transient expression phase in transfected cultured cells and also when stably integrated into the chromosome (see Box 10.1).

BOX 10.1: TRANSIENT EXPRESSION AND STABLE INTEGRATION

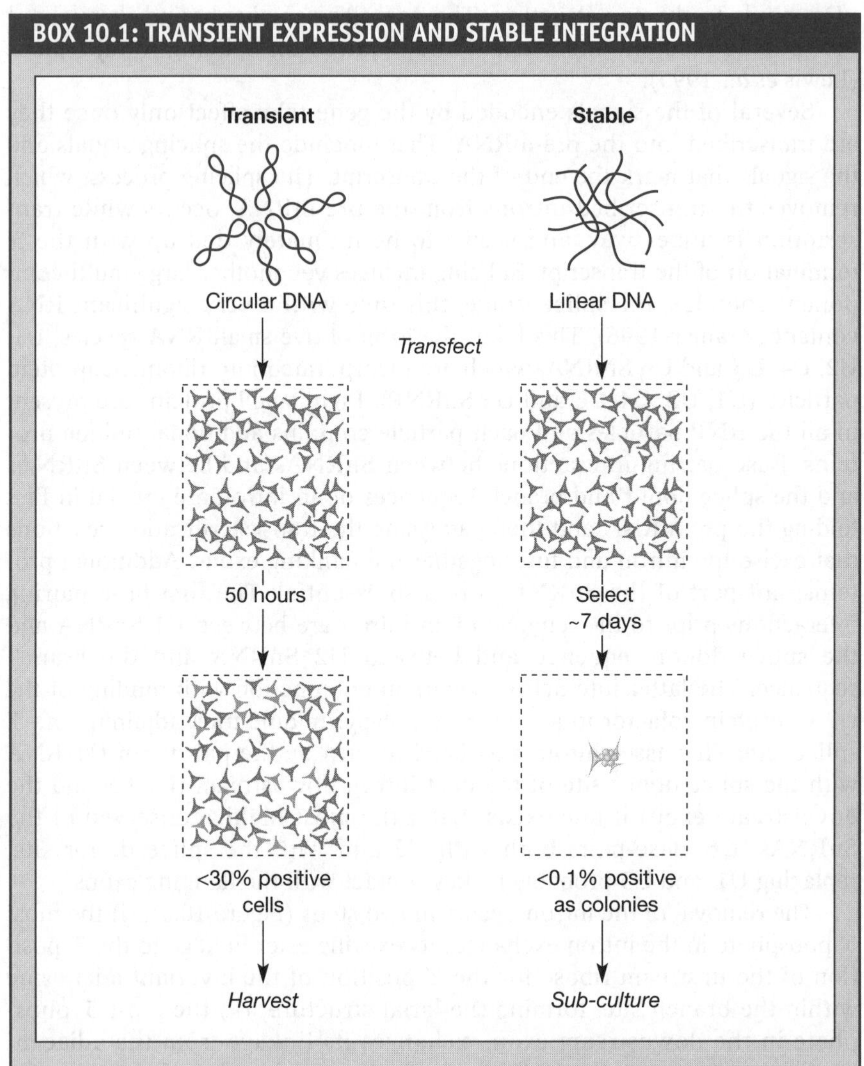

Transient expression experiments are initiated by transfecting cultured cells with covalently closed circular plasmid DNA. Between 40 and 60 hours later the expression of a competent gene carried by the plasmid is at a maximum. At this time the cells can be stained or harvested for the preparation of RNA, protein etc.

To produce stable transfectants, cultured cells are treated with linear DNA carrying a selective reporter and after allowing some time for expression selection is applied. Stably transfected cells develop into colonies while the others die off. The colonies can be subcultured and propagated indefinitely.

Shortly after its initiation, a cap structure (m^7G-5'-p-p-p-5'-N) is added to the 5' nucleotide (N) of a pre-mRNA transcript. This marks the molecule as a pre-mRNA. The cap-binding complex (CBC), two proteins of 80 and 20 kDa, binds to the cap structure. The complex remains bound throughout the splicing reaction and seems to be required for splicing to take place (Lewis *et al.* 1995).

Several of the signals encoded by the gene take effect only once they are transcribed into the pre-mRNA. These include the splicing signals and the signals that mark the end of the transcript. The splicing process, which removes the transcribed introns from the pre-mRNA, occurs while transcription is under way, and seems to be intimately tied up with the 3' termination of the transcript. Splicing involves yet another large multi-component complex, the spliceosome, this time with a very significant RNA content (Kramer 1996). This takes the form of five small RNA species, U1, U2, U4, U5 and U6 SnRNA, which are incorporated into ribonucleoprotein particles (U1, U2, U4/U6 and U5 SnRNP). Eight small proteins are present in all the RNP particles and each particle contains additional unique proteins. Base-pairing interactions between SnRNAs and between SnRNAs and the splice donor and branch sequences of an intron are crucial in first folding the pre-mRNA and then catalysing the transesterification reactions that excise the intron and fuse together the flanking exons. Additional proteins, not part of the SnRNPs, are also essential. The first base-pairing interactions prior to the removal of an intron are between U1 SnRNA and the splice donor sequence and between U2 SnRNA and the branch sequence. The latter interaction occurs in conjunction with binding of the U2AF protein cofactor to a conserved polypyrimidine tract adjoining the 3' splice site. This association is assisted in turn by the pairing of U1sRNA with the splice donor site of the next intron downstream (i.e. beyond the downstream exon) if one exists. Later the most highly conserved of the SnRNAs, U6, base-pairs both with U2 and with the splice donor site, replacing U1, and U5 probably makes contact with the flanking exons.

The removal of the intron occurs in two steps (Figure 10.2): (i) the most 5' phosphate in the intron exchanges its existing ester linkage to the 3' position of the upstream ribose for the 2' position of the invariant adenosine within the branch site, forming the lariat structure; (ii) the most 5' phosphate in the downstream exon exchanges its linkage from the adjacent intronic guanosine to the OH freed up by the previous transfer, linking the upstream and downstream exons together. The freed intron is released and subsequently degraded.

The 3' end of the pre-mRNA, which remains the 3' end of the molecule following splicing, is determined by a nucleolytic cleavage. This is signalled by two sequences, AAUAAA or AAUUAAA upstream of the cleavage site and the downstream sequence element (DSE) (Lewis *et al.* 1995). Cleavage is brought about by a small cluster of proteins focused on the cleavage and polyadenylation specificity factor (CPSF) which binds to the AAUAAA ele-

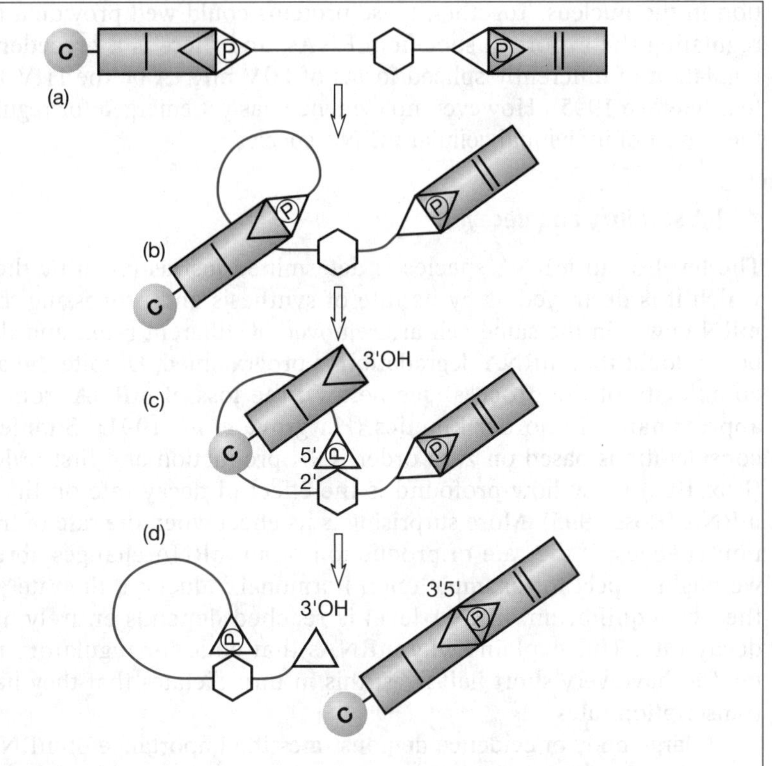

Figure 10.2 Simple geometry of pre-mRNA splicing. Exons are shown as blue rectangles, introns as single lines, splice sites as paired triangles, the branch site by a hexagon, the cap by a circled letter C and phosphate groups that participate in transesterification reactions by a circled P. (a) Pre-mRNA; (b) juxtaposition of the splice sites and the branch site; (c) first transesterification, between the 5' end of the intron and the branch site; (d) second transesterification, between the exons, releasing the lariat structure.

ment. The cleavage specificity factor (CStF) binds to the DSE and the CPSF and this stabilises the overall complex. Endonucleolytic cleavage is brought about by a cleavage factor and the addition of polyadenylic acid (poly-A) to the exposed 3' hydroxyl group is initiated by poly-A polymerase. Around 250 adenylic acid residues are added to complete the nuclear version of the mRNA. There is a link between the splicing out of the 3'-terminal intron and polyadenylation, mediated at least in part by U1snRNP (Lewis *et al.* 1995; Niwa and Berget 1991).

RNA is exported from the nucleus *via* the large nuclear pore complexes. The selection of an RNA molecule for export and its active transport through the pore are almost certainly mediated by proteins bound to it in a process closely akin to the process of protein export and employing similar peptide signal sequences (Izaurralde and Mattaj 1995). Both the 5' cap structure and the 3' poly-A tail stimulate export, presumably through associated proteins. While some of the proteins bound to RNA in the nucleus carry signals that promote export, others carry signals that promote retention in the nucleus. Together, these proteins could well provide a basis for regulating the export of specific mRNAs, and there is a precedent in the regulation of differently spliced forms of HIV mRNA by the HIV Rev protein (Gerace 1995). However, no evidence has yet emerged for regulation of the export of individual cellular mRNA species.

mRNA stability and decay

The level of an mRNA species is determined just as much by the rate at which it is destroyed as by its rate of synthesis and processing. Different mRNAs within the same cell are removed at different rates, and there can be no doubt that mRNA degradation is programmed. Despite the apparent complexity of the process (see below), the loss of mRNA from the cell approximates first order kinetics (Hargrove *et al.* 1991). Simple kinetic considerations based on zero order RNA production and first order decay (Box 10.2) show how profound is the effect of decay rate on the level of mRNA (Ross 1995). More surprising is its effect when the rate of transcription changes. If the rate of production of an mRNA changes abruptly, as we might expect for example upon hormonal induction, the rate at which the new equilibrium mRNA level is reached depends entirely upon the decay rate. This explains why mRNAs that code for regulatory proteins tend to have very short half-lives; this in turn dictates that they have high transcription rates.

A large body of evidence demonstrates the importance of mRNA decay rates in gene expression:

- The half-lives of individual mRNA species are known to differ over a wide range, testifying to the fact that mRNA decay is programmed. In a number of cases RNA sequence elements profoundly influence the stability of an mRNA and *trans*-acting proteins have been shown to interact with some of them.
- Changes in half-lives are known to accompany differentiation, some half-lives increasing and others decreasing within the same cell population. This indicates that at least some half-lives are regulated.
- A similar conclusion may be drawn from many observations of changes in half-life following hormonal stimulation or alterations in growth conditions. In a few cases the involvement of *trans*-acting proteins in such changes has been demonstrated.

Translation destabilises mRNA, or at any rate mRNA is stabilised when its translation is inhibited. The 5' cap structure and the 3' poly-A tail both promote translation. Paradoxically, when decapped or deadenylated, mRNA is destabilised; decapping and deadenylation probably allow nuclease access to the mRNA with an effect that outweighs the translation effect. Shortening of poly-A tails is progressive and is probably slowed by a specific binding protein, PABP. The 5' non-coding region can have a profound effect on the frequency with which an mRNA is translated, and different configurations in this region may contribute to programmed mRNA decay. The 3' non-coding region of many mRNAs contains elements based on the motif AUUUA, called AUREs (AU response elements). In general these have a destabilising effect which is ameliorated by the binding of members of a family of AU binding proteins.

Where they are known, elements that participate in the regulation of mRNA half-life are located in the 3' non-coding region or in the coding region. One of the clearest examples involves the 3' non-coding region of transferrin mRNA (Izaurralde and Mattaj 1995). This contains five stem-loop structures (iron response elements) to which an iron response protein (IRP) binds. Iron is bound by the IRP, destabilising its association with the response elements. This in turn decreases the half-life of the mRNA about 25-fold. As a result less transferrin will be synthesised and less iron will be delivered to cells. Elements that affect stability have been identified within the coding regions of *c-fos* and *c-myc* mRNAs, among others (Ross 1996; Kamakaka 1997).

Although a few more examples could be given, study of the regulation of mRNA stability is still in its infancy. Considering the impact that it can have on gene expression (Box 10.2), the subject probably merits more attention. It could assist, for example, in the production of transgenic bio-reactors with high expression levels (Chapter 11) and should also be considered when transgenes are being tailored to respond rapidly to an external stimulus (see below).

BOX 10.2: KINETICS OF RNA DECAY

The equations below are based on simple assumptions that the rate constant of RNA synthesis is zero order and that of RNA decay is first order.

$t_{\frac{1}{2}} = \ln 2/k_D$	$t_{\frac{1}{2}}$ is the RNA half-life; k_D is the rate constant of RNA decay
At equilibrium $k_S = k_D \cdot R_{eq}$ and $R_{eq} = k_S/k_D$	k_S is the zero order rate constant of RNA synthesis; R_{eq} is the steady-state level of RNA
$R_{eq}' = k_S/k_D'$ and $R_{eq}'/R_{eq} = k_D/k_D' = t_{\frac{1}{2}}'/t_{\frac{1}{2}}$	If k_D changes while k_S remains the same, the new value of R_{eq} (R_{eq}') is inversely proportional to k_D', the new value of k_D
$(R_t - R_0)/(R_{eq} - R_0) = 1 - e^{-k_D t}$	Describes the fact that when the transcription rate changes the *rate of approach* to the new equilibrium value R_{eq} is determined by the rate constant of RNA decay, k_D.

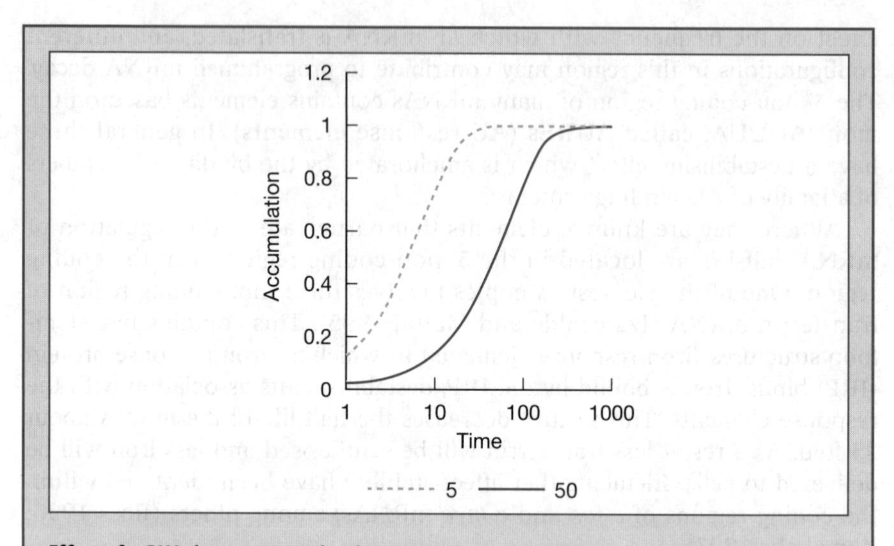

Effect of mRNA decay rate on the time to reach a new mRNA level when the rate of transcription increases. The initial mRNA level is arbitrarily set to 0 and the new level to 1. The time scale is also arbitrary. A difference in $t_{\frac{1}{2}}$ of 10-fold (5 or 50) alters the time to reach the new equilibrium 10-fold.

Protein synthesis

The important determinants in translation are the mRNA cap structure, the 5' untranslated region, the initiation codon and the sequence surrounding it, and the termination codon. The scanning model of eukaryote translation states that 40S ribosomal subunits attach to the 5' untranslated region of the mRNA at or near the cap site. The 40S ribosomal subunits are first withdrawn from association with 60S subunits and then activated by association with the initiator methionyl tRNA, with the participation of initiation factors (Figure 10.3). Attachment of the mRNA is mediated by a three-component complex of factors called eIF-4F (eukaryotic initiation factor 4F), which possesses affinity for the cap structure and RNA helicase activity. Its affinity for the cap is due to a phosphoprotein called cap binding protein (or eIF-4E). Phosphorylation of eIF-4E stimulates protein synthesis and conversely dephosphorylation inhibits it. Attached 40S ribosomal subunits move in a 3' direction, scanning for a methionine initiation codon, AUG. Having found one, a 40S subunit becomes associated with a 60S subunit to form a ribosome. Assisted by three elongation factors, the ribosome proceeds to assemble the polypeptide chain by sequential addition of amino acids. Each amino acid enters the reaction covalently linked to a cognate tRNA so that the growing polypeptide chain is serially transferred to these tRNAs, one after the other, while the mRNA passes synchronously, codon by codon, through the active site. Synthesis is terminated when any one of the three termina-

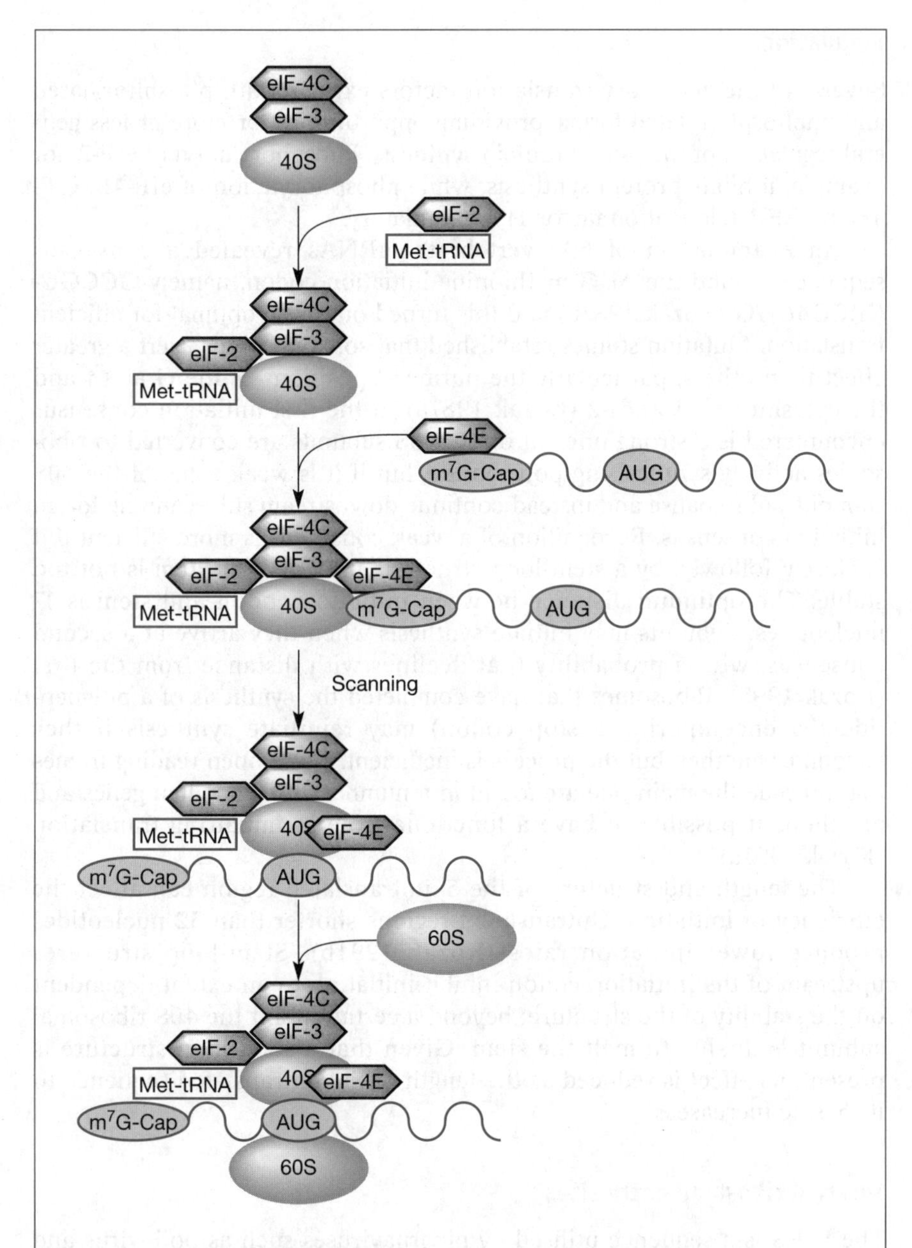

Figure 10.3 Initiation of protein synthesis. Blue ovals represent the 40S and 60S ribosomal subunits. Methionyl tRNA is shown as a blue box, and mRNA as a blue wavy line with the cap structure and the initiation codon marked. A selected few initiation factors are shown as black hexagons. After Merrick (1996).

tion codons UAA, UAG or UGA is encountered, at which point another factor, the release factor, severs the linkage between the last amino acid in the polypeptide chain and its cognate tRNA.

Several of the accessory translation factors exist in both phosphorylated and unphosphorylated forms, providing opportunities for more or less general regulation of the rate of protein synthesis. Phosphorylation of eIF-2, for example, inhibits protein synthesis, while phosphorylation of eIF-4B, eIF-4F and EF-1 (elongation factor 1) is stimulatory.

An examination of 699 vertebrate mRNAs revealed a consensus sequence around the AUG methionine initiation codon, namely GCCGCCRCC*AUG*G (Kozak 1987b) and this turned out to be optimal for efficient translation. Mutation studies established that some positions exert a greater effect than others, particularly the purine at position –3, the G at +4 and the cytosines at -1 and -2 (Kozak 1987a). If the first initiation consensus encountered is a strong one, all of the 40S subunits are converted to ribosomes actively synthesising polypeptide, but if it is weak some of the 40S subunits fail to pause and instead continue downstream still scanning for an initiation consensus. Recognition of a weak consensus is more efficient if it is closely followed by a stem-loop structure, provided the stem is not too stable. The optimum distance between initiation codon and stem is 17 nucleotides. Subunits may initiate synthesis when they arrive at a second consensus, with a probability that declines with distance from the first (Kozak 1995). Ribosomes that have completed the synthesis of a polypeptide (by encountering a stop codon) may reinitiate synthesis if they encounter another, but the process is inefficient. Short open reading frames that precede the main one are found in a number of mammalian genes and are thought possibly to have a functional role in impairing translation (Kozak 1991a).

The length and structure of the 5' untranslated region can affect the efficiency of initiation. Untranslated regions shorter than 32 nucleotides produce lower initiation rates (Kozak 1991b). Stem-loop structures upstream of the initiation codon inhibit initiation to an extent dependent on the stability of the structure; beyond a certain point the 40S ribosomal subunit is unable to melt the stem. Given that a stem-loop structure is present, its effect is reduced as the length of the untranslated sequence to its 5' side increases.

Internal ribosome entry sites

The IRES is a sequence utilised by picornaviruses such as poliovirus and encephalomyocarditis virus to allow the 'entry' of ribosomes at an internal position in an RNA sequence and the initiation of translation from a downstream AUG codon (Schmid and Wimmer 1994). This sequence can be incorporated into constructs, allowing bicistronic mRNAs to function efficiently in animal cells (Alexander *et al.* 1994; Mountford and Smith 1995; Mountford *et al.* 1994). This device removes the constraints on reinitiation of translation described above, allowing much more flexibility and therefore ease in the design of transgenes in some circumstances.

Measuring transgene expression

In measuring transgene expression it is important to appreciate that different estimates of expression may not agree and to decide what we want to know. The amount of each mRNA in the cell is determined by transcription and processing opposed by mRNA degradation; the amount of the corresponding protein is determined by the mRNA level and the rate of its translation opposed by protein degradation. All of these processes are known to be regulated both generally and in ways specific to particular genes and their corresponding pre-mRNA, mRNA and protein products. Since we seldom if ever have complete information about all these variables, it is usually unsafe to employ one measurement as an estimate of another. The case of a transgene that is expressed in the same tissue in different transgenic lines is a probable exception to this rule. Unless some crucial component is limiting, we can reasonably expect mRNA and protein stability and the rate of translation to be the same in the different lines. Relative transcription rates will then coincide with relative mRNA levels and levels of protein. At the opposite extreme, when the same gene is expressed in different tissues we cannot expect these measurements to agree.

Transcription rates cannot be measured with any confidence. The best method available is nuclear run-on analysis (Box 10.3). Its weakness is that observed transcription rates vary depending on the precise method by which the nuclear preparations are made. Thus, given tight control of the method, it could be expected to yield useful results in a comparison of the rate of transcription of the same gene in the same tissue in different transgenic lines, but other comparisons are more problematical. mRNA decay is also very difficult to measure with confidence, particularly *in vivo*. It requires an abrupt and substantial change in the rate of RNA production; either downregulation, after which the falling mRNA level is measured, or upregulation, taking advantage of the kinetic considerations outlined in Box 10.2. However, general inhibitors of RNA or mRNA synthesis such as actinomycin D and α-amanitin rapidly become general toxins as short-lived proteins and RNAs decay. Specific mRNAs can be studied if they are transcribed from a promoter that can be activated and deactivated by means of externally applied stimuli, provided of course that the stimuli do not have pleiotropic effects. Recent developments in the control of transgene expression (see below) are promising in this respect.

BOX 10.3: MEASURING TRANSCRIPTION BY NUCLEAR RUN-ON (Derman *et al.* 1981; Hofer *et al.* 1982)

Nuclei are isolated as free as possible from cytoplasmic tags and incubated with a labelled RNA precursor, for example α-[^{32}P]-CTP. The rate of incorporation into RNA decays rapidly and frequently ceases within 30 minutes. During this time the Pol II complexes extend the nascent incomplete pre-mRNA molecules by a few hundred nucleotides each (see Figure). The assumption is that all nascent RNA molecules will be extended to roughly the same extent. If so, the amount of RNA newly synthesised per unit length of DNA will be linearly proportional to the loading of the template (Pol II complexes per unit length).

The RNA is recovered from the nuclei and hybridised with cloned DNA attached to a small strip of nylon membrane. Several DNA clones derived from the same or different genes may be attached at different places on the same membrane. The cloned DNA should be in vast excess relative to the complementary RNA (radioactive and non-radioactive) in the preparation. Under these circumstances most of the complementary RNA will duplex with the membrane-bound DNA. In most cases more than 99% of the total radioactive RNA will not hybridise to the membrane bound DNA and must be washed away. The radioactivity hybridised with each bound DNA can then be measured. A correction must be made for the lengths of the DNA clones. If the assumptions are correct, the resulting figures will accurately reflect Pol II loading of the DNA.

The transcription rates of chosen genes can be estimated from their level of RNA polymerase 'loading' by limited extension of the nascent pre-mRNA chains *in vitro*. The DNA duplex is shown as a grey line, Pol II holoenzymes as oval beads, RNA molecules as black and blue lines; the direction of transcription is from left to right. I: a heavily loaded gene; II: a lightly loaded gene. (a) Before *in vitro* synthesis, showing Pol II arranged along the transcription unit with nascent RNA attached. The further the Pol II molecules have travelled the longer is the attached RNA. (b) During *in vitro* RNA synthesis each polymerase molecule translocates further along the duplex adding a small amount of radioactive RNA (blue) to each nascent chain. (c) The RNA molecules arranged to show that different parts of the duplex are equally represented in the newly transcribed RNA.

In contrast, mRNA levels can be measured quite reliably by a variety of methods. The two-fold problem is of course to pick out the mRNA species of interest from the entire cellular mRNA population and to do so quantitatively. Northern blotting and DNase and RNase protection methods are based on nucleic acid hybridisation. Primer extension assays involve the extension of a specific primer hybridised to the mRNA by retroviral reverse transcriptase. Given suitable controls, for instance RNA synthesised *in vitro* on a cloned DNA template, absolute measurements can be made by these methods. Reliable comparative measurements of abundant mRNAs have been made by translating the entire mRNA population using an mRNA-dependent *in vitro* translation system in the presence of radioactive amino acids; the amount of radioactivity incorporated into a specific protein is then measured after immunoprecipitation and/or gel electrophoresis.

The rate of synthesis of a protein product is normally not an issue in transgenic work, protein levels being more immediately important. Protein levels are most easily and reliably measured if the protein is an enzyme. Otherwise, if a suitable specific antibody is available, a protein can be quantified by standard immunological techniques such as ELISA, RIA or Western blotting.

The distribution of specific mRNAs within a tissue or a population of cells can be determined by *in situ* hybridisation and of proteins by enzymatic and antibody based methods. A wide variety of methods is available to make the resulting complexes visible.

Genetic background

By genetic background effect we mean the combined effect of all the other genes in the genome on the phenotypic expression of a gene that is under scrutiny. To detect such an effect we would have to study the gene against at least two backgrounds, and find a difference in its phenotypic expression. The effects of background on normal genes are well known, which is why in the mouse newly discovered allelic variants are regularly introduced into congenic strains by backcrossing. For general purposes the most widely accepted standard is the C57BL/6 strain.

Like normal genes, transgenes sometimes show different phenotypic effects in different genetic backgrounds (Threadgill *et al.* 1995; Bonyadi *et al.* 1997; Fawcett *et al.* 1995; Weng *et al.* 1995; Ishihara *et al.* 1995). Some inbred strains of mice are more closely related than others. We can take an average difference beween strains of 0.1% of base pairs, which corresponds to about two differences per mRNA species. This does not seem much, but consider that the mouse genome probably contains about 100 000 genes. If even 1 in 100 of these differences significantly affects the function of the gene in which it occurs, there will be on average 2000 genes that differ significantly between two inbred strains.

Variation in transgene expression due to genetic background can be expected to occur in many of the transgenic lines generated by either of the two methods most commonly employed. However, the problem has its origin in the history of the methods and can be avoided with relative ease.

Microinjection route

As a general rule, outbred mice and the offspring of F_1 crosses between inbred strains both perform better in reproduction than inbred mice. In consequence it is common practice to employ F_1 dams and sires to generate 1-cell embryos for microinjection. These are therefore F_2 embryos, i.e. as diverse a group of animals as it is possible to obtain from two inbred lines.

Transgenic lines can be derived from these genotypically F_2 founders in different ways. It is not desirable to maintain the line by full-sib mating, because this will generate a recombinant inbred line. A better solution would be to cross each generation to F_1 animals (obtained by crossing the same lines used to generate the parents of the 1-cell embryos). At each generation this will produce essentially the same spectrum of genotypes as is present in the F_2 generation. Although the transgenic phenotype may exhibit a high variance under these circumstances, requiring any measurement to be averaged over several animals, at least the mean and variance should remain constant from generation to generation since the range of genetic backgrounds should remain the same.

ES cell route

The most easily derived and maintained, and best performing ES cells are those obtained from strain 129 mice, and as a consequence these are by far the most widely employed. Unfortunately, strain 129 mice are poor breeders and are also poorly characterised relative to many other mouse strains. The temptation is therefore strong to cross the initial chimeras with a better-breeding mouse line, such as C57BL/6. This, of course, produces uniformly heterozygous individuals in the first instance. In many ways these F_1 animals are suitable for study, since they have the same uniform genotype and should exhibit low phenotypic variance due to 'hybrid vigour'. It is when these animals are bred, as they must be sooner or later, that genetic heterogeneity is introduced.

Backcrossing

Whichever method of generating transgenic mice is employed, the most efficient approach would probably be to employ an inbred strain such as C57BL/6 in the first instance. There are initial disadvantages in each case. C57BL/6 respond less uniformly to attempts to induce superovulation than C57BL×CBA F_1 animals, and although ES cells can be derived from them they are apparently less easily handled and maintained than the usual strain 129 ES cells. However, these disadvantages are almost certainly outweighed by the advantage of having the transgene ensconced from the outset in a very well defined and robust inbred mouse strain. Given that either of the less desirable methods outlined above has been adopted, one way forward is to transfer the transgene into the C57BL/6 background (for example) by repeated backcrossing. This is satisfactory as far as genes unlinked to the transgene are concerned. As shown in Figure 10.4, a plausibly congenic strain is produced with five generations of backcrossing (in about one year) and a convincing congenic strain in ten (two years). However, genes on the same chromosome as the transgene present a problem, because for the transgene to be transferred from the 129 chromosome (say) to a C57BL/6 chromosome requires recombination events to occur to both sides of it, and furthermore both will have to occur very close to the transgene if a significant amount of 129 chromosome is not to continue to segregate with it.

Figure 10.5 shows the amount of chromosome, in map units (cM), that will on average remain flanking the transgene after different numbers of backcrosses, and the percentage of the entire mouse genome that this represents. The 0.1% level reached after 100 generations (20 years!) represents on average two significant genetic polymorphisms (as assigned arbitrarily above). Of course where a transgene is introduced into C57BL/6×CBA F_1 embryos there is a 50% chance that it will be inserted into a C57BL chromosome and if so the question of nearby polymorphisms will not arise. However, mapping work would be required to identify the chromosome carrying the transgene.

In many cases whatever residue of 129 or CBA chromosome remains surrounding the transgene will have no effect on its expression and no relevant effect on the physiology of the mouse. However, it will generally not be known whether or not this is the case.

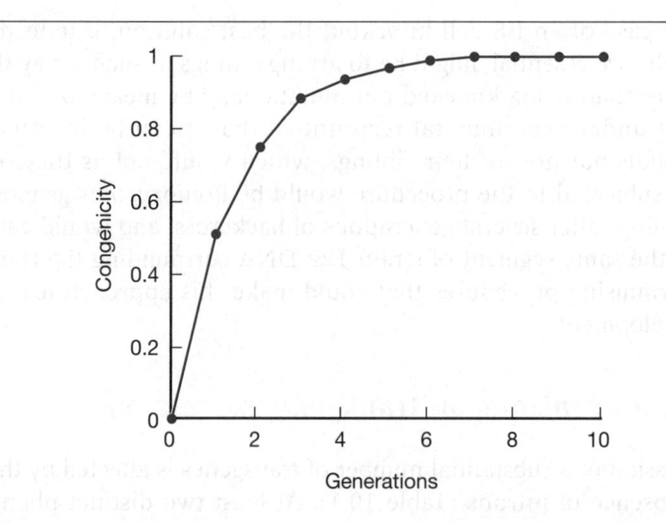

Figure 10.4 Introducing a transgene into a new genetic background by repeated backcrossing to a reference strain. The theoretical proportion of the genome that is homozygous for reference strain genes is shown as 'congenicity' against generations of backcrossing. Five generations of backcrosses occupy about 1 year in the mouse. This figure shows progress only insofar as genes not linked to the transgene are concerned.

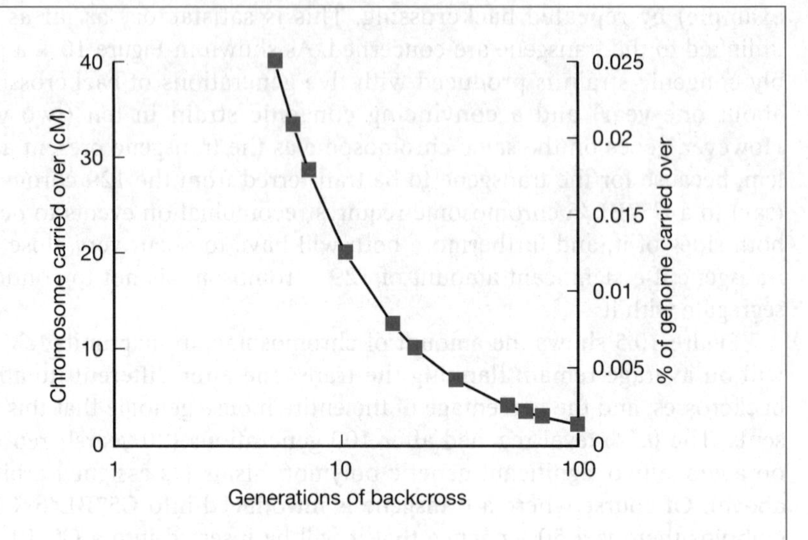

Figure 10.5 Showing that genes linked to a transgene are introduced only slowly into a reference genetic background by backcrossing. The left ordinate shows how many map units of the chromosome on which the transgene is located will on average remain associated with it, and the right ordinate what percentage of the total genome this represents. After 10 generations of backcross about 1.5% of the genome, representing significant differences in perhaps 30 genes, will remain heterozygous, on average. Even after backcrossing for 20 years some residual heterogeneity, which might or might not matter, remains.

In the case of an ES cell knockout the best solution, if indeed strain 129 ES cells are essential, might be to arrange things in such a way that the gene, rather than being knocked out mutationally by means of a deletion, is brought under experimental restraint so that it can be inactivated in some animals but not in their siblings, which would act as the controls. The mice subjected to the procedure would be homozygotes generated by mating siblings after several generations of backcross, and would carry two copies of the same segment of strain 129 DNA surrounding the transgene. Several promising procedures that could make this approach feasible are under development.

The effect of introns on transgene expression

The expression of a substantial number of transgenes is affected by the presence or absence of introns (Table 10.1). At least two distinct phenomena seem to be involved.

Table 10.1 Some genes with intronic enhancers that affect expression in transgenic mice.

Gene	*Product*	*Intron(s)*	*References*
h-ADA	Adenosine deaminase	1	Aronow *et al.* 1989
h-apoB	Apolipoprotein B	2	Brooks *et al.* 1994; Wang *et al.* 1996
h-CD8	CD8	1	Kieffer *et al.* 1996
COL1A1	$\alpha1_{(I)}$ collagen	1	Slack *et al.* 1991
FGFR3	FGF receptor 3	1	McEwen and Ornitz 1998
K18	Keratin 18	1	Umezawa *et al.* 1997
mMUP	Major urinary protein	1	Johnson *et al.* 1995
nestin	Intermediate filament protein	1 and 2	Zimmerman *et al.* 1994
TIE2	Receptor tyrosine kinase	1	Schlaeger *et al.* 1997

Intronic enhancers

Many genes possess what seem to be intronic enhancer sequences. As in the case of other transcriptional regulatory sequences, intronic enhancers frequently coincide with DNase hypersensitive sites, and the presence of a hypersensitive site in an intron is an indicator of the presence of an enhancer (Figure 10.6). Again, if the removal of one particular intron from a gene has a drastic effect on expression while removal of others has little or none, it is likely that the intron in question contains an enhancer. Quite a few intronic enhancers have been shown to contain putative DNA response elements. In some cases these have been shown by mutation to be an essential part of the enhancer (Karpinski *et al.* 1989; Oshima *et al.* 1990; Hengerer *et al.* 1990). In other cases cognate transcriptional regulatory proteins have been shown to bind to them (Karpinski *et al.* 1989; Bornstein *et al.* 1988) and in yet others to stimulate transcription in a cell-free system (Sorkin *et al.* 1993) or in co-transfected cells (Oshima *et al.* 1990). A notable number of the intronic enhancers contain response elements for general transcriptional activators of the Sp1 or AP-1 families. Unsurprisingly, at least two of these have been found to interact with heterologous promoters (Karpinski *et al.* 1989; Oshima *et al.* 1990), while some other intronic enhancers are known not to do so (Franklin *et al.* 1991).

Most intronic enhancers are located in the promoter proximal intron of the gene, although this is not invariably the case (Sorkin *et al.* 1993; Brooks *et al.* 1994), and genes with an enhancer in each of two introns have been reported (Ganguly *et al.* 1989; Zimmerman *et al.* 1994). The first intronic enhancer, or group of enhancers, to be detected was a rather special case (Gillies *et al.* 1983; Banerji *et al.* 1983). In immunoglobulin gene rearrangement one of a number of alternative variable (V) region gene segments is brought into proximity with a constant (C) region gene segment by the joining together of V and J or V, D and J regions. Since each V region gene has its own promoter, this brings the promoter of the rearranged one within the

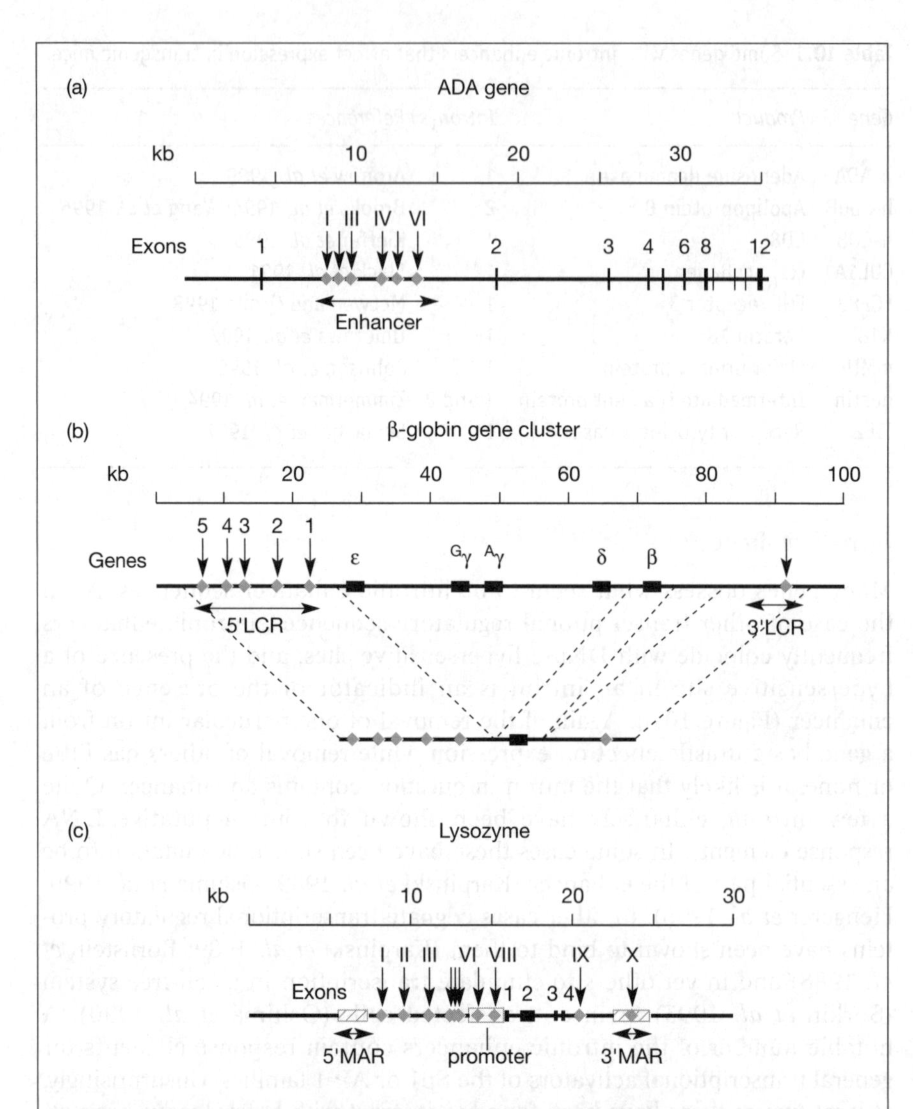

Figure 10.6 DNase hypersensitive sites. The hypersensitive sites are shown as blue diamonds. (a) The human adenosine deaminase (ADA) gene contains a thymus-specific enhancer, with 6 HS, within intron 1. (b) The human β-globin gene cluster contains the five genes given Greek letters, each of which is independently transcribed from its own promoter and contains two introns (not shown). The 5' LCR contains five HS and the 3' LCR one; the LCRs affects expression of each of the genes in the appropriate haematopoietic cell type. The bottom line of this part of the diagram shows the assembly of a minigene containing the β-globin gene and the two LCRs. (c) The chicken lysozyme gene has 10 HS, and two MARs at the boundaries of the transcription unit. HS I-VI are associated with enhancer sequences.

ambit of an enhancer located upstream of the C gene, within an intron of the rearranged immunoglobulin gene. What was thought at first to be a completely unique feature of rearranged genes of the immune system turns out to be a special case of a widespread phenomenon.

In common with 'classical' enhancers the effects of many, but not all, intronic enhancers are tissue-specific; enhancers in introns 1 and 2 of the nestin gene are muscle- and neural-specific respectively (Zimmerman *et al.* 1994). Similarly, the effect of an intronic enhancer on transcription can be negative rather than positive (Stred *et al.* 1990; McCready *et al.* 1997; Oskouian *et al.* 1997). Unlike classical enhancers, they are frequently position-dependent, orientation-dependent or both (Franklin *et al.* 1991; Magin *et al.* 1992; Sun and Means 1995).

Generic introns

Notwithstanding the existence of eukaryotic genes without introns, there is evidence that introns can increase gene expression in some general way unrelated to enhancer action (Hamer and Leder 1979; Villarreal and White 1983). Unfortunately, the most complete studies are of transgenes expressed in cultured cells, either infected with modified SV40 vectors or assessed during the transient expression phase following transfection with plasmid DNA (Buchman and Berg 1988; Huang and Gorman 1990a). Huang and Gorman (1990a) studied the effect of adding an intron to an otherwise intronless bacterial chloramphenicol acetyl transferase (CAT) gene under the transcriptional control of elements from simian virus 40. The addition of a hybrid intron, containing an adenovirus splice donor site and a splice acceptor site derived from an immunoglobulin gene, increased the level of cytoplasmic CAT mRNA about seven-fold. The intron did not affect the rate of transcription (measured by nuclear run-on analysis), the level of transcript in the nucleus or the stability of the CAT mRNA. However, the level of polyadenylated CAT RNA in the nucleus paralleled the increase in mRNA level. A lack of introns had previously been associated with a failure of polyadenylation (Villarreal and White 1983) and there is now more detailed evidence linking the splicing of an intron with the termination of the mRNA (see above).

The same hybrid intron had a similar stimulatory effect (ten-fold) on the CAT mRNA level in transgenic mice (Choi *et al.* 1991), but this study lacks direct evidence that the intron did not increase transcription. An investigation of the effect of an intron on the expression of the rat growth hormone gene in transgenic mice revealed a stimulation of 10- to 100-fold. This however was accompanied by an increase in the level of transcription (Brinster *et al.* 1988), suggesting that a transcriptional enhancer was involved. An extensive comparison of intronless cDNA coding regions with and without added introns from different sources revealed a variety of responses, but is also open to the possibility that enhancer sequences were present (Palmiter *et al.* 1991). Taken together, and with the evidence that the 3' proximal intron stimulates polyadenylation, the meagre evidence does strongly suggest the existence of a general intron effect on RNA processing.

From the viewpoint of transgenic work, it is obvious that the existence of intronic enhancers counsels caution in the design of constructs. It makes good sense to include a generic intron in any intronless construct, but it is worth noting that when added downstream of a coding sequence the SV40 small t antigen intron has a negative rather than a positive effect on expression; due to its small size it activates cryptic donor splice sites within the coding region, generating truncated mRNAs (Huang and Gorman 1990b). A generic intron will not of course be appropriate when the activity of the promoter is dependent on an intronic enhancer. Thus, if introns must be omitted from a transgene, a detailed knowledge of the implications for expression is invaluable.

Chromosomal position effect

The majority of transgenes exhibit two related effects that limit the value of some experimental approaches and pose particular problems for some commercial applications. These are chromosomal position effect and copy-number independence.

- Chromosomal position effect refers to the observation that different transgenic lines that carry the same transgene integrated into different chromosomal sites exhibit different levels of expression. Differences can be as much as 1000-fold. It can also affect the tissue distribution of expression and the time of the onset of expression during development.
- Copy-number independence refers to the fact that, although the number of transgene copies integrated at the different sites can differ widely (Chapter 8), the anticipated positive correlation between copy-number and expression level is not observed.

We are able to say that these effects are related because they can be reversed simultaneously (see below). This does not necessarily mean that they have the same cause.

Position effect is observed in cultured cells stably transfected with transgenes (see for example Feinstein *et al.* 1982) as well as in transgenic animals (Lacy *et al.* 1983; Jaenisch *et al.* 1981; Soriano *et al.* 1986). Transgenes that exhibited a putative position effect have been recovered both from cells (Butner and Lo 1986) and from transgenic mice (Al-Shawi *et al.* 1990) and reintroduced into an identical cellular or murine genome. In both cited cases the reintegrated transgenes failed to transmit their unusual patterns of expression, demonstrating that they were attributable to chromosomal position rather than mutation or rearrangement of the transgene. More recently two transgenes were each introduced several times into each of two chosen sites in the genome of CHO cells by site-specific recombination (Fukushige and Sauer 1992); the expression level of the transgenes at the two sites differed five- and eight-fold, while expression of each at each site was highly consistent. Together these results demonstrate formally the importance of chromosomal site of integration on transgene expression.

A program of DNA cytosine methylation begins in the early embryo and both methylation and demethylation continue throughout development (Chapter 6). Methylation of a gene predominantly correlates with transcriptional inactivity, although the opposite is sometimes the case. Methylation is a post-synthetic modification, but the preference of the DNA methyltransferase for hemimethylated DNA means that the methylation of specific cytosine residues is effectively inherited during cell proliferation. In contrast, the mechanism of *de novo* methylation is still obscure. In the same vein there is doubt as to whether or how commonly *de novo* methylation is a primary cause of gene inactivation, but general agreement that methylation contributes to the persistence of inactivation over generations of cells. For example demethylating agents, most commonly 5-azacytidine, bring about the activation of some inactive genes both in cell culture and *in vivo*.

The binding of some transcription factors to their cognate DNA response elements is weakened or eliminated when those elements are C-methylated (see Table 1 in Tate and Bird 1993). This opens a door to repression of transcription by the methylation of individual specific CpG sites. Since transcriptional activators and enhancers generally depend on the co-operation of several transcription factors, the fact that many response elements do not contain a CpG doublet may not greatly restrict the number of genes that can be regulated in this way. In addition, several proteins, including histone H1, are known to bind specifically or preferentially to methylated DNA. For example, MeCP1 binds to DNA that contains a certain critical density of methylated CpG sites (Boyes and Bird 1992), and repression *in vivo* may in some instances be similarly dependent on the density of methylation rather than on methylation of a few specific cytosines (Levine *et al.* 1991; Weng *et al.* 1995).

There are numerous examples of correlations between methylation and expression of transgenes. In some cases these parallel more or less closely the status of the corresponding resident gene. An example is the mouse apolipoprotein A1 gene and the corresponding human transgene, the former hypomethylated and expressed in both liver and intestine, the latter only in liver (Shemer *et al.* 1990).

There have long been suggestions that large transgene arrays attract methylation. In one study transgene copy number correlated positively with methylation and negatively with expression (Mehtali *et al.* 1990). Since the transgene arrays were integrated at different chromosomal sites, the possibility remained that the result was coincidentally dictated by position rather than copy number. More recently, by taking advantage of the Cre-l*oxP* site-specific recombination system (Chapter 9) arrays of different sizes were generated at the same site (Garrick *et al.* 1998). Two transgenic lines were first established carrying large (>100-copy) arrays of a transgene that contained a single internal *loxP* site. The arrays were then reduced in size by microinjecting a plasmid that expresses Cre recombinase into 1-cell embryos, which were then reintroduced into surrogate dams. New transgenic lines were established that carried reduced transgene arrays, having only one, one and five transgenes respectively. Two of these showed greatly increased levels of transcription and expression accompanied by reduced methylation.

This study showed a negative correlation between expression level and transgene copy number at the same locus. There are at least two ways in which this may be reconciled with several studies that show no correlation either negative or positive between the two parameters, and in which each insertion occupied a different chromosomal location. One is that the randomising effect of chromosomal location on expression overwhelms the negative effect due to copy number. The second is that the dependency of expression on copy number is non-linear, very large arrays having a disproportionate negative effect on expression. In view of the very large effect seen in this case, and in some earlier incidences of very large arrays, the latter explanation seems the more likely. Resolution of the question requires an experimental series in which arrays with moderate numbers of transgenes, in the range 1–20, are located at the same locus in different transgenic lines.

Variegation

The term heterochromatin refers to regions of chromatin that are condensed during interphase and relatively inactive transcriptionally, while the term euchromatin is applied to chromatin that is decondensed and transcriptionally relatively active. Constitutive heterochromatin is condensed in all cells and is mainly located close to the centromeres, while facultative heterochromatin is condensed in some cells but not in others.

A phenomenon called position effect variegation comes down to us from the classical period of *Drosophila* genetics. It was first observed as zonal variation in the colour of the facets of the eye that occurs when a pigmentation gene is brought into proximity with a region of constitutive heterochromatin by translocation, and was attributed to clonal variation in the 'spreading' of heterochromatinisation into the translocated chromosomal region. In this respect it resembles, at least superficially, the gradient of inactivation observed in parts of mammalian autosomes translocated to the inactivated X (Chapter 6). Similarly, variegation of transgene expression in the mouse has been shown to correlate with the integration of the transgene in proximity to pericentric heterochromatin (Festenstein *et al.* 1996; Dobie *et al.* 1996); in one case variegation coincided with a high intrastrain variance in the level of expression (Dobie *et al.* 1996). In the study quoted in the previous section (Garrick *et al.* 1998), the increased expression that accompanied the reduced numbers of transgenes in the arrays (and a reduced level of methylation) was mainly due to an increase in the number of expressing cells, i.e. to reduced variegation.

In a study of the erythroid expression of a globin-*lacZ* transgene in 11 transgenic lines Robertson *et al.* (1995) observed a correlation between the overall expression level and the proportion of expressing cells; expression per cell varied over an 80-fold range, whereas expression per expressing cell varied only three-fold. At the same time, there was no correlation between expression level and transgene copy number. Again variegation was the main cause of expressional variation, but the number of transgenic lines involved would seem to preclude the involvement of a pericentric location in this case. In contrast to those cases in which a pericentric location was implicated, there was little intra-strain variation in expression (Hsiao *et al.* 1994), and no correlation of expression with DNA methylation (Garrick *et al.* 1996).

It seems unarguable that variegation contributes to position effect in mammals, but it is not clear how common it is, or whether indeed it could be the main cause. Robertson *et al.* (1995) quote another nine studies in which variegated transgene expression is reported. In this context it may be recalled that the approximately 100-fold stimulation of extrachromosomal transgene expression in transfected cells (Box 10.1) by the SV40 enhancer was mainly due to an increase in the frequency of expressing cells from 0.1–1% to 10–30% (Weintraub 1988). More recently this effect of enhancers was confirmed and extended to stably transformed cells (Walters *et al.* 1995).

These various experiments give hints that chromosomal position effect may arise in a number of different ways:

- DNA methylation induced in some unknown way by high copy number
- inactivation due to proximity to heterochromatin
- other unknown cause(s)

As yet, however, not enough studies have been sufficiently comprehensive to allow firm conclusions to be drawn. The one common feature that runs through many of the extant studies is inaccessibility of the inactive or less active transgenes to DNase; this is hardly surprising, since it is a feature common to genes that are not being expressed, certainly when the cells in question are competent to express them. Site-specific recombination would now allow several transgene arrays of different sizes to be created at several different chromosomal locations, most efficiently in embryonic stem cells. A comprehensive study of transgenic lines generated from such a series of cell lines promises to answer many of the questions still outstanding.

Locus control regions and matrix attachment regions

Chromosomal position effect and copy-number independence can be simultaneously eliminated, or at least greatly reduced, through the agency of locus control regions (LCRs). LCRs share features with enhancer elements; they are marked by DNase hypersensitive sites (HS) and contain response elements to which recognised tissue-specific and ubiquitous transcription factors bind. Although an LCR tends to be situated at a greater distance from the transcription unit that it influences, this is not essential to its function; in transgene constructions an LCR brought close to the transcription unit will insulate it against position effect (Grosveld *et al.* 1987). LCRs contain multiple elements with distinguishable functions. One type in fact acts as an enhancer while others function in the remodelling of chromatin.

Known LCRs show tissue specificity to a high degree and relate to genes that are repressed in most tissues. The function of the LCR seems to be to release the gene from its generally repressed state, which may relate to chromatin condensation. The first LCR to be identified, and still the most closely studied, is associated with the human β-globin gene cluster. This is a cluster of five genes, spread across 32 kb of DNA (Figure 10.6), that fulfil the same basic function in three different cell types at three stages in development (yolk sac, ε gene; foetal liver, $^{G}\gamma$ and $^{A}\gamma$ genes; adult bone marrow, δ and β genes). Each gene has a promoter from which it is transcribed. LCRs lie at both ends of the cluster. The 5' LCR of the human gene cluster first drew attention because a set of overlapping deletions remote from the known location of the genes had a serious effect on their expression (Kioussis *et al.* 1983). It occupies the region between 6 and 22 kb upstream of the most 5' gene of the cluster, the ε-globin gene, and contains five sites marked by DNase hypersensitivity, HS1 to HS5 (Figure 10.6). Each of the five sites contains essentially the same response elements, but in different numbers and arrangements. In isolation the adult β-globin gene, for example, is poorly expressed as a transgene, but when coupled with the LCR it is expressed at a high level in haemopoietic tissues in a position-independent and copy-number-dependent manner (Grosveld *et al.* 1987; Ryan *et al.* 1989). The expression of other genes, linked to the LCR and stably integrated into cultured haemopoietic cells, is similarly stimulated (Blom van Assendelft *et al.* 1989). Parts of an LCR can be separately removed, for example from the resident mouse β-globin gene, without greatly impairing its function, and individual HS will act in isolation; HS2 acts as an enhancer and HS3 and HS4 confer copy-number dependence through chromatin remodelling.

In contrast to LCRs, MARs (matrix attachment regions, sometimes called scaffold attachment regions or SARs) are defined biochemically, indeed in two ways (Box 10.4). In several cases the same regions have been identified as MARs by both methods (Cockerill and Garrard 1986; Phi-Van and Stratling 1988). Again unlike LCRs, MARs have no well defined sequence motifs. They are several hundred base pairs in length and very AT-rich; the important feature seems to be the presence of quite long continuous regions of adenines paired of course with thymines on the other strand (Laemmli *et al.* 1992). They are not bound by known transcription factors, rather by topoisomerases I and II, histone H1 and a number of proteins not known in other contexts. An important characteristic of MARs is that they do not stimulate transient expression following cell transfection, although in identical constructs they greatly stimulate the expression of stably integrated transgenes (Stief *et al.* 1989). This is taken as evidence that they exert their effect by modifying *stable* chromatin structure (which is different in some way from the chromatin into which transfected DNA is

BOX 10.4: ISOLATION AND IDENTIFICATION OF MATRIX ATTACHMENT REGIONS

Two quite different methods of identifying MARs give essentially congruent results. One defines them as DNA elements that remain bound to the residue of nuclei that have been treated with a low concentration of lithium diiodosalicylate; the other defines them as DNA fragments that bind *in vitro* to the residue that remains after DNase-treated nuclei have been extracted with a high concentration of NaCl.

Isolation of DNA regions attached to the nuclear matrix *in vivo* (due to Mirkovitch *et al.* 1984)

Purified nuclei are extracted with lithium diiodosalicylate, which removes all histones and some other nuclear proteins. The residue is treated with restriction enzymes and both the DNA remaining bound to the matrix and the DNA released from it are purified to remove protein, size fractionated by gel electrophoresis, transferred to a membrane and hybridised with radioactive probes prepared from a cloned gene. Reactive fragments that are enriched in the pellet fraction contain matrix attachment regions.

In vitro identification (due to Cockerill & Garrard 1986)

Purified nuclei are treated with DNase and then extracted with 2M NaCl. The residue is the nuclear matrix. Radioactively labelled restriction enzyme fragments of cloned DNA are mixed with the matrix, in the presence of a 10 000-fold excess of bacterial DNA to eliminate non-specific binding. The fragments that bind, identified on the basis of their size by gel electrophoresis, contain matrix attachment regions.

rapidly incorporated). In agreement with this view, they are frequently located at the boundaries of a transcription unit as defined by sensitivity to nuclease attack (Figure 10.6). The two boundaries may lie at the base of a chromatin loop, binding it to the nuclear matrix.

Not surprisingly, confusion surrounds the use of the terms ‘enhancer’, ‘LCR’ and ‘MAR’. This is partly because of the complexity of the processes controlling gene expression in eukaryotes. There are at least two processes at work, the remodelling of ‘repressed chromatin’ into a potentially active *(accessible)* configuration and a more direct stimulation of transcription which operates on accessible chromatin. Enhancers can be operationally defined as elements that stimulate transcription both of genes embedded in chromosomes and of genes carried in extrachromosomal plasmids during the transient phase of expression. However enhancers may also promote chromatin remodelling. LCRs contain both enhancers and other elements with no enhancer activity that also promote chromatin remodelling. MARs can be defined biochemically as described above, but are frequently associated with enhancers which may be discovered only as a result of exhaustive analysis.

In the transgenic context LCRs are limited by their tissue specificity and there is some question as to whether or to what extent they are also specific to the promoter with which they are normally associated. While MARs are not tissue-specific, and the avian lysozyme MARs function in mammals, they do not confer convincing copy-number dependence (McKnight *et al.* 1992) and experience with them has not been uniform (Thompson *et al.* 1994).

Switching genes off or on

The most powerful experimental tool that transgenic technology can provide is a line of animals that differ from 'normal' or control animals in a single fundamental way. Under these circumstances differences detected in a pairwise comparison of the normal and abnormal animals can be attributed to the fundamental difference. One prime example of this approach is the addition of a foreign gene to the genome or the direction of gene expression to an ectopic site (Chapter 8). Another is the now routine gene 'knockout' (Chapter 9).

A single genetic change can have complicated ramifications and it can often be difficult or impossible to unravel primary from subsidiary effects. This complexity has often rendered futile the study of spontaneous or induced mouse mutants. An extreme, if obvious, example of an intractable situation is an embryonic lethal mutation, which can only be studied in heterozygous animals or by the use of chimeras. Exactly the same sort of difficulty besets many gene knockouts. Methods have accordingly been devised to induce change in a gene or its expression at will, for example to inactivate a function in the adult following normal development.

These methods are to a greater or lesser extent dependent on cellular specificity of transgene expression, which as we already know can be achieved by judicious selection of DNA regulatory elements, given enough understanding or some good fortune.

Antisense RNA and ribozymes

These RNA-based methods set out to inactivate a gene by neutralising its transcript in two different ways. Both depend upon base pairing between complementary sequences; antisense RNA acts by steric hindrance while ribozymes cleave the RNA with which they are paired. Both methods can be applied to animals in which gene targeting is not available. Another more far-reaching feature is that production of the RNA may be directed to a particular cell type at a particular time by the use of cell-specific DNA regulatory elements or by using one of the methods of temporal control described later in this chapter.

Antisense RNA has a close parallel in the use of natural and chemically modified oligodeoxynucleotides to inhibit gene expression. Interest in the latter has been stimulated by their possible therapeutic potential, and they have been quite carefully studied both by introduction into cultured cells and in cell-free translation systems. The evidence unsurprisingly shows that they are most effective when directed against the region of the RNA surrounding the initiation codon. Potentially, antisense RNA produced from a transgene *in vivo* could inhibit not only translation but also pre-mRNA processing and transport, or stimulate pre-mRNA or RNA degradation (Sokol and Murray 1996). Also, it is likely that intranuclear RNA duplexes are inactivated by the enzyme double strand-specific RNA adenosine deaminase which converts adenine to inosine (Bass and Weintraub 1988; Wagner *et al.* 1989).

We can anticipate that the ratio of antisense to sense RNA will be a very important factor in the efficacy of inhibition. Thus the most likely RNA targets will be those that have significant phenotypic effects although present in small amounts.

The first and arguably the most dramatic successful use of antisense RNA in mammals was the simulation of a known genetic defect, the 'shiverer' mutation in the mouse (Katsuki *et al.* 1988). In this study myelin basic protein antisense RNA reduced the mRNA level to about 20% of normal and produced variegated myelination in the central nervous system. Other informative examples include the exclusion of the Moloney leukemia virus RNA genome from virus particles by antisense RNA directed against the genomic packaging signal (Han *et al.* 1991) and the reduction of *Wnt-1* mRNA by 98% by antisense RNA under the control of the human testis-specific *PGK-2* promoter (Erickson *et al.* 1993). A reduction of receptor-mediated adenyl cyclase inhibition was obtained in fat tissue through a reduction of $G\alpha_{i2}$ to 5% brought about by antisense RNA under the control of the cAMP-inducible phosphoenolpyruvate carboxykinase gene promoter (Moxham *et al.* 1993b; Moxham *et al.* 1993a). Maternal tissue-type plasminogen activator synthesis directed by maternal mRNA was inhibited by antisense RNA controlled by the oocyte-specific promoter of the zona pellucida 3 (*Zp3*) gene (Richards *et al.* 1993). Dwarfism was induced in the rat by antisense growth hormone (GH) RNA directed to the pituitary somatotrophs by a remodelled GH gene promoter, despite a reduction of only 35–40% in GH mRNA and 25–40% in circulating GH (Matsumoto *et al.* 1993). On a cautionary note, no effect on enzyme levels was observed in transgenic lines in which *hprt* mRNA levels were reduced to 50% (Munir *et al.* 1990).

Although part of a ribozyme is complementary to its target RNA, another part is not and a ribozyme possesses the capacity to cleave its target RNA catalytically. Since the ribozyme identifies its target by base-pairing, it can be tailored to identify a chosen target and will do so, at least in principle, with a high degree of specificity. There are two basic ribozyme

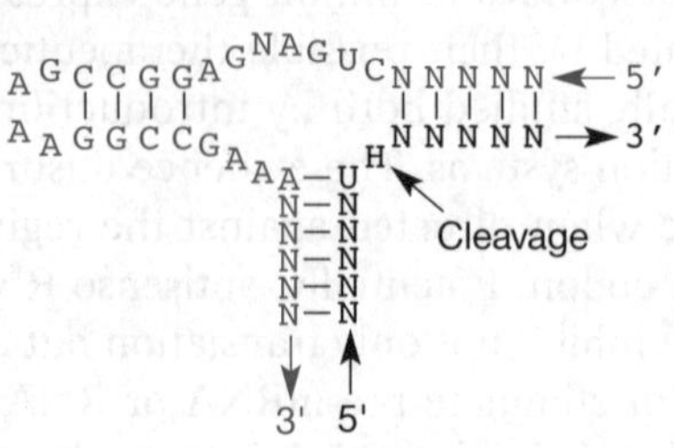

Figure 10.7 Hammerhead ribozyme (blue) base-paired with a target sequence (black). Paired nucleotides designated N are complementary. The sequences represented by black Ns are the normal sequences of the target RNA. The complementary blue Ns are built into the ribozyme.

conformations, the hammerhead and the hairpin. The hammerhead ribozyme (Hutchins *et al.* 1986; Buzayan *et al.* 1986) places a lesser constraint on the sequence of the RNA target, which is required to present only the sequence UH, where H may be U, C or A. To produce a hammerhead ribozyme that will cleave a new target, sequences complementary to those on either side of the UH doublet in the target are introduced at either end of the catalytic hammerhead as shown (Figure 10.7). The lengths of these should be chosen so as to strike a balance between selectivity for the target prior to cleavage and dissociation from it afterwards so that the ribozyme can act catalytically. The hammerhead ribozyme has been favoured, probably because a molecule can be built to attack any UH target, but this may be misguided; the longer target sequence of the hairpin ribozyme (BCUGNYR, where B represents C, U or G) confers greater selectivity with base pairing regions of a given length and this should favour turnover of the ribozyme. Finally, the ribozyme has to be introduced into a transcription unit; SnRNA transcription units have been favoured because of the stability of the RNA and its location predominantly in the nucleus.

In fact, very few mammalian transgenic studies with ribozymes have been published as yet, and none that suggest that ribozymes are likely to be more effective than antisense RNAs (Efrat *et al.* 1994; Larsson *et al.* 1994; L'Huillier *et al.* 1996). It is possible that the ionic conditions in the mammalian nucleus are unfavourable to ribozyme action (Pyle 1993). In one comparative study, four ribozymes were introduced into an antisense RNA. The antisense RNA was equally effective in inhibiting protein synthesis from the targeted RNA with and without the added ribozymes (Sokol *et al.* 1998), but the long duplex RNA that would be formed in this case may have precluded ribozyme turnover. The case for ribozymes remains open.

Site-specific recombination

Site-specific recombination, which is discussed in Chapter 9, provides a radical way of controlling gene expression. The recognition sites can be positioned in such a way that recombination between them activates or inactivates a gene (Figure 9.10). If the recombinase is under the control of a tissue-specific promoter, then we can expect the gene to be activated or inactivated solely in that tissue soon after the promoter becomes active.

The following example (Tsien *et al.* 1996a; Tsien *et al.* 1996b) is quite informative. Long-term potentiation (LTP) relates to a persistent increase in synaptic transmission that can be induced and measured in brain slices. Spatial learning is measured by observing the improvement in the response of an animal (a mouse in this case) in repetitions of a simple spatial test. The two are thought to be related functions of the CA1 region of the hippocampus, mediated by the N-methyl-D-aspartate (NMDA) receptor. Accordingly, two *loxP* sites were introduced into the NMDAR gene by homologous recombination, arranged in such a way that recombination between them would inactivate it. No CA1-specific promoter was available to direct Cre recombinase synthesis appropriately. The *Cre* gene was coupled to the promoter of the calcium-calmodulin-dependent kinase II (*CaMKIIα*) gene, which is expressed generally in forebrain neurons, and several transgenic mouse lines were derived by pronuclear microinjection. These lines were directly tested by mating each of them to a line carrying an inactive *lacZ* transgene that would be made active by Cre-mediated recombination *in vivo.* Surprisingly, the double heterozygotes derived by mating mice from one of the *CaMKIIα* lines to this tester line showed approximately the desired localisation of site-specific recombination to the CA1 region. Accordingly, this line was mated to the line carrying the *loxP*-modified NMDAR gene to produce mice heterozygous for the Cre gene and homozygous for the modified NMDAR gene. Both LTP and spatial learning were found to be impaired.

This study incorporates a good example of the opportunistic exploitation of chromosomal position effect. More than that, it demonstrates how site-specific recombination can be harnessed to a specific scientific objective.

Transgenic ablation

Several mammalian cell types are highly specialised to perform a single function *in vivo.* Some endocrine secretory cells are prominent examples of this. For example the anterior pituitary contains six distinct cell types each of which secretes a single hormone, or at most two hormones. The somatotrophs are the only source of growth hormone, the lactotrophs the only source of prolactin, and so on. Similarly, the thyroid contains two secretory cell types, the C-cells which secrete calcitonin and the thyrocytes (thyroid follicle cells) which are the only source of thyroid hormone.

Since such cells uniquely perform one essential function, it is possible to eliminate that function by eliminating (ablating) the cells. This can be done by directing to them the expression of a toxic protein which destroys them. An example is the diphtheria toxin A (Dt-A) protein. The toxin produced by *Corynebacterium diphtheriae* is cleaved into A and B proteins. The A protein is an ADP-ribosyltransferase which inactivates the translation factor EF-1 and brings protein synthesis to a stop. The B protein is required for entry of the A protein into cells. Consequently cells that produce only the A protein intracellularly from a transgene are killed, but any toxin released from dead cells cannot enter neighbouring cells. A suitable promoter can be the one exclusively concerned with producing the hormone that the cells synthesise, e.g. the growth hormone promoter for the ablation of somatotrophs. Early examples of this technique were the ablation of the pancreas by placing the *Dt-A* gene under the control of the elastase I promoter (Palmiter *et al.* 1987), the production of microphthalmic mice by directing Dt-A production to the lens with the γ_2-crystallin promoter (Breitman *et al.* 1987) and dwarfing of mice by the ablation of somatotrophs using the rat growth hormone promoter (Behringer *et al.* 1988).

Conditional ablation

Since the target cells are destroyed as soon as the promoter becomes active, usually during development, the transgenic mice develop abnormally. In order to avoid this and destroy the cells only after development is complete, a different strategy is required. Instead of a toxin, we can direct to the target cells a toxino*gen*, a protein which is itself harmless but which converts some substrate, also harmless, to a lethal product. The ablation of the cells occurs only when the substrate is administered; in other words the process is under temporal control. A further and not inconsiderable advantage of this approach is that the untreated mice are normal; consequently maintenance of the transgenic lines is simplified. An example of a toxinogen is the herpes simplex virus thymidine kinase, which efficiently phosphorylates a number of nucleoside analogues that are very poorly phosphorylated by cellular enzymes. Some analogues, such as Ganciclovir and FIAU, are harmless in themselves but become potently cytotoxic when phosphorylated. Expression of the viral kinase gene (*HSV-tk*) was first placed under the control of the immunoglobulin κ gene promoter combined with the μ gene intronic enhancer (Borrelli *et al.* 1988). Treatment with FIAU resulted in a deficit of B cells. As a second example, expression of HSV-tk was directed to the thyrocytes under the control of the bovine thyroglobulin promoter (Wallace *et al.* 1991; Wallace *et al.* 1994). The thyroid follicle cells were unharmed prior to the administration of the analogue and the mice developed normally. When Ganciclovir was administered to adult transgenic mice the follicle cells were rapidly destroyed by apoptosis (Wallace *et al.* 1996), despite the fact that they were not proliferating, and the mice could no longer produce thyroxine. The *HSV-tk* gene is expressed ectopically during spermatogenesis, independently of the promoter sequence, and causes male sterility (Al-Shawi *et al.* 1988; Al-Shawi *et al.* 1991; Cohen *et al.* 1998) but alterations to the coding region ameliorate this effect (Ellison *et al.* 1995). Conditional ablation has been used successfully to answer physiological questions (Wallace *et al.* 1995; Ahlgren *et al.* 1997).

Inducible gene expression

Ways have been devised to control the expression of a transgene by administering a non-toxic inducer to the mice. In some ways this method is analogous to conditional ablation, and carries the same requirement that the inducing compound should not be recognised by or have any effect upon a normal (non-transgenic) mouse, but in this case cell death is not a consequence, or at any rate a necessary consequence, of the effect of the inducer. The ideal is that it should be possible to switch the animal from a zero to a high level of expression and *vice versa*, that this should occur with minimal delay, and that intermediate levels of expression should be attainable, so that dose–response relationships can be assessed.

Two systems that employ elements of a bacterial tetracycline resistance gene and tetracycline or its analogue doxycycline have been developed (Gossen *et al.* 1993) (Figure 10.8). Each system is binary, i.e. it calls for two transgenes. One codes for a transcriptional activator and is under the control of either a general or a tissue-specific promoter, depending on circumstances. The other is under the control of a promoter that responds to the aforesaid activator and codes for the gene (gene X) that is to be regulated. In the first system, gene X is repressed in the presence of tetracycline or doxycycline, and activated when the drug is withdrawn (Gossen and Bujard 1992); in the second, gene X is activated in the presence of the drug and repressed when it is withdrawn (Gossen *et al.* 1995). Both methods have been tested in transgenic mice with model reporter genes (substituting *lacZ* or a luciferase gene for gene X in Figure 10.8) (Furth *et al.* 1994; Kistner *et al.* 1996). In each case the binary system was set up by making transgenic lines with each construct and mating selected lines to obtain double heterozygotes. Increases in expression of several orders of magnitude were observed upon induction in both systems. When the tTA transactivator was placed under the control of a liver-specific promoter, reporter gene expression in the liver was specifically induced by doxycycline withdrawal (Kistner *et al.* 1996). Mosaic expression was observed in these experiments together with a great deal of variation in the responses of different transgenic lines and also between individuals within a line. This is of little concern since no effort was made to avoid chromosomal position effects.

In a modification to the method the production of the activator protein was made subject to positive feedback by incorporating multiple copies of the *tetO* sequence into its promoter (Shockett *et al.* 1995). Interestingly, this greatly reduced the expressional mosaicism in transgenic mice.

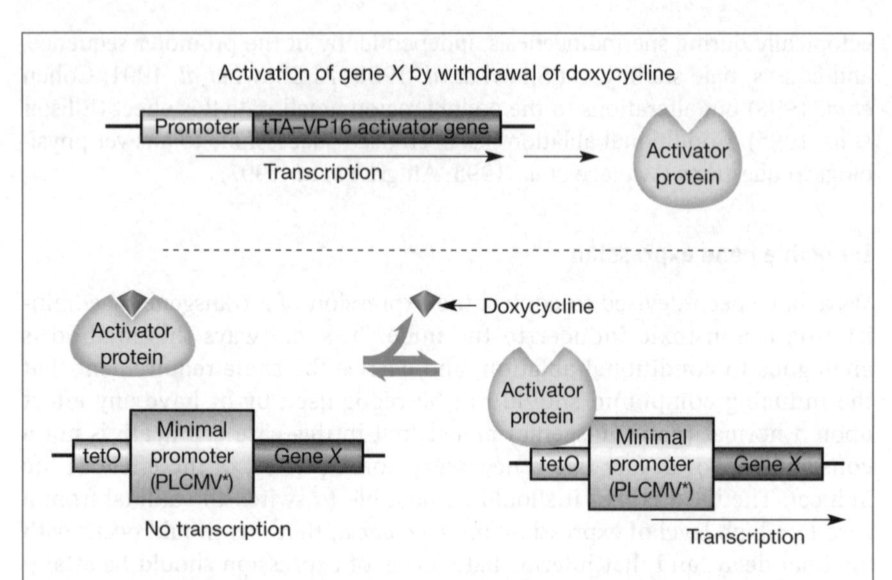

Figure 10.8 Two ways of controlling gene expression with the tetracycline analogue doxycycline (see also p.235). In both methods the expression of the conditionally expressed transgene is directed by a compound promoter sequence consisting of multiple copies of the bacterial *tet* operator (*tetO*) coupled to a minimal promoter derived from cytomegalovirus (PLCMV*). The activity of this promoter is regulated in the first case by a bacterial protein that binds to *tetO* in the absence but not the presence of doxycycline (tTA) and in the second by a mutant protein that binds in the presence but not the absence of doxycycline (rtTA). Each of these is coupled, as a hybrid protein transcription factor, with the transcription-activating domain of the herpes simplex virus VP16 transcriptional regulator protein.

This page – Gene activation by withdrawal of doxycycline. In the absence of doxycycline the hybrid tTA-VP16 protein binds to the compound tetO-PLCMV* promoter and activates the transcription of the gene [*X*]. Doxycycline binds to the tTA-VP16 protein and prevents its interaction with the tetO-PLCMV* promoter, so that in the presence of doxycycline gene *X* is transcribed only at a low basal level. Thus induction of gene *X* activity results from the withdrawal of doxycycline.

Other inducible systems are less well developed as yet. A ternary system based on the receptor for the *Drosophila melanogaster* steroid hormone ecdysone (EcR) has been described (Saez *et al.* 1997) (Figure 10.9). In common with other steroid hormones, the active form of EcR is a heterodimer with a second protein. A rather complicated hybrid receptor (VgEcR) was created incorporating parts of the EcR and the glucocorticoid receptor with the herpesvirus VP16 activation domain. This receptor forms an active heterodimer with RXR, the natural partner of the thyroid hormone receptor, and binds to a hybrid response element (AGGTCANAGAACA). VgEcR and RXR were introduced into the mouse genome in a tandem construct. A reporter gene controlled by four copies of the hybrid response element linked to a minimal promoter was introduced separately and double heterozygotes were produced by mating. Induction was observed upon

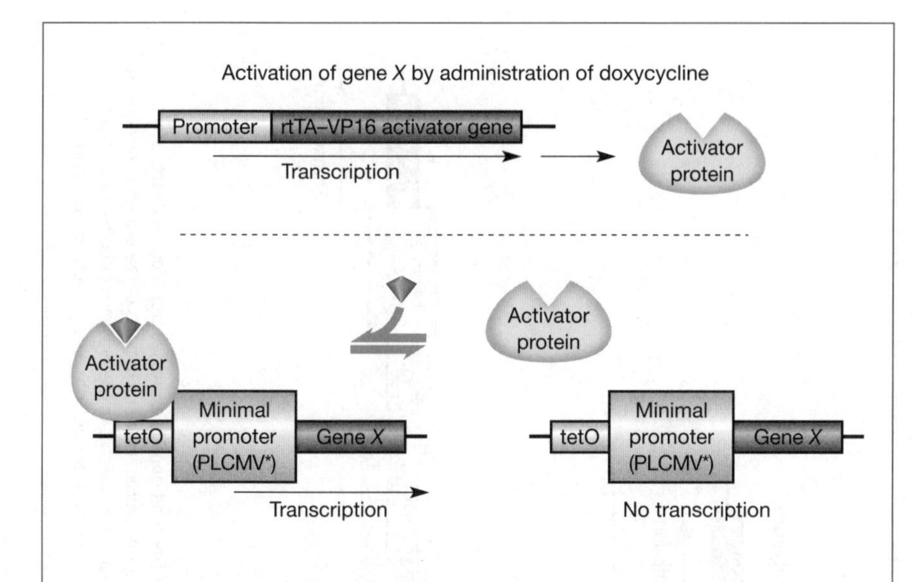

Figure 10.8 continued. Gene activation by administration of doxycycline. In the absence of doxycycline the hybrid rtTA-VP16 protein fails to bind to the compound tetO-PLCMV* promoter and gene *X* is transcribed only at a low basal level. Doxycycline binds to the rtTA-VP16 protein, and this makes the transcription factor competent to bind to the tetO-PLCMV* promoter, activating transcription of gene *X*.

administration of an ecdysone analogue, but not quantified. A similar system is based on a mutant human progesterone receptor ligand binding domain specific for a synthetic steroid, the VP16 activation domain and the yeast GAL4 DNA binding domain; the reporter gene contains four copies of the GAL4 response element (Wang *et al.* 1994).

Lastly, the promoter of a human cytochrome P450 1A1 gene, which is liver-specific and induced by aryl hydrocarbons, was used in combination with the coding region of the human apolipoprotein E gene. When the transgene was introduced into a line of hypercholesterolaemic mice and an inducer (β-naphthoflavone) was administered, the level of serum apolipoprotein E rose and the serum cholesterol level fell (Smith *et al.* 1995).

To a large extent these methods are in the developmental stage. An obvious way forward would be to take account of and eliminate chromosomal position effect and mosaic expression. For instance, the expression of activator proteins could probably be improved by introducing suitably modelled transgenes using homologous recombination in embryonic stem cells. Another improvement could be to put a site-specific recombinase under the control of an inducible promoter, gaining control of the timing of the reorganisation of the target gene.

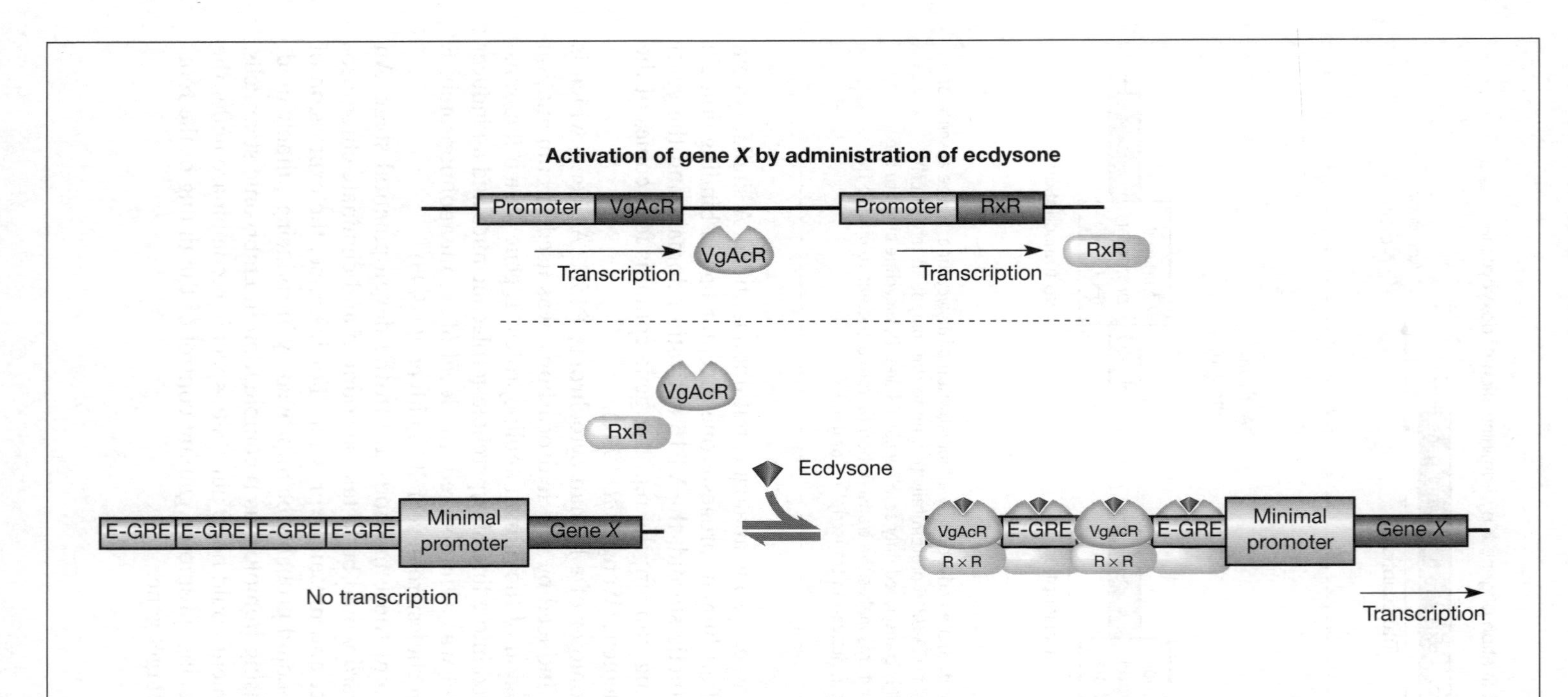

Figure 10.9 Regulating gene expression with the insect steroid hormone ecdysone. The expression of the conditionally expressed transgene is directed by a compound promoter sequence E-GRE (see text). The activity of this promoter is regulated by a hybrid receptor protein made up of parts of the ecdysone receptor and the related glucocorticoid receptor together with the activation domain of a herpesvirus transcriptional activator (VgEcR). In the presence of ecdysone the hybrid receptor forms a heterodimer with the RxR receptor and activates transcription of gene *X* from the compound promoter sequence. See the text for other details.

Ahlgren, S.C., Wallace, H., Bishop, J., Neophytou, C., and Raff, M.C. (1997) Effects of thyroid hormone on embryonic oligodendrocyte precursor cell development *in vivo* and *in vitro*. *Molecular and Cellular Neuroscience*, **9**, 420–432.

Al-Shawi, R., Burke, J., Jones, C.T., Simons, J.P., and Bishop, J.O. (1988) A Mup promoter-thymidine kinase reporter gene shows relaxed tissue-specific expression and confers male sterility upon transgenic mice. *Molecular & Cellular Biology*, **8**, 4821–4828.

Al-Shawi, R., Kinnaird, J., Burke, J., and Bishop, J.O. (1990) Expression of a foreign gene in a line of transgenic mice is modulated by a chromosomal position effect. *Molecular & Cellular Biology*, **10**, 1192–1198.

Al-Shawi, R., Burke, J., Wallace, H., Jones, C., Harrison, S., Buxton, D., Maley, S., Chandley, A., and Bishop, J.O. (1991) The herpes simplex virus type 1 thymidine kinase is expressed in the testes of transgenic mice under the control of a cryptic promoter. *Molecular & Cellular Biology*, **11**, 4207–4216.

Alexander, L., Lu, H.H., Gromeier, M., and Wimmer, E. (1994) Dicistronic polioviruses as expression vectors for foreign genes. *AIDS Research & Human Retroviruses*, **10** Suppl 2, S57–60.

Aronow, B., Lattier, D., Silbiger, R., Dusing, M., Hutton, J., Jones, G., Stock, J., McNeish, J., Potter, S., Witte, D., *et al.* (1989) Evidence for a complex regulatory array in the first intron of the human adenosine deaminase gene. *Genes & Development*, **3**, 1384–1400.

Banerji, J., Olson, L., and Schaffner, W. (1983) A lymphocyte-specific cellular enhancer is located downstream of the joining region in immunoglobulin heavy chain genes. *Cell*, **33**, 729–740.

Bass, B.L. and Weintraub, H. (1988) An unwinding activity that covalently modifies its double-stranded RNA substrate. *Cell*, **55**, 1089–1098.

Behringer, R.R., Mathews, L.S., Palmiter, R.D., and Brinster, R.L. (1988) Dwarf mice produced by genetic ablation of growth-hormone-expressing cells. *Genes & Development*, **2**, 453–461.

Blom van Assendelft, G., Hanscombe, O., Grosveld, F., and Greaves, D.R. (1989) The beta-globin dominant control region activates homologous and heterologous promoters in a tissue-specific manner. *Cell*, **56**, 969–977.

Bonyadi, M., Rusholme, S.A., Cousins, F.M., Su, H.C., Biron, C.A., Farrall, M., and Akhurst, R.J. (1997) Mapping of a major genetic modifier of embryonic lethality in TGF beta 1 knockout mice. *Nature Genetics*, **15**, 207–211.

Bornstein, P., McKay, J., Liska, D.J., Apone, S., and Devarayalu, S. (1988) Interactions between the promoter and first intron are involved in transcriptional control of alpha 1(I) collagen gene expression. *Molecular & Cellular Biology*, **8**, 4851–4857.

Borrelli, E., Heyman, R., Hsi, M., and Evans, R.M. (1988) Targeting of an inducible phenotype to mammalian cells. *Proceedings of the National Academy of Sciences of the United States of America*, **85**, 7572–7576.

Boyes, J. and Bird, A. (1992) Repression of genes by DNA methylation depends on CpG density and promoter strength: evidence for involvement of a methyl-CpG binding protein. *EMBO Journal*, **11**, 327–333.

Breitman, M.L., Clapoff, S., Rossant, J., Tsui, L.C., Glode, L.M., Maxwell, I.H., and Bernstein, A. (1987) Genetic ablation: targeted expression of a toxin gene causes microphthalmia in transgenic mice. *Science*, **238**, 1563–1565.

Brinster, R.L., Allen, J.M., Behringer, R.R., Gelinas, R.E., and Palmiter, R.D. (1988) Introns increase transcriptional efficiency in transgenic mice. *Proceedings of the National Academy of Sciences of the United States of America*, **85**, 836–840.

Brooks, A.R., Nagy, B.P., Taylor, S., Simonet, W.S., Taylor, J.M., and Levy-Wilson, B. (1994) Sequences containing the second-intron enhancer are essential for transcription of the human apolipoprotein B gene in the livers of transgenic mice. *Molecular & Cellular Biology*, **14**, 2243–2256.

Buchman, A.R. and Berg, P. (1988) Comparison of intron-dependent and intron-independent gene expression. *Molecular & Cellular Biology*, **8**, 4395–4405.

Butner, K. and Lo, C.W. (1986) Modulation of tk expression in mouse pericentromeric heterochromatin. *Molecular & Cellular Biology*, **6**, 4440–4449.

Buzayan, J.M., Hampel, A., and Bruening, G. (1986) Nucleotide sequence and newly formed phosphodiester bond of spontaneously ligated satellite tobacco ringspot virus RNA. *Nucleic Acids Research*, **14**, 9729–9743.

Cairns, B.R. (1998) Chromatin remodeling machines: similar motors, ulterior motives. *Trends in Biochemical Sciences*, **23**, 20–25.

Choi, T., Huang, M., Gorman, C., and Jaenisch, R. (1991) A generic intron increases gene expression in transgenic mice. *Molecular & Cellular Biology*, **11**, 3070–3074.

Cockerill, P.N. and Garrard, W.T. (1986) Chromosomal loop anchorage of the kappa immunoglobulin gene occurs next to the enhancer in a region containing topoisomerase II sites. *Cell*, **44**, 273–282.

Cohen, J.L., Boyer, O., Salomon, B., Onclerco, R., Depetris, D., Lejeune, L., Dubus-Bonnet, V., Bruel, S., Charlotte, F., Mattei, M.G., and Klatzmann, D. (1998) Fertile homozygous transgenic mice expressing a functional truncated Herpes simplex thymidine kinase delta TK gene. *Transgenic Research*, **7**, 321–330.

Cowell, I.G. (1994) Repression versus activation in the control of gene transcription. *Trends in Biochemical Sciences*, **19**, 38–42.

Derman, E., Krauter, K., Walling, L., Weinberger, C., Ray, M., and Darnell, J.E. (1981) Transcriptional control in the production of liver-specific mRNAs. *Cell,* **23**, 731–739.

Dobie, K.W., Lee, M., Fantes, J.A., Graham, E., Clark, A.J., Springbett, A., Lathe, R., and McClenaghan, M. (1996) Variegated transgene expression in mouse mammary gland is determined by the transgene integration locus. *Proceedings of the National Academy of Sciences of the United States of America,* **93**, 6659–6664.

Efrat, S., Leiser, M., Wu, Y.J., Fusco-DeMane, D., Emran, O.A., Surana, M., Jetton, T.L., Magnuson, M.A., Weir, G., and Fleischer, N. (1994) Ribozyme-mediated attenuation of pancreatic beta-cell glucokinase expression in transgenic mice results in impaired glucose-induced insulin secretion. *Proceedings of the National Academy of Sciences of the United States of America,* **91**, 2051–2055.

Ellison, A.R., Wallace, H., Al-Shawi, R., and Bishop, J.O. (1995) Different transmission rates of herpesvirus thymidine kinase reporter transgenes from founder male parents and male parents of subsequent generations. *Molecular Reproduction and Development,* **41**, 425–434.

Erickson, R.P., Lai, L.W., and Grimes, J. (1993) Creating a conditional mutation of Wnt-1 by antisense transgenesis provides evidence that Wnt-1 is not essential for spermatogenesis. *Developmental Genetics,* **14**, 274–281.

Fawcett, D., Pasceri, P., Fraser, R., Colbert, M., Rossant, J., and Giguere, V. (1995) Postaxial polydactyly in forelimbs of CRABP-II mutant mice. *Development,* **121**, 671–679.

Feinstein, S.C., Ross, S.R., and Yamamoto, K.R. (1982) Chromosomal position effects determine transcriptional potential of integrated mammary tumor virus DNA. *Journal of Molecular Biology,* **156**, 549–565.

Festenstein, R., Tolaini, M., Corbella, P., Mamalaki, C., Parrington, J., Fox, M., Miliou, A., Jones, M., and Kioussis, D. (1996) Locus control region function and heterochromatin-induced position effect variegation. *Science,* **271**, 1123–1125.

Franklin, G.C., Donovan, M., Adam, G.I., Holmgren, L., Pfeifer Ohlsson, S., and Ohlsson, R. (1991) Expression of the human PDGF-B gene is regulated by both positively and negatively acting cell type-specific regulatory elements located in the first intron. *EMBO Journal,* **10**, 1365–1373.

Fukushige, S. and Sauer, B. (1992) Genomic targeting with a positive-selection lox integration vector allows highly reproducible gene expression in mammalian cells. *Proceedings of the National Academy of Sciences of the United States of America,* **89**, 7905–7909.

Furth, P.A., St Onge, L., Boger, H., Gruss, P., Gossen, M., Kistner, A., Bujard, H., and Hennighausen, L. (1994) Temporal control of gene expression in transgenic mice by a tetracycline-responsive promoter. *Proceedings of the National Academy of Sciences of the United States of America,* **91**, 9302–9306.

Ganguly, S., Vasavada, H.A., and Weissman, S.M. (1989) Multiple enhancer-like sequences in the HLA-B7 gene. *Proceedings of the National Academy of Sciences of the United States of America,* **86**, 5247–5251.

Garrick, D., Sutherland, H., Robertson, G., and Whitelaw, E. (1996) Variegated expression of a globin transgene correlates with chromatin accessibility but not methylation status. *Nucleic Acids Research,* **24**, 4902–4909.

Garrick, D., Fiering, S., Martin, D.I., and Whitelaw, E. (1998) Repeat-induced gene silencing in mammals. *Nature Genetics,* **18**, 56–59.

Gerace, L. (1995) Nuclear export signals and the fast track to the cytoplasm. *Cell,* **82**, 341–344.

Gillies, S.D., Morrison, S.L., Oi, V.T., and Tonegawa, S. (1983) A tissue-specific transcription enhancer element is located in the major intron of a rearranged immunoglobulin heavy chain gene. *Cell,* **33**, 717–728.

Gossen, M., Bonin, A.L., and Bujard, H. (1993) Control of gene activity in higher eukaryotic cells by prokaryotic regulatory elements. *Trends in Biochemical Sciences,* **18**, 471–475.

Gossen, M., Freundlieb, S., Bender, G., Muller, G., Hillen, W., and Bujard, H. (1995) Transcriptional activation by tetracyclines in mammalian cells. *Science,* **268**, 1766–1769.

Gossen, M. and Bujard, H. (1992) Tight control of gene expression in mammalian cells by tetracycline-responsive promoters. *Proceedings of the National Academy of Sciences of the United States of America,* **89**, 5547–5551.

Grosveld, F., van Assendelft, G.B., Greaves, D.R., and Kollias, G. (1987) Position-independent, high-level expression of the human beta-globin gene in transgenic mice. *Cell,* **51**, 975–985.

Halle, J.P. and Meisterernst, M. (1996) Gene expression: increasing evidence for a transcriptosome. *Trends in Genetics,* **12**, 161–163.

Hamer, D.H. and Leder, P. (1979) Splicing and the formation of stable RNA. *Cell,* **18**, 1299–1302.

Han, L., Yun, J.S., and Wagner, T.E. (1991) Inhibition of Moloney murine leukemia virus-induced leukemia in transgenic mice expressing antisense RNA complementary to the retroviral packaging sequences. *Proceedings of the National Academy of Sciences of the United States of America,* **88**, 4313–4317.

Hargrove, J.L., Hulsey, M.G., and Beale, E.G. (1991) The kinetics of mammalian gene expression. *Bioessays,* **13**, 667–674.

Hengerer, B., Lindholm, D., Heumann, R., Ruther, U., Wagner, E.F., and Thoenen, H. (1990) Lesion-induced increase in nerve growth factor mRNA is mediated by c-fos. *Proceedings of the National Academy of Sciences of the United States of America*, **87**, 3899–3903.

Hofer, E., Hofer-Warbinek, R., and Darnell, J.E., Jr. (1982) Globin RNA transcription: a possible termination site and demonstration of transcriptional control correlated with altered chromatin structure. *Cell*, **29**, 887–893.

Hoffmann, A., Oelgeschlager, T., and Roeder, R.G. (1997) Considerations of transcriptional control mechanisms: do TFIID-core promoter complexes recapitulate nucleosome-like functions? *Proceedings of the National Academy of Sciences of the United States of America*, **94**, 8928–8935.

Hsiao, K.K., Groth, D., Scott, M., Yang, S.L., Serban, H., Rapp, D., Foster, D., Torchia, M., DeArmond, S.J., and Prusiner, S.B. (1994) Serial transmission in rodents of neurodegeneration from transgenic mice expressing mutant prion protein. *Proceedings of the National Academy of Sciences of the United States of America*, **91**, 9126–9130.

Huang, M.T. and Gorman, C.M. (1990a) Intervening sequences increase efficiency of RNA 3' processing and accumulation of cytoplasmic RNA. *Nucleic Acids Research*, **18**, 937–947.

Huang, M.T. and Gorman, C.M. (1990b) The simian virus 40 small-t intron, present in many common expression vectors, leads to aberrant splicing. *Molecular & Cellular Biology*, **10**, 1805–1810.

Hutchins, C.J., Rathjen, P.D., Forster, A.C., and Symons, R.H. (1986) Self-cleavage of plus and minus RNA transcripts of avocado sunblotch viroid. *Nucleic Acids Research*, **14**, 3627–3640.

Ince, T.A. and Scotto, K.W. (1995) A conserved downstream element defines a new class of RNA polymerase II promoters. *Journal of Biological Chemistry*, **270**, 30249–30252.

Ishihara, H., Tashiro, F., Ikuta, K., Asano, T., Katagiri, H., Inukai, K., Kikuchi, M., Yazaki, Y., Oka, Y., and Miyazaki, J. (1995) Inhibition of pancreatic beta-cell glucokinase by antisense RNA expression in transgenic mice: mouse strain-dependent alteration of glucose tolerance. *FEBS Letters*, **371**, 329–332.

Izaurralde, E. and Mattaj, W. (1995) RNA export. *Cell*, **81**, 153–159.

Jaenisch, R., Jahner, D., Nobis, P., Simon, I., Lohler, J., Harbers, K., and Grotkopp, D. (1981) Chromosomal position and activation of retroviral genomes inserted into the germ line of mice. *Cell*, **24**, 519–529.

Johnson, D., Harrison, S., Pineda, N., Heinlein, C., Al-Shawi, R., and Bishop, J.O. (1995) Localization of the response elements of a gene induced by intermittent growth hormone stimulation. *Journal of Molecular Endocrinology*, **14**, 35–49.

Kamakaka, R.T. (1997) Silencers and locus control regions: opposite sides of the same coin. *Trends in Biochemical Sciences,* **22**, 124–128.

Karpinski, B.A., Yang, L.H., Cacheris, P., Morle, G.D., and Leiden, J.M. (1989) The first intron of the 4F2 heavy-chain gene contains a transcriptional enhancer element that binds multiple nuclear proteins. *Molecular & Cellular Biology,* **9**, 2588–2597.

Katsuki, M., Sato, M., Kimura, M., Yokoyama, M., Kobayashi, K., and Nomura, T. (1988) Conversion of normal behavior to shiverer by myelin basic protein antisense cDNA in transgenic mice. *Science,* **241**, 593–595.

Kieffer, L.J., Bennett, J.A., Cunningham, A.C., Gladue, R.P., McNeish, J., Kavathas, P.B., and Hanke, J.H. (1996) Human CD8 alpha expression in NK cells but not cytotoxic T cells of transgenic mice. *International Immunology,* **8**, 1617–1626.

Kioussis, D., Vanin, E., deLange, T., Flavell, R.A., and Grosveld, F.G. (1983) Beta-globin gene inactivation by DNA translocation in gamma beta-thalassaemia. *Nature,* **306**, 662–666.

Kistner, A., Gossen, M., Zimmermann, F., Jerecic, J., Ullmer, C., Lubbert, H., and Bujard, H. (1996) Doxycycline-mediated quantitative and tissue-specific control of gene expression in transgenic mice. *Proceedings of the National Academy of Sciences of the United States of America,* **93**, 10933–10938.

Koleske, A.J. and Young, R.A. (1995) The RNA polymerase II holoenzyme and its implications for gene regulation. *Trends in Biochemical Sciences,* **20**, 113–116.

Kornberg, R.D. and Lorch, Y. (1995) Interplay between chromatin structure and transcription. *Current Opinion in Cell Biology,* **7**, 371–375.

Kozak, M. (1987a) At least six nucleotides preceding the AUG initiator codon enhance translation in mammalian cells. *Journal of Molecular Biology,* **196**, 947–950.

Kozak, M. (1987b) An analysis of 5'-noncoding sequences from 699 vertebrate messenger RNAs. *Nucleic Acids Research,* **15**, 8125–8148.

Kozak, M. (1991a) An analysis of vertebrate mRNA sequences: intimations of translational control. *Journal of Cell Biology,* **115**, 887–903.

Kozak, M. (1991b) A short leader sequence impairs the fidelity of initiation by eukaryotic ribosomes. *Gene Expression,* **1**, 111–115.

Kozak, M. (1995) Adherence to the first-AUG rule when a second AUG codon follows closely upon the first. *Proceedings of the National Academy of Sciences of the United States of America,* **92**, 2662–2666.

Kramer, A. (1996) The structure and function of proteins involved in mammalian pre-mRNA splicing. *Annual Review of Biochemistry,* **65**, 367–409.

L'Huillier, P.J., Soulier, S., Stinnakre, M.G., Lepourry, L., Davis, S.R., Mercier, J.C., and Vilotte, J.L. (1996) Efficient and specific ribozyme-mediated reduction of bovine alpha-lactalbumin expression in double transgenic mice. *Proceedings of the National Academy of Sciences of the United States of America,* **93**, 6698–6703.

Lacy, E., Roberts, S., Evans, E.P., Burtenshaw, M.D., and Costantini, F.D. (1983) A foreign beta-globin gene in transgenic mice: integration at abnormal chromosomal positions and expression in inappropriate tissues. *Cell,* **34**, 343–358.

Laemmli, U.K., Kas, E., Poljak, L., and Adachi, Y. (1992) Scaffold-associated regions: *cis*-acting determinants of chromatin structural loops and functional domains. *Current Opinion in Genetics & Development,* **2**, 275–285.

Larsson, S., Hotchkiss, G., Andang, M., Nyholm, T., Inzunza, J., Jansson, I., and Ahrlund-Richter, L. (1994) Reduced beta 2-microglobulin mRNA levels in transgenic mice expressing a designed hammerhead ribozyme. *Nucleic Acids Research,* **22**, 2242–2248.

Levine, A., Cantoni, G.L., and Razin, A. (1991) Inhibition of promoter activity by methylation: possible involvement of protein mediators. *Proceedings of the National Academy of Sciences of the United States of America,* **88**, 6515–6518.

Lewis, J.D., Gunderson, S.I., and Mattaj, I.W. (1995) The influence of 5' and 3' end structures on pre-mRNA metabolism. *Journal of Cell Science – Supplement,* **19**, 13–19.

Magin, T.M., McEwan, C., Milne, M., Pow, A.M., Selfridge, J., and Melton, D.W. (1992) A position- and orientation-dependent element in the first intron is required for expression of the mouse hprt gene in embryonic stem cells. *Gene,* **122**, 289–296.

Matsumoto, K., Kakidani, H., Takahashi, A., Nakagata, N., Anzai, M., Matsuzaki, Y., Takahashi, Y., Miyata, K., Utsumi, K., and Iritani, A. (1993) Growth retardation in rats whose growth hormone gene expression was suppressed by antisense RNA transgene. *Molecular Reproduction & Development,* **36**, 53–58.

McCready, P.M., Hansen, R.K., Burke, S.L., and Sands, J.F. (1997) Multiple negative and positive cis-acting elements control the expression of the murine CD4 gene. *Biochimica et Biophysica Acta,* **1351**, 181–191.

McEwen, D.G. and Ornitz, D.M. (1998) Regulation of the fibroblast growth factor receptor 3 promoter and intron I enhancer by Sp1 family transcription factors. *Journal of Biological Chemistry,* **273**, 5349–5357.

McKnight, R.A., Shamay, A., Sankaran, L., Wall, R.J., and Hennighausen, L. (1992) Matrix-attachment regions can impart position-independent regulation of a tissue-specific gene in transgenic mice. *Proceedings of the National Academy of Sciences of the United States of America,* **89**, 6943–6947.

Mehtali, M., LeMeur, M., and Lathe, R. (1990) The methylation-free status of a housekeeping transgene is lost at high copy number. *Gene,* **91**, 179–184.

Mirkovitch, J., Mirault, M.E., and Laemmli, U.K. (1984) Organization of the higher-order chromatin loop: specific DNA attachment sites on nuclear scaffold. *Cell,* **39**, 223–232.

Mountford, P., Zevnik, B., Duwel, A., Nichols, J., Li, M., Dani, C., Robertson, M., Chambers, I., and Smith, A. (1994) Dicistronic targeting constructs: reporters and modifiers of mammalian gene expression. *Proceedings of the National Academy of Sciences of the United States of America,* **91**, 4303–4307.

Mountford, P.S. and Smith, A.G. (1995) Internal ribosome entry sites and dicistronic RNAs in mammalian transgenesis. *Trends in Genetics,* **11**, 179–184.

Moxham, C.M., Hod, Y., and Malbon, C.C. (1993a) Gi alpha 2 mediates the inhibitory regulation of adenylylcyclase in vivo: analysis in transgenic mice with Gi alpha 2 suppressed by inducible antisense RNA. *Developmental Genetics,* **14**, 266–273.

Moxham, C.M., Hod, Y., and Malbon, C.C. (1993b) Induction of G alpha i2-specific antisense RNA in vivo inhibits neonatal growth. *Science,* **260**, 991–995.

Munir, M.I., Rossiter, B.J., and Caskey, C.T. (1990) Antisense RNA production in transgenic mice. *Somatic Cell and Molecular Genetics,* **16**, 383–394.

Niwa, M. and Berget, S.M. (1991) Mutation of the AAUAAA polyadenylation signal depresses in vitro splicing of proximal but not distal introns. *Genes & Development,* **5**, 2086–2095.

Novina, C.D. and Roy, A.L. (1996) Core promoters and transcriptional control. *Trends in Genetics,* **12**, 351–355.

Oshima, R.G., Abrams, L., and Kulesh, D. (1990) Activation of an intron enhancer within the keratin 18 gene by expression of c-fos and c-jun in undifferentiated F9 embryonal carcinoma cells. *Genes & Development,* **4**, 835–848.

Oskouian, B., Rangan, V.S., and Smith, S. (1997) Regulatory elements in the first intron of the rat fatty acid synthase gene. *Biochemical Journal,* **324**, 113–121.

Palmiter, R.D., Behringer, R.R., Quaife, C.J., Maxwell, F., Maxwell, I.H., and Brinster, R.L. (1987) Cell lineage ablation in transgenic mice by cell-specific expression of a toxin gene. *Cell,* **50**, 435–443.

Palmiter, R.D., Sandgren, E.P., Avarbock, M.R., Allen, D.D., and Brinster, R.L. (1991) Heterologous introns can enhance expression of transgenes in mice. *Proceedings of the National Academy of Sciences of the United States of America,* **88**, 478–482.

Phi-Van, L. and Stratling, W.H. (1988) The matrix attachment regions of the chicken lysozyme gene co-map with the boundaries of the chromatin domain. *EMBO Journal,* **7**, 655–664.

Pyle, A.M. (1993) Ribozymes: a distinct class of metalloenzymes. *Science,* **261**, 709–714.

Richards, W.G., Carroll, P.M., Kinloch, R.A., Wassarman, P.M., and Strickland, S. (1993) Creating maternal effect mutations in transgenic mice: antisense inhibition of an oocyte gene product. *Developmental Biology,* **160**, 543–553.

Robertson, G., Garrick, D., Wu, W., Kearns, M., Martin, D., and Whitelaw, E. (1995) Position-dependent variegation of globin transgene expression in mice. *Proceedings of the National Academy of Sciences of the United States of America,* **92**, 5371–5375.

Roeder, R.G. (1996) The role of general initiation factors in transcription by RNA polymerase II. *Trends in Biochemical Sciences,* **21**, 327–335.

Ross, J. (1995) mRNA stability in mammalian cells. *Microbiological Reviews,* **59**, 423–450.

Ross, J. (1996) Control of messenger RNA stability in higher eukaryotes. *Trends in Genetics,* **12**, 171–175.

Ryan, T.M., Behringer, R.R., Martin, N.C., Townes, T.M., Palmiter, R.D., and Brinster, R.L. (1989) A single erythroid-specific DNase I super-hypersensitive site activates high levels of human beta-globin gene expression in transgenic mice. *Genes & Development,* **3**, 314–323.

Saez, E., No, D., West, A., and Evans, R.M. (1997) Inducible gene expression in mammalian cells and transgenic mice. *Current Opinion in Biotechnology,* **8**, 608–616.

Schlaeger, T.M., Bartunkova, S., Lawitts, J.A., Teichmann, G., Risau, W., Deutsch, U., and Sato, T.N. (1997) Uniform vascular-endothelial-cell-specific gene expression in both embryonic and adult transgenic mice. *Proceedings of the National Academy of Sciences of the United States of America,* **94**, 3058–3063.

Schmid, M. and Wimmer, E. (1994) IRES-controlled protein synthesis and genome replication of poliovirus. *Archives of Virology – Supplementum,* **9**, 279–289.

Shemer, R., Walsh, A., Eisenberg, S., Breslow, J.L., and Razin, A. (1990) Tissue-specific methylation patterns and expression of the human apolipoprotein AI gene. *Journal of Biological Chemistry,* **265**, 1010–1015.

Shockett, P., Difilippantonio, M., Hellman, N., and Schatz, D.G. (1995) A modified tetracycline-regulated system provides autoregulatory, inducible gene expression in cultured cells and transgenic mice. *Proceedings of the National Academy of Sciences of the United States of America,* **92**, 6522–6526.

Slack, J.L., Liska, D.J., and Bornstein, P. (1991) An upstream regulatory region mediates high-level, tissue-specific expression of the human alpha 1(I) collagen gene in transgenic mice. *Molecular & Cellular Biology,* **11**, 2066–2074.

Smith, J.D., Wong, E., and Ginsberg, M. (1995) Cytochrome P450 1A1 promoter as a genetic switch for the regulatable and physiological expression of a plasma protein in transgenic mice. *Proceedings of the National Academy of Sciences of the United States of America,* **92**, 11926–11930.

Sokol, D.L., Passey, R.J., MacKinlay, A.G., and Murray, J.D. (1998) Regulation of CAT protein by ribozyme and antisense mRNA in transgenic mice. *Transgenic Research,* **7**, 41–50.

Sokol, D.L. and Murray, J.D. (1996) Antisense and ribozyme constructs in transgenic animals. *Transgenic Research,* **5**, 363–371.

Soriano, P., Cone, R.D., Mulligan, R.C., and Jaenisch, R. (1986) Tissue-specific and ectopic expression of genes introduced into transgenic mice by retroviruses. *Science,* **234**, 1409–1413.

Sorkin, B.C., Jones, F.S., Cunningham, B.A., and Edelman, G.M. (1993) Identification of the promoter and a transcriptional enhancer of the gene encoding L-CAM, a calcium-dependent cell adhesion molecule. *Proceedings of the National Academy of Sciences of the United States of America,* **90**, 11356–11360.

Steinmetz, E.J. and Brow, D.A. (1996) Repression of gene expression by an exogenous sequence element acting in concert with a heterogeneous nuclear ribonucleoprotein-like protein, Nrd1, and the putative helicase Sen1. *Molecular & Cellular Biology,* **16**, 6993–7003.

Stief, A., Winter, D.M., Stratling, W.H., and Sippel, A.E. (1989) A nuclear DNA attachment element mediates elevated and position-independent gene activity. *Nature,* **341**, 343–345.

Stred, S.E., Stubbart, J.R., Argetsinger, L.S., Shafer, J.A., and Carter Su, C. (1990) Demonstration of growth hormone (GH) receptor-associated tyrosine kinase activity in multiple GH-responsive cell types. *Endocrinology,* **127**, 2506–2516.

Sun, Z. and Means, A.R. (1995) An intron facilitates activation of the calspermin gene by the testis-specific transcription factor CREM tau. *Journal of Biological Chemistry,* **270**, 20962–20967.

Svejstrup, J.Q., Vichi, P., and Egly, J.M. (1996) The multiple roles of transcription/repair factor TFIIH. *Trends in Biochemical Sciences,* **21**, 346–350.

Tate, P.H. and Bird, A.P. (1993) Effects of DNA methylation on DNA-binding proteins and gene expression. *Current Opinion in Genetics and Development,* **3**, 226–231.

Thompson, E.M., Christians, E., Stinnakre, M.G., and Renard, J.P. (1994) Scaffold attachment regions stimulate HSP70.1 expression in mouse preimplantation embryos but not in differentiated tissues. *Molecular & Cellular Biology,* **14**, 4694–4703.

Threadgill, D.W., Dlugosz, A.A., Hansen, L.A., Tennenbaum, T., Lichti, U., Yee, D., LaMantia, C., Mourton, T., Herrup, K., Harris, R.C., *et al.* (1995) Targeted disruption of mouse EGF receptor: effect of genetic background on mutant phenotype. *Science,* **269**, 230–234.

Tsien, J.Z., Chen, D.F., Gerber, D., Tom, C., Mercer, E.H., Anderson, D.J., Mayford, M., Kandel, E.R., and Tonegawa, S. (1996a) Subregion- and cell type-restricted gene knockout in mouse brain. *Cell,* **87**, 1317–1326.

Tsien, J.Z., Huerta, P.T., and Tonegawa, S. (1996b) The essential role of hippocampal CA1 NMDA receptor-dependent synaptic plasticity in spatial memory. *Cell,* **87**, 1327–1338.

Umezawa, A., Yamamoto, H., Rhodes, K., Klemsz, M.J., Maki, R.A., and Oshima, R.G. (1997) Methylation of an ETS site in the intron enhancer of the keratin 18 gene participates in tissue-specific repression. *Molecular & Cellular Biology,* **17**, 4885–4894.

Verrijzer, C.P. and Tjian, R. (1996) TAFs mediate transcriptional activation and promoter selectivity. *Trends in Biochemical Sciences,* **21**, 338–342.

Villarreal, L.P. and White, R.T. (1983) A splice junction deletion deficient in the transport of RNA does not polyadenylate nuclear RNA. *Molecular & Cellular Biology,* **3**, 1381–1388.

Wagner, R.W., Smith, J.E., Cooperman, B.S., and Nishikura, K. (1989) A double-stranded RNA unwinding activity introduces structural alterations by means of adenosine to inosine conversions in mammalian cells and Xenopus eggs. *Proceedings of the National Academy of Sciences of the United States of America,* **86**, 2647–2651.

Wallace, H., Ledent, C., Vassart, G., Bishop, J.O., and Al-Shawi, R. (1991) Specific ablation of thyroid follicle cells in adult transgenic mice. *Endocrinology,* **129**, 3217–3226.

Wallace, H., McLaren, K., Al-Shawi, R., and Bishop, J.O. (1994) Consequences of thyroid hormone deficiency induced by the specific ablation of thyroid follicle cells in adult transgenic mice. *Journal of Endocrinology,* **143**, 107–120.

Wallace, H., Pate, A., and Bishop, J.O. (1995) Effects of perinatal thyroid hormone deprivation on the growth and behaviour of newborn mice. *Journal of Endocrinology,* **145**, 251–262.

Wallace, H., Clarke, A.R., Harrison, D.J., Hooper, M.L., and Bishop, J.O. (1996) Ganciclovir-induced ablation non-proliferating thyrocytes expressing herpesvirus thymidine kinase occurs by p53-independent apoptosis. *Oncogene,* **13**, 55–61.

Walters, M.C., Fiering, S., Eidemiller, J., Magis, W., Groudine, M., and Martin, D.I. (1995) Enhancers increase the probability but not the level of gene expression. *Proceedings of the National Academy of Sciences of the United States of America,* **92**, 7125–7129.

Wang, D.M., Taylor, S., and Levy-Wilson, B. (1996) Evaluation of the function of the human apolipoprotein B gene nuclear matrix association regions in transgenic mice. *Journal of Lipid Research,* **37**, 2117–2124.

Wang, Y., O'Malley, B.W., Jr., Tsai, S.Y., and O'Malley, B.W. (1994) A regulatory system for use in gene transfer. *Proceedings of the National Academy of Sciences of the United States of America,* **91**, 8180–8184.

Weintraub, H. (1988) Formation of stable transcription complexes as assayed by analysis of individual templates. *Proceedings of the National Academy of Sciences of the United States of America,* **85**, 5819–5823.

Weng, A., Engler, P., and Storb, U. (1995) The bulk chromatin structure of a murine transgene does not vary with its transcriptional or DNA methylation status. *Molecular & Cellular Biology,* **15**, 572–579.

Wolffe, A.P. (1996) Chromatin and gene regulation at the onset of embryonic development. *Reproduction, Nutrition, Development,* **36**, 581–606.

Zimmerman, L., Parr, B., Lendahl, U., Cunningham, M., McKay, R., Gavin, B., Mann, J., Vassileva, G., and McMahon, A. (1994) Independent regulatory elements in the nestin gene direct transgene expression to neural stem cells or muscle precursors. *Neuron,* **12**, 11–24.

CHAPTER ELEVEN

Transgenic livestock

Many of the scientific and medical issues that are addressed by transgenic methods are in the mainstream of enquiries that were under way before the new technology became available. The transgenic approach adds a new dimension in these fields, often making it possible to answer old questions in a more meaningful way and opening up previously inaccessible areas of investigation.

Truly novel possibilities have been opened up by the advent of methods of generating transgenic livestock, by which we mean here rabbits, sheep, goats, pigs and cows. Many of these closely parallel the new developments in transgenic plants, and in common with them tend to have commercial applications. They include:

- Qualitative changes in animal products, such as wool, meat and milk. Some of these are improvements in the suitability of the product for well established purposes, for instance altering the casein composition of milk to simplify cheese-making. Others are intended to address new markets. As an example, attempts are being made to produce bovine milk that would be an adequate substitute for human milk.
- The use of animals as bioreactors for the production of human proteins, mainly for medical purposes.
  - There are potential advantages both of scale and in cost. Important target products include proteins that require post-translational modifications not available in microorganisms. The expectation is that such proteins will be more cheaply and reliably produced by animal bioreactors than by large-scale cell culture methods. In some cases the animal bioreactor potentially offers a less costly way of producing a product of established efficacy. In others, the increased scale of production that may be achievable could make available products that cannot be produced at affordable cost by other means.

  - Many of the products in question have in the past been isolated from human blood or from cadavers, with the well known risk of contamination by human infectious agents. Recipients of human products have contracted hepatitis (B and C), AIDS and CJD, and there is no guarantee that a new infective agent will not appear. While animal products eliminate this problem, they do introduce a hypothetical risk of zoonotic disease (disease caused by an animal infectious agent crossing the species barrier and infecting humans).
  - The long list of products under development includes α1-antitrypsin, blood clotting Factors VIII and IX, tissue plasminogen activator, fibrinogen and protein C.
  - Although the products mentioned could be most easily produced in the blood plasma, they are all being produced in the milk. There are good reasons for this. Transgenic animals could tolerate only limited concentrations of blood products in their bodies, whereas milk is effectively extracorporeal. Also, the harvesting of milk is non-invasive. Thirdly, milk is perceived by consumers to be benign.

- Radical changes in the dependency of livestock on nutrients by introducing biochemical pathways from other organisms, even from other phyla. An example is the attempt to introduce a prokaryotic cysteine biosynthetic pathway into sheep (Ward *et al.* 1994; Bawden *et al.* 1995). This approach impinges on the cost-effectiveness of husbandry. In relation to cost, it may be possible to extend the range of livestock into less favourable environments.
- Radical changes in the resistance of animals to diseases, for instance by modifying the structure of a protein that acts as a receptor for a virus.

Compared with mice livestock are costly, roughly in relation to their size. Caring for them is costly in terms of space requirements, housing, labour and feed. The necessary surgical procedures require expensive facilities. All this places transgenic livestock out of reach of most academics so that most publicly funded work is carried out in national agricultural research institutes. More, however, is carried out by commercial organisations. As a result of the involvement of companies, much of the work is not in the public domain, and for that reason the account given here may be out-of-date or inaccurate in some or even many respects.

Up to now virtually all transgenic livestock animals have been generated by pronuclear microinjection. Valiant attempts to develop functional ES cells have not yet resulted in reports of germ line transmission from chimeras, so this transgenic route is still unavailable. A very significant development, of course, is nuclear transfer technology (Wilmut *et al.* 1997; Schnieke *et al.* 1997; Cibelli *et al.* 1998), which promises to provide a substitute for the ES cell route and may even be superior to it in some ways.

Pronuclear microinjection

Pronuclear microinjection technique is essentially the same for livestock as for the mouse, but different protocols for preparing donor and recipient dams are dictated by their different patterns of reproductive physiology. Short articles on livestock protocols can be found in Houdebine (Various Authors 1997). The optimal time interval between fertilisation and injection varies according to the different timing of the formation of pronuclear membranes. Immediate post-injection events will also differ to some extent because of differences between mammals in the immediate post-fertilisation and early cleavage stages, for instance in the time of onset of DNA replication and in the stage at which RNA synthesis is initiated in the embryo (Fulka, Jr. *et al.* 1996).

The main problems encountered relate to gene expression. This is most obvious when the objective of the experiment is production, e.g. of a pharmaceutical protein in milk, where the importance of both tissue-specificity and a high level of expression are self-evident. Construct design is of course vital to transgene expression. Tissue-specific promoters are available to give appropriate ectopic expression, but a range of effects discussed in Chapter 10 needs to be taken into account, generally against a background of inadequate information. In maximising expression the two main issues are the rate of transcription and processing and the rate of mRNA decay. Chromosomal position effect is the most serious potential problem. Transgenes susceptible to position effect are transcribed at different rates in different transgenic lines and fail to show copy-number dependence. Also, there is a tendency for many of the lines to be transcriptionally inert. These features are particularly important when the cost of each founder animal is high. Thus it is very important, if possible, that the transgene should not be susceptible to position effect and should display copy-number dependence. As a general rule this is more likely to be the case when more of the gene and its flanking regions are introduced. Thus in the absence of complete information as to what DNA sequences are required for full expression, BAC and YAC fragments may be favoured (see for example Schedl *et al.* 1993). An alternative is to add in LCRs, if available, or MARs.

At the same time, many native genes are not transcribed at the maximum possible rate. This can be due to many features of the structure of the gene and also of its normal chromosomal environment. Thus a transgene (or promoter) that perfectly replicates the transcriptional regulation of the normal gene will usually not be transcribed maximally, and potential opportunities will often exist to increase the transcription rate by modifying gene structure. To do this intelligently requires a very detailed knowledge of the functional aspects of the structure, and unfortunately every gene seems to be unique in this respect.

As discussed in Chapter 10, the level of cytoplasmic mRNA depends as much on mRNA stability as on the rate of transcription. Since the processes that determine RNA stability are not at all well understood, it is not easy to decide on the best course of action. An obvious step would be to remove known destabilising elements from the 3' non-coding region. Another might be to adopt into the gene the 3' end of an mRNA species known to have a long half-life.

Expression level is likely also to be important in attempts to improve natural animal products, where in some instances it may be very difficult to determine the optimum level, which is unlikely to be the maximum attainable level. It is simple enough to determine optimum levels of secretion when humanising bovine milk, but modifying metabolic processes so as to approximate the best levels is likely to be difficult and it may be even more difficult to strike an optimum balance between positive direct effects and negative side effects. An obvious and simple example is the serious deleterious effects that occurred in a range of livestock engineered to produce growth hormone ectopically (see for example Pursel *et al.* 1990, Rexroad *et al.* 1990). These problems stemmed from the fact that GH has multiple (pleiotropic) effects that affect many organs. It seems likely that introducing any new pathway that uses an existing metabolite as its initial substrate will cause widespread perturbations in linked pathways, with the danger that some of these could be deleterious. This is a well-worked problem in the field of fermentation. There, pilot experiments are available to test modified organisms and nutritional conditions can be rather easily controlled. But there is no reliable pilot for a transgenic cow or pig, the mouse being altogether too different from any farm animal to serve as a reliable predictor of transgene action at the physiological level.

The problems are further compounded when, as is often the case, the objective is to obtain the ectopic expression of a protein. This requires the coding region of one gene (B) to be combined with the promoter and other elements of another gene (A) that will direct expression to the desired cells. The principles are in general similar to those already discussed, with the added complication that the activity of some promoters is strongly affected by an intronic enhancer that is usually located downstream of the translation initiation codon. Since this codon would initiate the translation of A rather than B it requires to be eliminated or neutralised. Various alternative strategies suggest themselves. Coding region B might be substituted for A downstream of the enhancer, which will usually be in intron 1. Then the start codon of A, and any other potential translation initiation codons upstream of the intron, might be deleted or altered, or an internal ribosome entry site might be added at the junction of the intron and the coding region of B. A more conventional arrangement is to make the junction between A and B in the 5' non-coding region of both genes; in this case the intron containing the enhancer might be substituted for the equivalent intron of B, or if it is sufficiently well defined the enhancer might simply be inserted into an intron of B.

Gene targeting

With the development of the nuclear transfer technique (Chapter 8), gene targeting (Chapter 9) is likely to become a major route to creating transgenic livestock. Concepts can be tested in mice using ES cell technology to attain the same goals. Two alternative strategies would be appropriate to different objectives:

- Where the objective is to maximise expression, it may be possible to identify chromosomal sites that favour expression, perhaps sites remote from heterochromatic regions. If so, different genes could be targeted to those sites by homologous or site-directed recombination.
- Gene targeting should prove very powerful in permitting the modification of existing natural products, since the new gene is substituted for the old. If only the coding region of a gene is altered, the transcriptional and processing signals are essentially unchanged and the probability is that RNA stability will not be greatly affected. In this case the outcome should be more predictable, although protein stability would remain to be accounted for.

For discussions of specific objectives, the reader is referred to review articles, some of which are cited below.

References

Bawden, C.S., Sivaprasad, A.V., Verma, P.J., Walker, S.K., and Rogers, G.E. (1995) Expression of bacterial cysteine biosynthesis genes in transgenic mice and sheep: toward a new in vivo amino acid biosynthesis pathway and improved wool growth. *Transgenic Research*, **4**, 87–104.

Cibelli, J.B., Stice, S.L., Golueke, J., Kane, J.J., Jerry, J., Blackwell, C., Ponce de Leon, F.A., and Robl, J.M. (1998) Cloned transgenic calves produced from non-quiescent foetal fibroblasts. *Science*, **280**, 1256–1258.

Fulka, J., Jr., First, N.L., and Moor, R.M. (1996) Nuclear transplantation in mammals: remodelling of transplanted nuclei under the influence of maturation promoting factor. *Bioessays*, **18**, 835–840.

Pursel, V.G., Bolt, D.J., Miller, K.F., Pinkert, C.A., Hammer, R.E., Palmiter, R.D., and Brinster, R.L. (1990) Expression and performance in transgenic pigs. *Journal of Reproduction & Fertility – Supplement*, **40**, 235–245.

Rexroad, C.E., Jr., Hammer, R.E., Behringer, R.R., Palmiter, R.D., and Brinster, R.L. (1990) Insertion, expression and physiology of growth-regulating genes in ruminants. *Journal of Reproduction & Fertility – Supplement*, **41**, 119–124.

Schedl, A., Montoliu, L., Kelsey, G., and Schütz, G. (1993) A yeast artificial chromosome covering the tyrosinase gene confers copy number-dependent expression in transgenic mice. *Nature*, **362**, 258–261.

Schnieke, A.E., Kind, A.J., Ritchie, W.A., Mycock, K., Scott, A.R., Ritchie, M., Wilmut, I., Colman, A., and Campbell, K.H. (1997) Human factor IX transgenic sheep produced by transfer of nuclei from transfected fetal fibroblasts. *Science*, **278**, 2130–2133.

Various Authors. (1997) Part II. Section A. Gene transfer into mammal embryos. In *Transgenic animals, generation and use*, L.M. Houdebine (ed), Harwood Academic Press, Amsterdam, pp. 7–44.

Ward, K.A., Leish, Z., Bonsing, J., Nishimura, N., Cam, G.R., Brownlee, A.G., and Nancarrow, C.D. (1994) Preventing hair loss in mice. *Nature*, **371**, 563–564.

Wilmut, I., Schnieke, A.E., McWhir, J., Kind, A.J., and Campbell, K.H. (1997) Viable offspring derived from fetal and adult mammalian cells. *Nature*, **385**, 810–813.

Further reading

General

Cameron, E.R. (1997) Recent advances in transgenic technology. *Molecular Biotechnology*, **7**, 253-265.

Velander, W.H., Lubon, H., and Drohan, W.N. (1997) Transgenic livestock as drug factories. *Scientific American*, **276**, 70–74.

Ward, K.A. (1997) The transfer of isolated genes having known functions. In *Transgenic animals, generation and use*, L.M. Houdebine (ed), Harwood Academic Press, Amsterdam, pp. 511–518.

Bioreactors

Colman, A. (1996) Production of proteins in the milk of transgenic livestock: problems, solutions, and successes. *American Journal of Clinical Nutrition*, **63**, 639S–45S.

Colman, A. (1998) Production of therapeutic proteins in the milk of transgenic livestock. *Biochemical Society Symposia*, **63**, 141–147.

Drohan, W.N. (1997) The past, present and future of transgenic bioreactors. *Thrombosis & Haemostasis*, **78**, 543–547.

Wall, R.J., Kerr, D.E., and Bondioli, K.R. (1997) Transgenic dairy cattle: genetic engineering on a large scale. *Journal of Dairy Science*, **80**, 2213–2224.

Milk composition

Martin, P. and Grosclaude, F. (1993) Improvement of milk protein quality by gene technology. *Livestock Production Science*, **35**, 95–115.

Mercier, J.C. and Vilotte, J.L. (1997) The modification of milk protein composition through transgenesis: progress and problems. In *Transgenic animals, generation and use*, L.M. Houdebine (ed), Harwood Academic Press, Amsterdam, pp. 473–482.

Wool

Bullock, D.W., Damak, S., Jay, N.P., Su, H.Y., and Barrell, G.K. (1997) Improved wool production from insulin-like growth factor 1 targeted to the wool follicle in transgenic sheep. In *Transgenic animals, generation and use*, L.M. Houdebine (ed), Harwood Academic Press, Amsterdam, pp. 507–510.

Powell, B.C., Walker, S.K., Bawden, C.S., Sivaprasad, A.V., and Rogers, G.E. (1994) Transgenic sheep and wool growth: possibilities and current status. *Reproduction, Fertility, & Development*, **6**, 615–623.

Xenotransplantation

Michler, R.E., Minanov, O.P., Artrip, J.H., and Itescu, S. (1997) Immature host for xenotransplantation. *World Journal of Surgery*, **21**, 924–931.

Platt, J.L. and Logan, J.S. (1997) The generation and use of transgenic animals for xenotransplantation. In *Transgenic animals, generation and use*, L.M. Houdebine (ed), Harwood Academic Press, Amsterdam, pp. 455–460.

Glossary

α-amanitin	Specific inhibitor of RNA polymerase II.
α-fetoprotein	Foetal homologue of serum albumin.
animal bioreactor	An animal that produces a product, usually a human protein.
acentric	Description of a chromosome fragment without a centromere.
acetylase	An enzyme that adds an acetyl group (–CH_2COOH).
acrosome	Vesicle at the tip of the spermatozoon containing enzymes required for successful fertilisation.
acrylamide	Used as a matrix to support electrophoresis gels, e.g. for sequencing.
actinomycin	Inhibitor of RNA synthesis.
ADA	Adenosine deaminase.
adenovirus	Animal virus used as an expression vector.
agarose	Used as a matrix to support electrophoresis gels.
allophenic	Chimeric.
AMH	*See* anti-Müllerian hormone.
aminopterin	Inhibitor of *de novo* AMP, GMP and TMP synthesis.
aneuploid	Having an abnormal chromosome complement.
anti-Müllerian hormone	Also Müllerian inhibitory substance (MIS). A polypeptide hormone produced by the testis Sertoli cells early in development, it promotes the regression of the proto-Müllerian ducts.
antipolar	Of opposite polarity; used to describe the orientation of the complementary strands of DNA.

antisense	Describes a DNA or RNA molecule complementary to the sense or reading strand.
antrum	The fluid-filled space that opens up in the maturing oocyte follicle.
apoB	Apolipoprotein B.
apoptosis	Programmed cell death (as opposed to necrosis).
aprt	The adenosine phosphoribosyltransferase gene.
ATG	The translation initiation codon (sense-strand DNA form).
autosome	Chromosome other than X and Y.
avidin	Protein that binds biotin very firmly.
β-galactosidase	Enzyme that cleaves a galactoside residue from a sugar.
BAC	Literally bacterial artificial chromosome. In fact, a cloning vector based on the replication origin and other necessary parts of a large bacterial plasmid of the resistance transfer factor type.
backcross	Mating between offspring of a cross and one of the parental strains.
bacteriophage	Bacterial virus.
*Bam*HI	Restriction enzyme, recognises the sequence GGATCC.
bFGF	Basic fibroblast growth factor, a mitogen.
biallelic	Pertaining to both alleles of a gene, as in 'biallelic expression'.
biparental	Pertaining to both parents, as in 'biparental expression of the X chromosome'.
blastocoel	Fluid-filled cavity of the blastocyst.
blastocyst	In mammals, the stage of development between morula and gastrula.
blastomere	Cell of an early embryo, actually *before* the blastocyst stage.
CaMKII	Calcium calmodulin-dependent kinase.
cAMP	3',5' cyclic adenosine monophosphate.
capsid	The core structure of an envelope virus, e.g. a retrovirus.
CBA	Inbred mouse strain.
CDK	*See* cyclin-dependent kinases.
cDNA	*See* copy DNA.
centimorgan	Unit of map distance, from recombination frequency.
centromere	Specialised part of the chromosome, attaches to the spindle at mitosis and meiosis.
centromeric	Pertaining to centromeres, as in centromeric heterochromatin.

c-fos	Protooncogene.
chimera	Animal made up of two or more cell types differing in origin.
chloramphenicol	An antibiotic.
CHO	Chinese hamster ovary cell line.
chromatid	Daughter chromosome, before separation from sister.
chromogenic substrate	A colourless substrate that is enzymatically converted to a coloured product.
cis	On the same side, e.g. genes in *cis* configuration are on the same chromosome. *Cis*-acting means acting on an element on the same chromosome or DNA fragment, e.g. an enhancer usually acts on the cognate promoter in *cis*. See also *trans*.
CJD	Creutzfeld-Jakob disease, form of spongiform encephalopathy.
CKI	Inhibitor of cyclin-dependent kinase.
clonal	A genetically identical group (e.g. of cells), all descendants of one individual.
cM	*See* centimorgan.
c-myc	Protooncogene.
codon	Three RNA nucleotides that specify an amino acid, or alternatively transcriptional initiation or termination, or the corresponding three DNA nucleotides.
cointegrate	A circular plasmid molecule created by the joining together of two smaller circular plasmids.
compaction	The early developmental process whereby the loosely aggregated cells (usually between 8 and 16) are transformed into a compact sphere made up of inner and outer cells with different properties.
concatemer	A number of DNA molecules joined end-to-end.
congenic strains	Strains that differ genetically only at a single locus.
constitutive	Invariably present, not susceptible to alteration, as in 'constitutive synthesis', 'constitutive heterochromatin'.
contig	A genetic region defined by overlapping cloned segments.
copy DNA	Prepared by copying an RNA template (or mixture of templates) with the enzyme reverse transcriptase and an oligodeoxynucleotide primer.

cosmid	Cloning vector with the replication origin of a multicopy plasmid and the bacteriophage l cohesive-end site, can be packaged into bacteriophage particles *in vitro*.
CpA	Dinucleotide.
CpG	Dinucleotide that attracts C-methylation.
c-ras	Protooncogene.
Cre	Enzyme that brings about site-specific recombination between *loxP* sites.
cryopreservation	Preservation by storage in liquid nitrogen or at a similar low temperature.
cycler	A machine that runs automatically through a cycle of predetermined temperature changes, used in PCR protocols.
cyclin-dependent kinase	Cyclin-dependent kinases are involved in the regulation of the cell cycle.
cyclins	Cofactors of cyclin-dependent kinases.
cytochalasin	Drug that prevents microtubule formation.
cytochrome	Haem-containing protein involved in electron transport.
cytokines	Regulatory proteins secreted by a variety of cell types; have numerous effects on other cells.
cytomegalovirus	Human pathogen, member of the herpesvirus group.
cytotoxic	Toxic to cells.
Dalton, Da	The unit of molecular mass; the mass of a hydrogen atom is about 1 Da.
deacetylases	Enzymes that remove acetyl (CH_2COOH) groups.
deadenylation	Removal of the 3' poly-A sequence from mRNA.
deaminase	An enzyme that removes amino (NH_3) groups.
decapping	Removal of the 5' cap structure from mRNA.
demethylation	Removal of methyl (CH_3) groups.
dephosphorylation	Removal of phosphate (H_2PO_4) groups.
DIA	Another name for LIF, *q.v.*
dicentric	Describes a chromosome with two centromeres.
Dictyostelium	Slime mould (*Dictyostelium discoideum*).
dihydrotestosterone	Derivative of testosterone, affects development of male genitalia.
dimer	Two joined molecules.
dinucleotide	Two joined nucleotides, e.g. CpGp.
DNase	Deoxyribonuclease, usually deoxyribonuclease I, an endonuclease.
Dnmt	Locus of the gene that codes for DNA-cytosine-5-methytransferase.
doxycycline	Tetracycline analogue.

dpc	Days *post coitum.*
Drosophila	Fruit fly (usually *Drosophila melanogaster*).
DSB	Double-strand break in DNA.
Dt-A	Diphtheria toxin A protein.
E.coli	*Escherichia coli*, gut bacterium, principal tool of *in vitro* DNA manipulation.
ecdysone	Insect steroid hormone.
*Eco*RI	Restriction enzyme, recognises the sequence AAGCTT.
ectopic	Not in the usual place, as in 'ectopic pregnancy'.
EC cells	Embryonal carcinoma cells, derived from blastocysts implanted ectopically *in vivo*; similar to ES cells but usually not totipotent.
EG cells	Embryonic germ cells, modified PGCs (*q.v.*) that share properties, such as the ability to contribute to the embryo in chimeras, with ES cells.
elastase	A protease.
electrofusion	Fusion together of cells by applying a voltage potential that destabilises cell membranes.
electroporation	Introduction of substances into cells by applying a membrane-destabilising voltage potential.
ELISA	Enzyme-linked immuno-sorbent assay. An antibody-based quantitative assay for specific antigens.
embryogenesis	Development of the embryo from the zygote.
embryoid body	Cluster of cells that develops from EC or ES cells in suspension structure bearing a superficial resemblance to a morula.
encephalomyocarditis virus	Picornavirus with internal ribosome entry site.
endonuclease	A nuclease that cleaves an RNA or DNA polynucleotide internally.
endonucleolytic	Classifies the cleavage of a polynucleotide as internal.
enhancer	Cluster of DNA response elements to which regulatory protein factors bind, resulting in increased transcriptional initiation from a nearby promoter.
enucleated	With nucleus removed, as in 'enucleated oocyte'.
env	Envelope gene of a retrovirus.
ENV	Envelope protein.
epiblast	Part of the blastocyst, derived from the inner cell mass, destined to become the embryo proper.
epistasis	Describes the situation in which a homozygous recessive mutation at one locus is overridden or concealed by particular allelic combinations at another.

erythroid	Pertaining to blood cells.
erythropoiesis	Blood cell production.
erythropoietin	Polypeptide hormone that stimulates erythropoiesis.
ES cells	Embryonic stem cells, derived from epiblast cells of blastocysts explanted *in vitro*; ES cell lines contribute to the embryo in chimeras and many contribute to the germ line.
EST	*See* expressed sequence tag.
ethidium bromide	Dye that binds to RNA and particularly strongly to DNA and fluoresces under UV illumination.
eukaryotes	Organisms with nuclei in their cells.
exons	Those parts of a gene not removed from the RNA transcript by pre-mRNA splicing.
exonuclease	Nuclease that degrades a nucleic acid progressively from the end(s).
explant	Piece of tissue placed in a dish for culture.
expressed sequence tag	Nucleotide sequence derived from a cDNA (*q.v.*) molecule randomly extracted from a library (*q.v.*) of cDNA sequences prepared using as template a mixed population of mRNA molecules.
extrachromosomal	Nuclear, but not incorporated into a chromosome.
extracorporeal	Outside the body.
extraembryonic	Accessory tissues outside the embryo proper.
FACS	Fluorescence-activated cell sorter, also used to sort chromosomes.
feminisation	Making female or more like a female.
FGF	Fibroblast growth factor, a mitogen.
fetus	*See* foetus.
FIAU	Antiherpetic analogue of thymidine, phosphorylated by some herpesvirus thymidine kinases but not by cellular kinases and lethal to cells when phosphorylated.
fibroblast	Cell type found in all tissues.
FISH	Fluorescence *in situ* hybridisation, used to locate intracellular macromolecules.
FLP	Yeast recombinase that brings about site-specific recombination specifically between two FRT sites.
foetus (fetus)	Stage of intrauterine development after the embryo.
FRT	*See* FLP.
FSH	Follicle stimulating hormone, a pituitary gonadotrophin.
galactose	A hexose sugar.
galactoside	Galactose residue within a disaccharide or polysaccharide.

gamete	Mature germ cell.
gametogenesis	Maturation of germ cells.
Ganciclovir	Antiherpetic analogue of thymidine, phosphorylated by some herpesvirus thymidine kinases but not by cellular kinases and lethal to cells when phosphorylated.
gastrula	Early embryo with ectoderm, mesoderm and endoderm.
gastrulation	Process of cell migration and rearrangement by which the blastula is transformed into the gastrula during early development.
genome	All the genetic material (DNA) that makes up one haploid set of chromosomes.
genotype	All the genes of an organism (both copies of each).
germ line	Cells designated at an early stage of development to form the germ cells.
GFP	Green fluorescent protein. A reporter gene product that fluoresces under UV light.
GH	Growth hormone; GH genes are used as reporters in some circumstances.
globin	Polypeptide part of haemoglobin.
glucocorticoid	Steroid hormones such as cortisone.
glycoprotein	Any protein with covalently attached carbohydrate.
gonadal	Pertaining to the gonads.
gonadotrophins	Polypeptide sex hormones.
gp130	Transmembrane protein that forms part of several cytokine receptors, including LIF receptor and oncostatin M receptor.
granulosa cells	Cells surrounding the early oocyte that develop into follicle cells.
haematopoiesis	Process of blood cell production.
HCG	Human chorionic gonadotrophin, used to stimulate ovulation in mice.
helicase	Enzyme that unwinds a polynucleotide double helix.
hemimethylase	Enzyme that methylates the C residue of a CpG doublet on one DNA strand when the CpG doublet on the other strand is already methylated.
hepatocyte	Principal liver cell type, liver parenchyma cell.
heterochromatin	Condensed chromatin.
heterodimer	Dimer composed of two different molecules. Antonym – homodimer.
heteroduplex	Duplex of different polypeptides. Antonym – homoduplex.

heterokaryon	Cell with multiple nuclei of at least two types.
heterozygote	Animal with two different chromosome sets.
hexaparental	Chimera of three cell types (six chromosome sets and 'parents').
*Hha*I	Restriction endonuclease, sequence GCGC, sensitive to CpG methylation.
histocompatibility loci	Loci of alleles that cause rejection of grafted tissues.
histones	Protein components of chromatin; H2A, H2B, H3 and H4 make up the nucleosomes while H1 is the 'linker' histone.
Holliday junction	Junction formed by strand exchange between paired matched DNA duplexes followed by repair and ligation.
holoenzyme	An enzyme or enzyme complex with all its parts present.
homodimer	A dimer of two like molecules.
homopolymer	A polymer made up of many copies of the same unit.
homozygote	Animal with two identical chromosome sets.
*Hpa*II	Restriction endonuclease, sequence CCGG, sensitive to CpG methylation.
hph	Bacterial gene conferring resistance to hygromycin, used as a selective reporter gene in animal cells.
hprt	Hypoxanthine-guanine phosphoribosyltransferase gene; the enzyme converts guanine and hypoxanthine to GMP and inosine monophosphate respectively so that they can be recycled; a 'scavenging' enzyme.
HSV	Herpes simplex virus.
HSVtk	Herpes simplex virus thymidine kinase gene.
hybridase	Enzyme that catalyses duplex formation between complementary DNA strands.
hygromycin	Antibiotic, *see hph*.
hypercholesterolaemic	The condition of having an elevated level of circulating cholesterol.
hypomethylated	Not fully methylated.
hypophosphorylated	Not fully phosphorylated.
hypoxanthine	Intermediate base in purine degradation.
ICM	Inner cell mass, totipotential cells of an early blastocyst.
immunofluorescence	Technique that combines the detection of a macromolecule with a specific antibody and visualisation of the bound antibody by means of a fluorescent tag, used on histological preparations to localise the macromolecule.

immunoprecipitation	Precipitation of a particular protein from a mixture of proteins with a specific antibody. The immunoprecipitate can be analysed by gel electrophoresis for example.
immunosurgery	Isolation of a cell type by using specific immobilised antibodies to bind to it, or alternatively to unwanted cells.
imprint	Signal attached to a gene during gametogenesis that persists and marks it as paternal or maternal when the gamete contributes to a zygote.
inducer	Chemical that switches a gene on by binding to a gene-specific regulatory protein.
inosine	The nucleoside of hypoxanthine.
Inr	Transcription initiation signal.
integrase	Enzyme that brings about the integration of one DNA molecule into another, usually through site-specific recombination.
intergenic	Describes the DNA regions between genes.
interleukins	Many cytokines are referred to as interleukins, e.g. IL-1, IL-8.
internal ribosome entry site	Region found in genomes of small RNA viruses, among others, that allows ribosomes to initiate polypeptide synthesis efficiently from an internal initiation codon.
internucleosomal	Describes the DNA that lies between nucleosomes.
interphase	The phase of the cell cycle between mitoses, comprised of G_1, S and G_2 phases.
interspecific cross	Cross between interfertile species.
intron	Those parts of the primary transcript that are excised by pre-mRNA splicing.
IRES	*See* internal ribosome entry site.
isogenic	Genetically identical mice are isogenic. Used to describe cloned DNA isolated from cells into which it is to be introduced.
isotypes	Different proteins that perform the same function in the same organism. As isozyme, but not confined to enzymes.
isozymes	Different enzymes that perform the same function in the same organism; often, but not always, stage- or tissue-specific.
JAK	*See* Janus kinases.
Janus kinases	Family of protein kinases that associate with hormone and cytokine receptors, such as growth hormone receptor and LIF receptor.
karyokinesis	Separation of daughter nuclei at meiosis or mitosis.

karyotype	The chromosomal complement of an animal or cell line, usually as visualised during chromosome condensation prior to mitosis.
kDa	One thousand Daltons.
kilobase	One thousand bases (of DNA or RNA).
lactoglobulin	A milk protein.
lactotrophs	Pituitary cells that synthesise and secrete prolactin.
*lac*Z	Bacterial gene that codes for β-galactosidase.
LCR	*See* locus control region.
LDH	The enzyme lactate dehydrogenase.
Leukemia inhibitory factor	Polypeptide cytokine that maintains embryonic stem cells in the undifferentiated state *in vitro*. However, not essential for early development *in vivo*.
Leydig cells	Extratubular cells in the testis that synthesise testosterone.
LH	*See* luteinising hormone.
LIF	*See* leukemia inhibitory factor.
LIF receptor	Transmembrane receptor related to erythropoietin and growth hormone receptors. The active receptor complex also contains gp130.
LINEs	*See* long interspersed repeat elements.
lipofection	Introduction of macromolecules into cells by means of complexes between the macromolecules and charged lipid vesicles, positively charged in the case of DNA.
locus control region	Chromosomal region containing response elements for *trans*-acting regulatory proteins that stimulates transcription from a nearby gene or group of genes. Tissue-specific and normally associated with genes that are expressed at a high level in one or few tissues.
long interspersed repeat elements	Families of dispersed longer repetitive DNA elements such as L1 in humans and IAP in the mouse.
long-term potentiation	A prolonged increase in hippocampal synaptic transmission induced by stimulation.
long terminal repeat	Region that is duplicated directly at the ends of a retroviral provirus. The LTR at the 5' end provides the promoter for proviral transcription, which is initiated within the LTR. The 3' LTR provides the cleavage and polyadenylation signal. The LTR is compounded from sequences at the 5' and 3' ends of the viral RNA.
loxP	Recognition site for the Cre site-specific recombinase of bacteriophage P1.

LTP	*See* long-term potentiation.
LTR	*See* long terminal repeat.
luciferase	Firefly protein that produces light with the consumption of ATP. The *luc* gene is used as a very sensitive reporter gene.
luteinising hormone	Pituitary gonadotrophin. A pulse of elevated circulating LH provides the stimulus for ovulation by stimulating progesterone production by the follicle cells.
lysozyme	An enzyme that attacks bacterial cell walls.
MARs	*See* matrix attachment regions.
masculinisation	Making male or more like a male.
matrix attachment regions	DNA regions operationally defined as the anchor points for attachment of the chromosomal DNA to the nuclear matrix or scaffold; alternatively scaffold attachment regions, SARs.
maturation promoting factor	M-phase promoting factor; complex of cyclin B and a cyclin-dependent kinase that controls the entry into mitosis; initiates germinal vesicle breakdown in oocytes.
MDa	One million Daltons.
MeCpG	Methylated CpG doublet (in DNA).
megabase	One million bases (of DNA).
melanocyte	Pigment cell.
methylase	An enzyme that adds methyl (CH_3) groups.
methylation	Addition of methyl groups.
methyltransferase	As methylase.
microcell	Cell fragments containing fragmented nuclei (micronuclei) each with only one or a few chromosomes; produced by treatment with colcemid and cytochalasin B, which inhibit the polymerisation of tubulin and actin respectively.
microphthalmia	Condition of having very small eyes.
microsatellite	Simple sequence DNA based on a very small repeat unit (e.g. CAA).
midpiece	Part of the spermatozoon between the head and the tail that contains the mitochondria.
minigene	A gene reduced in size, usually by removal of some or all of its introns.
minisatellite	Simple sequence DNA based on a small (*ca.* 10 bases) repeat unit.
MIS	Müllerian duct inhibitory substance. *See* anti-Müllerian hormone.
mitomycin c	Drug that intercalates into DNA and inhibits its replication.
mitosis	Normal cell division, with segregation of daughter chromosomes to daughter cells.

monoallelic expression	Expression of only one of the two alleles of a gene, due to an epigenetic process and not related to mutational inactivation. Thus, X-inactivation leads to monoallelic expression from the active X and uniparental expression is monoallelic.
monolayer	Layer one molecule thick.
morula	Early embryo, 8–32 cells, after compaction but before the blastocoel develops.
mosaic	An animal made up of at least two genetically different cell types that have developed from one zygote.
MPF	*See* maturation promoting factor.
mRNA	Messenger RNA.
*Msp*I	Restriction endonuclease, recognition sequence CCGG, *not* methylation-sensitive.
Müllerian ducts	Female genital primordia, regress under the action of AMH.
multicopy	Bacterial plasmid that exists in the cells in many copies per bacterial chromosome copy.
MutS	Bacterial protein that identifies mismatched base pairs in DNA.
neo	Bacterial gene used as a reporter in animal cells; confers resistance to G418.
neonatal	Soon after birth.
neurotrophic	Affecting neurons.
nonpermissive conditions	Culture conditions that preclude conditional expression.
*Not*I	Restriction endonuclease, recognition sequence GCGGCCGC.
nucleolytic	Refers to the process of nucleic acid degradation.
nucleosome	Ball-like structure of eight histones of four types (H2A, H2B, H3 and H4) around which DNA is wrapped in chromatin.
oestradiol	Female steroid hormone.
oligo-dT	A short polymer of deoxythymidylate residues.
oligodeoxynucleotide	A short deoxynucleotide polymer.
oligonucleotide	A short nucleotide polymer.
oncostatin M	A cytokine.
oocyte	The female gamete before fertilisation, including premeiotic stages.
oogenesis	The process of oocyte development.
osteoclasts	Cells responsible for depositing bone.
ovariectomy	Surgical removal of the ovaries.
PAC	Cloning vector based on bacteriophage P1.

paracrine stimulation	Hormonal stimulation originating in nearby cells (as opposed to systemic endocrine stimulation).
passage	Subculture of animal cells, essentially to reduce their density and at the same time provide fresh growth medium.
PCR	Polymerase chain reaction
penetrance	Used when not all individuals carrying the same altered gene show the corresponding phenotype, generally due to differences in genetic background. Low penetrance means that few exhibit the phenotype.
pericentric	Around the centromere, as in 'pericentric heterochromatin'.
pericentromeric	Same as pericentric.
perinatal	Around the time of birth, i.e. just before, at, or just after birth.
PGCs	*See* primordial germ cells.
pgk	Phosphoglycerate kinase gene.
phagocytosis	Internalisation of extracellular matter by engulfment (by pseudopodia).
phosphatase	An enzyme that removes phosphate (H_2PO_4) groups.
phosphodiester	Linkage of two phosphates via an oxygen atom.
phosphoprotein	A phosphorylated protein, may be phosphorylated on serine, threonine or tyrosine residues.
phosphotransferase	An enzyme that transfers a phosphate residue from one molecule to another.
picornavirus	Small RNA virus.
pinocytosis	Internalisation of materials external to the cell by vacuolation.
plasmid	Circular extrachromosomal DNA molecule with an origin of replication that replicates along with proliferation of the host.
PLCMV	Cytomegalovirus late promoter.
pleiotropic	Having multiple effects, as in 'pleiotropic mutation'.
pluripotent cell type	A cell type able to contribute to many cell lineages; sometimes used to mean that a cell line is able to contribute to both somatic cells and the germ line.
poliovirus	Small RNA virus that has an internal ribosome entry site.
polyadenylated RNA	RNA with a poly-A tail.
polymerase	An enzyme that catalyses polymer formation; polynucleotides are linear polymers.

polymorphism	Any difference between macromolecules dependent on a difference in primary DNA sequence, or a difference in the DNA sequence itself, between two interbreeding individuals. Mainly amino acid and nucleotide substitutions, deletions and insertions, but also includes large deletions and insertions and inversions of DNA.
polynucleotide	Linearly polymerised nucleotides; longer than an oligonucleotide, but the boundary size between one and the other is not defined.
polypeptide	Linearly polymerised amino acids.
polyploid	Having more than two haploid sets of chromosomes.
polyprotein	Precursor to several different proteins that are released from it by proteolysis.
polypyrimidine	A string of contiguous pyrimidine nucleotides (thymidine and cytidine).
polyribosome	Several ribosomes simultaneously translating the same mRNA molecule.
pRb	The retinoblastoma protein.
primordial germ cells	Immature germ cells prior to sexual commitment.
prolactin	Pituitary hormone.
promoter	Region of a gene required for the initiation of transcription; contains response elements bound by the necessary *trans*-acting proteins.
pronuclei	Haploid nuclei of a fertilised oocyte (1-cell embryo).
protamine	Basic protein bound to DNA in spermatozoa.
protooncogene	Normal cellular gene that can become oncogenic by mutation, overexpression or unregulated expression.
provirus	DNA copied from the retrovirus RNA genome and integrated into the chromosome.
pseudoautosomal genes	Genes located on the homologous regions of the X and Y chromosomes.
pseudohermaphrodite	Animal with non-functional male and female sex organs.
puromycin	Protein synthesis inhibitor. The *pur* gene, which confers resistance to puromycin, is used as a selective reporter.
radioimmune assay	An assay for a single component of a mixture of proteins based on competition with a radioactive seed for specific antibodies.
radiomimetic agent	A chemical that has an effect similar to that of radiation.

recombinant inbred strains	A set of inbred strains descended in parallel from a cross between two established inbred strains; each RI strain of the set is homozygous for a different combination of alleles from the parental strains.
recA	Gene coding for a multifunctional bacterial enzyme required for normal DNA repair and recombination processes.
recombinogenic	Competent to recombine.
reporter	Part of a transgene construct; frequently attached to a promoter so as to reveal the time and place of its activation; also used to select cells carrying integrated or altered transgenes in various different experimental contexts.
resection	Process of cutting back one strand at the end of a DNA duplex by exonuclease action.
response element	Short DNA sequence recognised and bound to by a *trans*-acting protein.
restriction fragment length polymorphism	Polymorphism based on differences in the lengths of restriction enzyme fragments identified by hybridisation with a specific probe.
retinoic acid	Chemical inducer; influences the differentiation of EC and ES cells.
retroposon	Retrotransposon; transposable DNA element with a sequence organisation resembling a retroviral provirus.
retrovirus	Type of RNA virus with a replication cycle centred on reverse transcriptase.
reverse transcriptase	Polymerase that makes the DNA complement of an RNA template, utilising an RNA or DNA primer.
RFLP	*See* restriction fragment length polymorphism.
RI strains	*See* recombinant inbred strains.
RIA	*See* radioimmune assay.
ribonuclease	An enzyme that degrades RNA.
ribonucleoprotein	A complex molecule containing RNA and protein, usually not bound covalently.
ribosome	Large ribonucleoprotein complex responsible for protein synthesis.
ribozyme	A catalytic RNA species.
RNP	*See* ribonucleoprotein.
*Sal*I	Restriction endonuclease, recognition sequence GTCGAC.
SAR	*See* matrix attachment region.
*Sce*I	Yeast endonuclease, cleaves a particular 18 base pair sequence.

seminiferous tubules	Tubules of the testes within which spermatogenesis takes place.
Sendai virus	Virus sometimes used to bring about fusion between cells.
Sertoli cells	Supporting cells of the testis, male homologue of follicle cells.
short interspersed repeat elements	Families of shorter dispersed repetitive DNA elements, such as Alu in humans and B1 in the mouse.
simple sequence length polymorphism	Polymorphism due to variation in the number of identical or near-identical short sequences in an array.
SINEs	*See* short interspersed repeat elements.
SnRNA	Small nuclear RNA species.
SnRNP	Complex of snRNA with proteins.
somatotroph	Pituitary cell that synthesises and secretes growth hormone.
spermatids	Post-meiotic maturing spermatozoa.
spermatocytes	Meiotic precursor of spermatozoa.
spermatogonia	Precursor cells of spermatozoa; some are cycling stem cells while others are committed to differentiation.
spliceosome	Large ribonucleoprotein complex responsible for pre-mRNA splicing.
Sry	Sex determining gene on the Y chromosome.
SSLP	*See* simple sequence length polymorphism.
superfamily	Of proteins; proteins with sequence similarities that indicate descent from a common ancestral protein.
superovulation	Treatment to increase the number of oocytes ovulated.
Sxr	Mouse locus defined by sex-reversal; due to translocation of *Sry*.
syncytium	A group of cells, each with a nucleus but an incomplete (or non-existent) cellular boundary.
syntenic	Describes parts of chromosomes of different species in which the gene order is the same.
TAA	Translational stop codon (DNA form).
tamoxifen	Oestrogen analogue.
TATA	Sequence present in the minimal promoter of many genes, contacted by TATA binding protein during the initiation of transcription.
TDF	Region of the human Y chromosome determining sex; includes *SRY*.
Tdy	Region of the mouse Y chromosome determining sex, includes *Sry*.

telomere	End of a chromosome.
teratocarcinoma	Malignant outgrowth from an embryo transplanted to an ectopic site and containing both differentiated and undifferentiated cells.
teratoma	Non-malignant outgrowth from an embryo transplanted to an ectopic site and containing only differentiated cells.
tet	Bacterial gene conferring resistance to tetracycline.
tetO	Operator of the *tet* gene.
tetraploid	Having four haploid sets of chromosomes during the period of the cell cycle between mitosis and S phase.
Tfm	Gene coding for the androgen receptor; an inactive receptor causes testicular feminisation.
TGA	Translational stop codon (DNA form).
thioguanine	Guanine analogue; HPRT phosphorylates the nucleoside and this causes cell death.
thymidine	Nucleoside of thymine.
thyrocyte	Thyroid cell that synthesises thyroglobulin and thyroxin; thyroid follicle cell.
thyroglobulin	Protein precursor molecule; contains tyrosine residues that are converted to thyroxin and cleaved off.
thyroxin	Tyrosine derivative tetra-iodo-thyronine; a hormone.
tk	Thymidine kinase gene, used as a selective reporter.
topoisomerases	Enzymes that alter the degree of DNA supercoiling.
totipotency	Property of being able to contribute to both somatic and germ lines, as do embryonic stem cells; sometimes used to mean the capacity to contribute also to extraembryonic structures, which excludes ES cells.
toxinogen	An enzyme that converts a harmless substrate to a lethal toxin.
trans	On the other side. Different mutations on chromosomal homologues, e.g. two different alleles of the same gene, are said to be in *trans* configuration. '*Trans*-acting' now signifies the action of any gene product on another gene, irrespective of whether it lies on the same chromosome in *cis* or *trans* configuration or on another chromosome altogether. Thus '*trans*-acting transcription factor'. See also '*cis*'.

transactivator	*Trans*-acting protein that binds to an activation site and stimulates transcription from an adjacent minimal promoter.
transcription	RNA synthesis.
transdifferentiate	Differentiation of already differentiated cells in a different direction.
transduction	Infection of bacteria with a bacteriophage preparation that carries over DNA from its previous bacterial host.
transesterification	Exchange of one phosphoester linkage for another.
transfection	Introduction of foreign DNA into cells.
transferrin	Iron-binding protein of serum.
transgene	Any DNA integrated into the chromosomes after deliberate introduction into a cell.
transition mutation	Single-base substitution of pyrimidine for pyrimidine or purine for purine.
translation	Synthesis of proteins.
translocation	Transfer of a segment of one chromosome to another.
transmembrane receptor	A receptor that spans the cell membrane; usually with extracellular ligand-binding, transmembrane and cytoplasmic signal-transducing domains.
transversion mutation	Single-base substitution of pyrimidine for purine or *vice versa*.
tRNA	Transfer RNA.
trophectoderm	Trophoblast.
trophoblast	Outer cell layer of the blastula or blastocyst.
UAA, UAG, UGA	Translation termination codons.
unimolecular	Of reaction rates; spontaneous transformation of a single molecular species.
uniparental expression	Expression of the allele that came from one parent while the other remains silent.
vasectomy	Sterilisation of the male by severing the *vas deferens*.
virion	Virus particle.
Wolffian duct	Male sexual primordium, stimulated to develop by testosterone.
Xa	The active X chromosome.
Xce	X controlling element; as for *Xic* below.
xenotransplant	Transplant of tissue between species.
Xi	The inactive X chromosome.
Xic	X-inactivation centre; region of the X chromosome controlling X-inactivation.

Xist	Gene controlling X-inactivation; active on the inactive X.
YAC	Yeast artificial chromosome.
zona pellucida	The glycoprotein layer surrounding the oocyte.
zoonotic	Infectious disease transmitted from an animal to humans.

Index

References to **Figures** are shown in **bold**, to *Tables* in *italic*, and to ***Boxes*** in ***bold italic***. Chapter 1 is not indexed.